For all Zimbabweans,
but especially in memory of
Theresa Chimhondo,
a mother deeply missed,
who I hardly knew,
but to whom I am greatly indebted.

THE SILENCE OF GREAT ZIMBABWE

THE SILENCE OF GREAT ZIMBABWE

Contested Landscapes and the Power of Heritage

Joost Fontein

University College London Institute of Archaeology Publications,
Vol. 37

LONDON AND NEW YORK

First published 2009 by Left Coast Press, Inc.

First published by UCL Press in 2006.
Distributed by Left Coast Press, Inc. as of 2009.

Published 2016 by Routledge
2 Park Square, Milton Park, Abingdon, Oxon OX14 4RN
711 Third Avenue, New York, NY 10017, USA

Routledge is an imprint of the Taylor & Francis Group, an informa business

Library of Congress Cataloguing-in-Publication Data available from the
publisher

ISBN 978-1-59874-220-6 hardback
ISBN 978-1-59874-221-3 paperback

Contents

List of Illustrations

List of Acronyms

AAEC	Association of African Earth-keeping Churches
AZTREC	Association of Zimbabwean Traditional Ecologists
BSACo	British South Africa Company
CID	Central Investigations Department
CIO	Central Intelligence Organisation
DA	District Administrator
ICOMOS	International Council on Monuments and Sites
IUCN	The World Conservation Union
MDC	Movement for Democratic change
NMMR	National Museums and Monuments of Rhodesia
NMMZ	National Museums and Monuments of Zimbabwe
ODA	Overseas Development Authority
PA	Provincial Administrator
SAREC	Swedish Agency for Research Co-operation
UNESCO	United Nations Educational, Scientific and Cultural Organisation
WHC	World Heritage Centre
ZANLA	Zimbabwe African National Liberation Army
ZANU PF	Zimbabwe African National Union Patriotic Front
ZAPU	Zimbabwe African Peoples' Union
ZBC	Zimbabwe Broadcasting Corporation
ZINATHA	Zimbabwe National Traditional Healers Association
ZIPRA	Zimbabwe Peoples' Revolutionary Army
ZIRRCOM	Zimbabwe Institute of Religious Research and Ecological Conservation
ZRP	Zimbabwe Republic Police

Acknowledgements

There are many people to whom I owe a debt of gratitude. First of all I must acknowledge the financial support of the ESRC, the British Academy and the Munro Fund at the University of Edinburgh. Each provided funding at different times in the long process from initial proposal, through archival research and fieldwork to the writing, rewriting and publication of this book. The Carnegie Trust for the Universities of Scotland deserves special mention for their grant which has enabled the co-publication of this book in Zimbabwe by Weaver Press.

In 2000, as Chairman of the History Department at the University of Zimbabwe, Dr Innocent Pikirayi assisted me greatly by offering me a position as a Research Associate (2000–01). Beyond merely being instrumental for gaining permission to conduct research, this link with historians and archaeologists at the University of Zimbabwe proved very helpful as I got to grips with the impressive array of historical and archaeological work being carried out in Zimbabwe. Special thanks go to Gerald Mazarire who gave me numerous pointers to files at the National Archives, and off whom I frequently bounced ideas as they emerged from my research. I feel privileged to be able to call him my friend. Similarly I extend my gratitude to Terence Ranger, for inviting me as a fledgling postgraduate to present a paper at the 'A View of the Land' conference in Bulawayo in July 2000, for his support throughout my period of research in Zimbabwe and since, and for the memorable discussion we had about this study in October 2003.

The National Museums and Monuments of Zimbabwe (NMMZ) deserves special appreciation for their enthusiasm and support of this research. At NMMZ head office in Harare, both Dawson Munjeri, and Dr G. Mahachi made time available for remarkably frank and engaging interviews. In Masvingo, Edward Matenga provided access to NMMZ files, which had a great impact on my research. He also made time available for interviews and conversations, and a very informative and enjoyable tour of Great Zimbabwe one Sunday in June 2001. I got to know many members of the conservation team at Great Zimbabwe very well, and was constantly impressed by their openness and willingness to discuss. I hope that they will recognise that any critique of NMMZ contained in this book is made in the spirit of co-operation, and I hope they find my limited insights of some assistance for the course that NMMZ has embarked upon.

I should also acknowledge the help of members of the World Heritage Centre, UNESCO in Paris, with whom I worked as an intern in the spring of 1999. I am grateful for their intellectual engagement and encouragement during that time. In Harare, I am grateful to staff at the National Archives of Zimbabwe who were consistently helpful and provided every assistance I needed as I attempted to plough through the extensive archives they hold. I should also mention Ranche House College in Harare with whom I did an intensive Shona course during the early months after my arrival in Zimbabwe. Whilst I was in Harare I stayed with Mike Zimonte – 'VaChedgelow' – who has also become a close friend. When I first arrived in Zimbabwe he helped me get to grips with spoken and written Shona, and later he assisted me at the National Archives by going through many years of Zimbabwean newspapers.

My greatest debt must be to the communities around Great Zimbabwe among whom I carried out extensive fieldwork, particularly members of the Nemanwa, Charumbira and Mugabe clans whose perspectives on Great Zimbabwe were the focus of my research. Their co-operation, enthusiasm and hospitality was unrivalled, and a key to the success of this project. Whilst they were always keen to emphasise the correctness of their particular versions of the past, and their group or individual claims over Great Zimbabwe, it was always understood and accepted without question, that I was undertaking this research among all sides of the dispute. I only hope that I have done justice to their efforts to help me with this research, and that I have managed to present them in a sensitive and balanced light. They are far too numerous to mention all by name here, but the contributions of particular individuals are apparent throughout the book.

Sadly, since this book was written several of these people have passed away. They include Aiden Nemanwa, Chief Nemanwa, Chief Murinye, VaMututuvari, VaMhike and Eddison Zvobgo, among others. Given their general enthusiasm for this research it is particularly regrettable that they were unable to witness the published results. I hope I have done justice to their efforts.

My research among these communities was considerably aided by my three 'research assistants', Timothy Madende, Dzingarai Mubayiwa and Pardon Masunganye, with whom I travelled around Masvingo District during different periods of my fieldwork. Special mention must be made of the Rukasha family, near Nemanwa Growth Point, at whose household I stayed for the duration of my field work. I have known them since my first visit to the area in 1995, and a subsequent visit in 1997, and their continued kindness, hospitality and generosity of spirit continue to inspire me. We have developed a deep friendship that is akin to 'genuine' kinship, and I doubt if I will ever be able to repay the kindness they have shown me.

At Edinburgh University I have continued to benefit greatly from the support of my colleagues in Social Anthropology, and at the Centre of Africa Studies. Among many others, Paul Nugent, Sara Rich Dorman, Alan Barnard and Richard Whitecross have provided support, comments and advice. Charles Jedrej, in particular, has been a constant source of moral support since my undergraduate days, and through a long series of proposals and funding applications. He gave me the confidence to take this path, and I hope I have done justice to his efforts. To Jeanne Cannizzo, who supervised the PhD that was an early version of this book, I also owe a great deal. Throughout the arduous experience of writing up my research, she inspired me with confidence and thoughtful insight. Always generous with praise, and subtle and constructive in her criticism, I was very lucky to have had her supervise this work. I could not have done this without her help. Many thanks. More recently, this book has also benefited from the careful and insightful comments of Peter Ucko of the Institute of Archaeology at UCL, Murray McCartney of Weaver Press, and two anonymous reviewers selected by UCL Press. Their efforts and comments are acknowledged and appreciated.

Finally, to my parents, Hans and Marianne Fontein, I am grateful for the continued support you have offered me along my chosen path. I hope I have made you proud. Leaving for last my most important debt of gratitude, my deepest thanks go to my wife Barbara Chimhondo. She has been a pillar of strength through sometimes difficult times of profound change. Living with someone who is writing a book is perhaps more difficult than writing one, and I am grateful for your infinite patience and support.

Some Notes on Fieldwork, Language and Sources

This book is based predominantly on research carried out between 2000 and 2004. The main period of research in Zimbabwe was from June 2000 to December 2001, although material for the last chapter was collected during a follow up visit in July 2004. Most of this research took the form of ethnographic fieldwork among the communities who live around Great Zimbabwe, but I also spent almost four months at the National Archives of Zimbabwe, and several weeks working through NMMZ files at the Conservation Centre at Great Zimbabwe. I also carried out some interviews in and around Harare with war veterans, *masvikiro* belonging to ZINATHA, and members of NMMZ head office. In total I conducted 'formal' interviews with more than 70 people, of whom about 20 were established *masvikiro* or claimed to have spirits that occasionally possessed them.

After having spent several months on an intensive Shona language course at Ranche House College in Harare, and at the National Archives, I finally moved to Masvingo District in October 2000. I stayed in a rural household near Nemanwa Growth Point, close to Great Zimbabwe, with people who I had got to know very well during previous visits in 1995 and 1997. Whilst I was staying in an area that falls under the jurisdiction of Headman Nemanwa (himself under Chief Charumbira), the people with whom I lived are 'incomers' who arrived there after Independence. This was not deliberate (I was invited to stay there because we were already well acquainted), but it was helpful in that I was able to move freely and unhindered between the different communities of Nemanwa, Charumbira and Mugabe, without being pushed into any particular direction. Maintaining a balanced and sensitive view of the disputes between these clans was therefore facilitated by living in a household that is not intricately involved in these inter-clan disputes.

I began by visiting each of the chiefs to ask for permission to conduct research and from there widened my search for different informants. I got to know key actors from all three clans, though I hasten to add that I was, inevitably, unable to interview or meet with *all* the elders and *masvikiro* from these clans. I also carried out some interviews with people from the VaDuma clans of Murinye and Shumba, and the unrelated Mapanzure clan. Apart from those with a direct interest in, or claim to, Great Zimbabwe, I also carried out interviews with other influential *masvikiro* across Masvingo Province, as well as local politicians, and employees of NMMZ. I spent a considerable amount of time seeking out war veterans though with limited success. I was only able to track down one war veteran who had operated in the district during the war, though I did carry out interviews with a number of different members of the War Veterans Association, both in Harare and Masvingo, many of whom were involved in the 'fast track' land reform programme being implemented at the time.

Nearly all the interviews were recorded on tape, and in most cases they were conducted in *Chikaranga*, the local dialect of Shona. Exceptions to this were interviews with members of NMMZ, local politicians, and people who chose to speak in English.

Only one or two people refused the use of recording equipment, and most were very keen to have their chance to be heard. Most interviews were carried out with the assistance of a 'translator' (research assistant), who I employed from the local communities. This tended to help maintain a flow of conversation during interviews, especially in the early months of fieldwork when my Shona was still very infantile. All the transcriptions were written in English and done by myself. Often with the benefit of being able to rewind and re-listen, I was able to improve the translations given by my research assistants. As I found my Shona improving, I continued to work with a research assistant, even as the translation work became less important, because they often provided helpful insights and assistance, especially in the process of seeking out and meeting new informants. In the interests of protecting informants I have kept all the interview tapes myself, and my transcriptions of them. I would be willing, at my discretion, to make them available for interested researchers.

I deliberately did not seek to formally interview any *masvikiro* when they were possessed because I considered it invasive. I did, however, witness several possession ceremonies, which were extensively recorded in my field notes. On these occasions I found working with a research assistant particularly useful, as we would later compare our notes of the complicated events taking place. All my field notes and interview transcriptions are sorted by date, and direct references to 'interview with' refer to those that were recorded on tape, while references to 'field notes', and 'notes of interview' refer to un-taped conversations and interviews. All quotes from recorded interviews appear in italics, while quotes from my field notes, like those from printed sources, are in normal font.

In the vast majority of instances I found people across Masvingo District and Province to be very enthusiastic and keen to assist me with my research. For those clans immediately surrounding Great Zimbabwe there was a sense in which I became a resource that could offer the opportunity to air concerns and interests, though I was careful to emphasise that I was working with people from *all* clans, and that I was unrelated to any government authorities or NMMZ. As for war veterans, I found them, for the most part, equally welcoming and enthusiastic about my research. Perhaps the most difficult people to gain access to were national politicians and members of the government, though I was able to interview the late veteran nationalist and Masvingo M.P., Dr E. Zvobgo, as well as the Governor of Masvingo Province, Josiah Hungwe and the Provincial Administrator.

I found NMMZ staff, both at their Regional Office at Great Zimbabwe and at Head Office in Harare, to be very helpful and I was very fortunate to be allowed access to NMMZ files at the Conservation Centre at Great Zimbabwe. This is a rich resource, including some files that date back to the last days of the liberation struggle, which I was only able, due to limitations of space and time, to make partial use of here in this book. Nevertheless, they provided an important link between information gathered from the National Archives in Harare and my own ethnographic fieldwork in the area. Through NMMZ files I was able to glimpse its internal workings, and in particular, its development since independence into the organisation it is today. One particular file, (NMMZ G1 I and II) is particularly interesting as it contains correspondence between local chiefs and NMMZ since the early 1980s, revealing the changing relationship between NMMZ and local communities during this period. In my references I have used NMMZ's own reference numbers for their files, and, where possible, a date for the particular letters and documents cited. They are all held at the Conservation Centre at Great Zimbabwe.

Not surprisingly there is a great wealth of material, published and unpublished, relating to Great Zimbabwe and Masvingo District at the National Archives of Zimbabwe. In my four months at the National Archives (September/October 2000, March and October 2001) I was really only able to skim the surface of these resources. I chose to focus most of my attention on unpublished manuscripts and Rhodesian Government documents, only a fraction of which are cited in this book. Great potential exists for further historical research on the relationship between how Great Zimbabwe was viewed and how it was managed during the first part of the twentieth century, particularly under Wallace, during whose 'reign' at Great Zimbabwe some of the most profound transformations were made. From the 1960s onwards the archival record is less consistent, though some files do exist that document the discussions that occurred between ministries over the transfer of the entire estate to NMMR in the late 1970s (the National Archives of Zimbabwe file H15/10/1/3, 10, 20). There is also a file containing Colonel Hartley's correspondence in the late 1960s (the National Archives of Zimbabwe file HA 17/1/2) which indicates the extreme reaction of some Rhodesian apologists against growing nationalist use of the site. Unfortunately I have not, through lack of space, been able to fully develop some of these themes here. I have provided the National Archives reference numbers for all the files cited and referred to, though these by no means represent all the files that I worked through or that are available at the National Archives.

Apart from archival records, the National Archives of Zimbabwe also holds a huge collection of published material on Great Zimbabwe from the end of the nineteenth century to the present. The size of this collection in itself illustrates the extent to which the site has captured the imagination of so many for so long, and in this sense my study is merely another addition to the ever-expanding literature on the site. The National Archives also contains a full collection of the national and regional newspapers published in Rhodesia and Zimbabwe, and while my references to newspaper articles are not all cited to the National Archives here, they are accessible for research there.

The UNESCO material referred to in Chapter 9 was largely collected during research carried out in 1999. In March/April that year I spent a month working as an intern with the World Heritage Centre at UNESCO in Paris, and I returned in June to conduct interviews with members of the Centre. This research formed the basis of an M.Sc. dissertation, entitled 'UNESCO, Heritage and Africa: an anthropological critique of World Heritage', which was later published by the Centre of African Studies as an Occasional Paper (Fontein 2000). Most of the UNESCO literature referred to is available on the World Heritage Centre's website at http://whc.unesco.org/reql.asp. Copies of my notes of interviews with World Heritage Centre staff are in my possession, and can be viewed on request.

Figure I.1: Plan of Great Zimbabwe, at Great Zimbabwe. (Author 2000)

PART I: THE SILENCE OF UNREPRESENTED PASTS AT GREAT ZIMBABWE

CHAPTER 1

THE 'ZIMBABWE CONTROVERSY': THE POWER OF 'FACT' OVER 'FICTION'

RUMOURS OF GOLD, SHEBA AND THE LAND OF OPHIR

Today I dare to close this account with:

The Queen of Seba is the Queen of Simbaöe,
Psalm 72, 10 – The Seba mentioned there is Simbaöe,
Math. 2, 11 – Of the three kings the one was from here, the others from Arabia and India
The reported pot is possibly an Ark of the Covenant.
The ruins are copies of Salomo's temple and palace
(Carl Mauch, Wednesday 6 March 1872, in Burke 1969: 191)

Carl Mauch, a German explorer, is often attributed with the 'discovery' of Great Zimbabwe in 1871, and whilst he was probably the first European to publish an account of visits there, it is clear from his published journals (Burke E. 1969 *The Journals of Carl Mauch 1869–1872*) that he did not stumble upon them, rather he was looking for them. His journal entry for Friday 1 September 1871 – four days before he first 'discovered' the ruins – makes this clear.[1]

Some more people were questioned about this pot and they, too, told curious stories about it. There was one among these people who came forth with still more important news, namely of the presence of quite large ruins which could never have been made by blacks. Could these be the ruins of the Banyai for which I've been looking? (Carl Mauch Friday 1 Sept. 1871 in Burke 1969: 139)

Mauch already held faith in the fanciful theories of the 'ancient' and 'white' origins of the stone walls, even before hearing of the ruins themselves from locals. This would certainly explain his delight – evidenced in his journal entry for the previous day – when 'natives' told him white people once lived in the country.

This most exciting news was that, according to the natives, white people had once lived in this country and that when they [ie the 'natives'] took possession of these parts (about 40–50 years ago), they had, now and again, picked up tools while working in their gardens; for instance, once a piece of iron which, according to their description, could have been a miner's pick. They affirmed with conviction that they would not have been able to make such things. Remnants of furnaces were still numerous and, as they do not keep it secret but, on the contrary, would like to see white people living here, the ancestors of whom once owned this country, I started early to look for traces on a higher-lying terrace. (C. Mauch Thursday 31 Aug. 1871 in Burke 1969:137)

The search for ruins of an 'ancient' and 'foreign', preferably white, civilisation were linked in the explorer's imagination to the search for gold – the discovery of which he was already credited with at what became known as the Hartley Hills, during previous explorations (Burke 1969: 3). Indeed, prior to this expedition, Carl Much had come under the influence of a certain Rev. A. Merensky of the Berlin Mission in the

Transvaal, who 'was of the conviction that in the country Northeast and east of Mosilikatse the ancient Ophir of Solomon is to be found and that in the times of the Ptoleymies Egyptian trade penetrated to our coasts'[2] (Burke 1969: 4). Garlake (1973: 62) has also emphasised the role of Rev. Merensky in the formation of Mauch's expectations of, and subsequent explanation for, the ruins. Describing Mauch as a 'young man of courage and great tenacity, but certainly no thinker', Garlake (1973: 62) suggested that,

> In fact Mauch was, of course, not conducting an investigation but giving unquestioning acceptance to someone else's ideas. His opinions reflect nothing more than the sources, selected, channelled and coloured by Merensky, that had stimulated him. Thus Mauch, the first certain foreign visitor to the Ruins and the first person to describe them to the outside world set the final touches to Muslim tales that had reached the Portuguese over three centuries before. (Garlake 1973: 64)

The 'Muslim tales' that Garlake referred to are detailed in a variety of Portuguese documents from the sixteenth century, alongside the Portuguese writers' own accounts of encounters with the kingdom of Mwene Mutapa, on the plateau of northern Mashonaland. While the Portuguese archives of their explorations between 1506 and 1890 are extensive, there are only a few references to stone buildings (Garlake 1973: 51; Beach 1998: 48). Some of these accounts refer to stone buildings within the 'Mwene Mutapa's own Karanga Kingdom' (Garlake 1973: 51) in the north of Zimbabwe, but one in particular (that of João de Barros published in the first decade of his *Da Asia* in 1552) seems to describe Great Zimbabwe's geographical position, surroundings and architectural features very closely (Garlake 1973: 51–52).

Garlake has argued convincingly that the ideas of Great Zimbabwe's 'foreign origins' that emerged in the Carl Mauch's descriptions, and later sparked the 'Zimbabwe Controversy', originated from the tales of Swahili traders and Portuguese writers who ignored 'the most obvious assumption ... that they were the work of the local people' (Garlake 1973: 54). In particular, he laid blame on the Portuguese writers De Barros and Dos Santos, who 'with a completely uncritical acceptance of Swahili tales and with generalisations based on incomplete knowledge, ruled out an indigenous origin for the stone buildings' (Garlake 1973: 54). He continues:

> It is much more certain that the Swahili did not build the Ruins. De Barros' 'Moorish merchants' admitted that they knew nothing of the builders and Dos Santos' 'aged Moors' were indicating much the same thing in naming Solomon or Sheba as the builders for these were popular figures in Muslim folklore and two names from the remote past with which every Muslim was familiar ... .

> Having thus eliminated the Karanga and Swahili, de Barros and dos Santos were left with no concrete evidence at all. They therefore drew from their own resources. They could scarcely conceive of any area of major human achievement that had gone completely un-recorded and recognised in the Bible as the most precise historical account of the Human past, elaborated perhaps by some classical authors. Their view of Africa was coloured by memories of the hopes once raised by the great lost Christian Kingdom of Prestor John, and by their faith, however much it was dwindling, in an enormous wealth in gold in the Sofalan interior. These diverse inspirations were reinforced by an awareness that Ethiopia contained definite ruins of Biblical kingdoms, coupled with a very hazy and exaggerated idea of how far Ethiopia extended. With these

premises, it now looks nearly inevitable that they should have suggested Prestor John, Solomon and Sheba as the instigators of Mwene Mutapa's stone buildings. (Garlake 1973: 54–5).

By the end of the nineteenth century, these accounts were feeding a frenzy of European imperial and capitalist discourses and activities in southern Africa. Mauch's 'discovery' of Great Zimbabwe 1871, and the almost inevitable comparison with Solomon's Temple, and the Queen of Sheba only fuelled a fire that was already burning with fury. Set in the context of what Pakenham (1991) described as 'the scramble for Africa', the effects of the reports that Carl Mauch had found the mythical ruins of the 'land of Ophir' were exponential. By 1890 'historical legend had inspired the colonisation of Zimbabwe' (Kuklick 1991: 139).

The territory occupied by the British South Africa Company was, as Kuklick noted, the 'only British Colonial preserve acquired for the explicit purpose of exploiting its mineral resources' (1991: 138). Furthermore, 'Rhodes and his kind in southern Africa were accustomed to invoking their version of the history of the area to demonstrate that African polities had no stronger claim to land than they did' (1991: 138–39). Therefore the rumours of ancient civilisations, and the land of Ophir[3] 'helped fire his imagination and shape his policy' (Colquhoun 1914: 485). The link between Great Zimbabwe, rumours of gold and ancient mythical/biblical 'civilisations' was used to encourage early setters to join the British South Africa Company's (BSACo) 'pioneer column' in 1890. Beyond attracting colonial settlers with the lure of abundant and exploitable gold resources, Great Zimbabwe 'also quickly became a symbol of the essential rightness and justice of colonisation and gave the subservience of the Shona an age-old precedent if not Biblical sanction' (Garlake 1973: 65). When Cecil Rhodes came to visit for the first time, 'local Karanga chiefs were told that the "Great Master" had come to see the ancient temple which once upon a time belonged to white men' (Garlake 1973:66).

Clearly theories of 'ancient' and 'exotic' origins for Great Zimbabwe were of great political use for the British South Africa Company, and particularly Cecil Rhodes. In 1891 he began financing a variety of research projects, including a search among European archives and libraries for descriptions of Zimbabwe by Alexander Wilmot (1896). With the Royal Geographical Society and the British Association for the Advancement of Science, he commissioned an expedition to, and investigation of, the ruins led by an explorer called Theodore Bent, a man with 'antiquarian inclinations but no formal archaeological training' (Garlake 1973: 66). This was to be the first in a series of 'officially sanctioned' excavations at Great Zimbabwe which signalled the systematic, and institutionalised appropriation of both its past and the site itself.

In terms of artefacts, the process of appropriation had already begun when Carl Mauch took samples of what he thought was cedarwood[4] from the lintel of the Great Enclosure's north entrance on Wednesday 6 March 1872 (Matenga 1998: 21). Much more significant than Mauch's splinters of wood, was Willie Posselt's dubious acquisition of one of the – soon to be famous – 'Zimbabwe Birds' in 1889, which was later sold to Cecil Rhodes (Matenga 1998: 22). While Great Zimbabwe was apparently spared from the relic hunting of the Ancient Ruins Company, authorised by the British

South Africa Company in 1895 to 'plunder for profit in all the ruins save Great Zimbabwe' (Kuklick 1991: 142; see also Ndoro and Pwiti 2001: 23), it still fell victim to destruction and pillaging by unauthorised excavators. Most destructive of all were Richard Hall's vast clearances of the Great Enclosure between 1902 and1904, done under the auspices of 'preservation work' (Garlake 1973: 72). Further destruction of archaeological deposits occurred under the authority of the Public Works department in the second decade of twentieth century, in a misguided attempt to prevent the collapse of the western wall of the Hill Complex (Ndoro 2001: 41). But the appropriation of the site did not only take the form of relic hunting, or the destruction of archaeological deposits. It also refers to the process by which local communities were increasingly distanced and alienated from the site. At the end of the nineteenth century members of the Mugabe clan occupied parts of the Hill Complex, and much of what is now the Great Zimbabwe estate was grazed by cattle, hunted on, and used for gathering fruits, and collecting thatching grass. More importantly perhaps, the site was considered sacred by different local communities who were deeply involved in a contest over the ownership of the site. Today a fence exists around the estate, access is tightly controlled, and entrance fees are charged. In short, Great Zimbabwe has become a *heritage* site, and a tourist destination.

Theodore Bent's excavations at Great Zimbabwe began in the Great Enclosure in June 1891. The site was already receiving a great deal of European visitors inspired by the rumours of King Solomon and the Queen of Sheba, that had been widely proliferated through popular novels such as Rider Haggard's *King Solomon's Mines* (1885). Bent was sceptical of these ideas.

> The names of King Solomon and the Queen of Sheba were on everybody's lips, and have become so distasteful to us that we never expect to hear them again without an involuntary shudder. (Bent 1896: 64)

Indeed, his excavations in the Great Enclosure came across very little that was not of much more recent, local African origin. Bent wrote, 'We found but little depth of soil, very little *débris*, and indications of a Kaffir occupation of the place up to a very recent date, and no remains like those we afterwards discovered in the Fortress' (Bent 1896: 118). As Garlake noted, Bent's guide, C.C. Meredith, said that 'on one occasion at this time Bent "looking rather depressed" and confided to him "I have not much faith in the antiquity of these ruins I think they are native ... . Everything we have so far is native" ' (Garlake 1973: 66). But Bent's preconceived ideas of 'exotic', 'foreign' builders held strong and his finds indicating 'kaffir occupation' were ignored as failing 'to bring any definite records of the past' (Bent 1896: 121). Rather, he suggested that perhaps,

> a spot situated on the shady side of the hill behind the great rock might possibly be free from Kaffir desecration; and the results of our excavations on this spot proved this to be the case, for here, and here only, did we come across relics of the past in our digging. In fact, the ancient builders seemed to have originally chosen the most shady spots for their buildings. Undoubtedly the oldest portions of the Zimbabwe ruins are those running along the sunless side of the hill fortress; on the other side, where now the Kaffir village is, we found hardly any trace of ancient structures. Our difficulty was to get the shivering Kaffirs to work there, for whenever our backs were turned they would hurry off to bask in the rays of their beloved sun. (Bent 1896: 122)

It seems extraordinary that Bent would have used the reasoning that his African workers disliked the shade and preferred to work in the sun, as a basis upon which to choose a site 'free from Kaffir desecration', in order to 'come across relics of the past'. What he came across there were more material remains that seemed 'indistinguishable from contemporary Karanga articles', as well as a few less obviously dateable objects he never-the-less considered African, and some 'clearly identifiable and dateable ... pieces of Arabian, Persian and Chinese glass and ceramics ... no more than a few centuries old' (Garlake 1973: 67). These objects were of little help to bolster his conviction that the ruins were of ancient and exotic origins. However, he did come across various soapstone objects, including four 'Zimbabwe birds' on carved monoliths, as well as other decorated monoliths, soapstone bowls, figurines and 'phalli'. Unlike anything else that had been found in sub-Saharan Africa before, these objects were 'assumed by Bent to be the only clues to the origins of the builders and he started to look for parallels' (Garlake 1973: 67). His parallels of choice were from various origins in the ancient middle east (Assyria, Mycenae, Phonecian Cyprus, Egypt and Sudan), and combined with a somewhat idiosyncratic analysis of the architecture of the ruins, he arrived at the rather imprecise conclusion that there was 'little room for doubt that the builders and workers of Great Zimbabwe came from the Arabian peninsula' (Bent 1896: 288–89; Garlake 1973: 68). However sceptical he might have been of the King Solomon and Queen of Sheba myths, his conviction that ancient foreign builders were responsible was clearly based on the prevailing racial attitudes of the time. This is explicit in the foreword to the third edition of his *The Ruined Cities of Mashonaland* (1896), where he referred to surveying work carried out by a Mr. R.M.W. Swan[5] at another ruin on the Lundi River.

> It is, however, very valuable evidence when taken with the other points, that the builders were of a Semitic race and of Arabian origin, and quite excludes the possibility that any negroid race having had more to do with their construction than as slaves of a higher race of cultivation; for it is a well-accepted fact that the negroid brain never would be capable of taking the initiative in work of such intricate nature. (Bent 1896: xiv)

It was this racialised perspective that lay behind all of the 'foreign origins' theories (for example, Wilmot 1896; Hall and Neal 1902; Hall 1905, 1909) that together formed one side of what became known as the 'Zimbabwe Controversy'. It was essential for the moral justification of the colonial enterprise to be able to show that 'Africans were incapable of great achievements unless forcefully supervised by a superior race' (Kuklick 1991: 147). Therefore, it is not surprising that when the 'foreign origins' theories of Rhodesian apologists first came to be challenged in 1905 by the work of a 'professional' archaeologist from Britain David Randal-MacIver (1906), most Rhodesian settlers shared the opinion of the explorer and colonial official Sir Harry Johnstone, that the views of a 'supposed expert' from abroad could never be as reliable as those of people 'intimately acquainted with the Bantu negroes of Africa' (Johnstone 1909: 564). This was the argument most strongly made by Richard Hall, in his attempt (1909) to refute David Randal-MacIver's case for the African origins of Great Zimbabwe.

Thus the 'battle lines' for the first stage of the 'Zimbabwe Controversy' were set. Richard N. Hall, a leading Rhodesian protagonist and author of several books on the subject (1902, 1905, 1909), emphasised the superiority of Rhodesian knowledge of the 'natives' to argue for the Semitic origin of Great Zimbabwe. On the other side, David Randal-MacIver challenged the 'amateur methods' of the Rhodesian protagonists, and claimed the authority of 'science' as the basis for his argument that 'the people who inhabited the elliptical temple' belonged to 'tribes whose arts and manufacture were indistinguishable from those of the modern Makalanga' (Randal-MacIver 1906: 63). While Randal-MacIver became a champion of professional British archaeologists, Hall had the support of Rhodesian public opinion until his death in 1914.

The debate remained unresolved, and in 1929 the British Association for the Advancement of Science sent Gertrude Caton-Thompson to conduct the 'examination of the ruins at Zimbabwe or any other monument or monuments of the same kind in Rhodesia, which seem likely to reveal the character, date and source of the culture of the builders' (Caton-Thompson 1931: 1). 'Already recognised as one of the outstanding archaeologists of her generation' (Kuklick 1991: 152), it is not surprising that she concurred with Randal-MacIver's view of the African origins of the site, though she made some concessions to Rhodesian public opinion by arguing that 'the architecture at Zimbabwe ... strikes me as essentially the product of an infantile mind, a pre-logical mind' (Caton-Thompson 1931: 103). Kuklick has used this to illustrate how even 'sound archaeological evidence can document the legitimacy of political regimes as effectively as can fanciful accounts' (1991: 164).

Caton-Thompson's expedition was deliberately timed so that she would be able to deliver a report to the British Association meeting in Cape Town and Johannesburg in July–August 1929. There she faced strong opposition from the supporters of recent versions of the 'foreign origins' theories, in the form of Professor Leo Frobenius, President of the Frankfurt Research Institute, and Raymond Dart. The British Association meetings provided 'sensational material' for the South African press, and Raymod Dart's very noisy, indignant protests 'delighted the press' (Kuklick 1991: 152).

Clearly the 'Zimbabwe Controversy' continued to be as emotive an issue in 1929 as it had been during Hall and Randal-MacIver's time 20 years earlier. In fact, the 'ancient', 'exotic' origins of Great Zimbabwe remained 'an article of faith for Southern Rhodesian settlers and officials' (Ranger 1987: 159), right up until Zimbabwean independence in 1980. But in the years following Caton-Thompson's expedition in 1929, and especially with the introduction of carbon dating in the 1950s, Great Zimbabwe's African origins gradually became widely accepted beyond the narrow archaeological circles that had originally supported Randal-MacIver in the first decade of the twentieth century. Although 'foreign origins' theories did see a revival in the late 1960s and 1970s in response to growing African nationalist use of Great Zimbabwe, they no longer carried much credence among audiences outside of Rhodesia. 'Professional archaeology' had triumphed over the 'amateur' approaches of Rhodesia settlers and established its own authority to represent the past through claims to 'scientific objectivity' and 'professionalism'.

THE 'ZIMBABWE CONTROVERSY': 'A CAVALCADE OF FACT AND FANTASY'

> It is a cavalcade of fact and fantasy. A blending of history and hypothesis against a background of migrations, mines and monuments whose histories have yet to be unveiled. So this story of the land men have named Ophir, Punt, Sofala, Zambesia, and Rhodesia is a tale written in blood and salt sea spray, in gold and ink, recorded in the literature bearing on Rhodesia's past.
>
> A hundred books have played their part, now the cavalcade departs, yet much remains untold. There is a tale of bleeding sculptures which vanished from Zimbabwe. Stories of spiritualism and séance. A fertility pattern on Bantu pottery and Zanzibar doors. A tale of clay Zimbabwe, of caves below the Acropolis. A story of the chevron pattern on the Temple Wall, and its supposed link with the summer and winter solstice. There is a story of a great cavern in the Inyanga mountains with the face of the rock tool-marked by man, and of strange cuttings in the hillsides for roads leading nowhere. The supposed purpose of the parallel passage. The untold tale of soapstone Zimbabwe birds transported to the Cape Province; a hill near Zimbabwe known as Inyoni – the bird, and a tribe whose totem is the fish eagle.
>
> (Paver [1950] 1957: 197)

Half a century later, Paver's romantic, though apt, description of the 'Zimbabwe Controversy' as a cavalcade (Paver [1950] 1957) continues to resonate. True to his time, Paver considered 'the black man's past [to] be speechless, and the words of the present-day Bantu valueless' (Paver 1957: 156) despite his deliberate, if unsuccessful, attempt to position his work on neither side of the polarised debate over the origins of Great Zimbabwe. He neither favoured the arguments of Rhodesian apologists and 'amateur' antiquarians (Bent 1896; Hall 1905; Dart 1925) that sought an ancient, exotic and non-African origin for Great Zimbabwe, nor the opposing 'medieval school of thought' of British archaeologists that stressed its African origins (Randal-MacIver 1906; Caton-Thompson 1931). Delaying judgement, he sought rather to assess and consider arguments on both sides, while maintaining faith that 'a key to Rhodesia's mysteries will yet be found' (Paver 1957: 156).

The 'cavalcade' has continued, and now, at the beginning of a new century, the archaeological discourses that grew to dominate debates about Great Zimbabwe's past have moved on. Ancient, exotic origins are no longer sought after – its African origins are rarely disputed. The introduction of carbon dating in the early 1950s established Great Zimbabwe's 'medieval' date (Summers 1955) and excavations in 1958 (Robinson, Summers and Whitty 1961) revealed 'a complete stratified sequence of deposits on the hill, spanning the entire occupation of the site' (Garlake 1983: 4). Combined with Robinson's (1961) work on pottery sequences, and Whitty's (1961) architectural analysis, a chronology of the site was established that suggested continued occupation until the nineteenth century. After investigations by Garlake (1968 and 1973), this chronology was slightly modified, and 1450 AD became the accepted date for the 'demise' of Great Zimbabwe. To say the city 'flourished' (Garlake 1983: 1; Beach 1998: 47; Matenga 1998: 6) for a period of 300 years between 1150 and 1450

AD has became standard, in effect these dates have now been *black-boxed* (Hodder et al., 1995: 8).

Archaeologists and historians have searched for the causes of the emergence of the 'Zimbabwe State' and 'Culture'. Some emphasised the role of external trade in gold (Huffman 1972), while others (Garlake 1973, 1978; Connah 1987; Sinclair 1987; Pwiti 1996b) 'found stimuli in the organisation of agriculture, the management of cattle, and the propagation of religion, as well as control of trade' (Pikirayi 2001: 21). More recently, archaeological discourses on Great Zimbabwe have been dominated by a vigorous debate on the role of cognitive and symbolic archaeological approaches to the interpretation of space and power, that has been developed by Thomas Huffman in various works dating back to the early 1980s (1981, 1984a, 1984b, 1987, 1996). The robust and intense critiques of Huffman's work delivered by Beach and others, which appeared in the pages of *Current Anthropology* in 1998 (see Beach 1998), demonstrate the continued vitality of interest and intrigue that surrounds the interpretation of Great Zimbabwe's past today. The 'cavalcade' has moved far beyond the fascination with the origins that characterised the 'Zimbabwe Controversy', and continues to grow with Pikirayi's *The Zimbabwe Culture* (2001) being the most recent addition to the archaeological literature on the site, alongside an emerging academic discourse on heritage management and community participation (Ndoro 1996; Pwiti 1996; Pwiti and Muvenge 1996; Pwiti and Ndoro 1999; Ndoro and Pwiti 2001; Sinamai 2003).

Since Paver's work was published in 1950, the overt political use (or abuse) of the past has become widely recognised as a defining feature of the 'Zimbabwe Controversy' and is the subject of much recent literature (Garlake 1983; Hall 1984, 1995; Kuklick 1991; Pikirayi 2001: 1–24). Kuklick's paper (1991) in particular, highlighted how Rhodesian ideas about Great Zimbabwe's ancient, foreign origins both inspired and provided historical and moral legitimacy for the colonisation of what became Rhodesia in 1890.

> There is no novelty in the observation that history often serves to rationalise national purpose, and that archaeology is a form of historical inquiry. But the history espoused by officialdom conventionally serves to sanction present practices by arguing that they have always been thus. Colonial ventures are *ruptures* of what has always, or at least previously, been. In this context, *archaeology has been a vehicle for explaining away the obvious*, for transforming a decisive break with the past into an inconsequential moment, and it can do this because its esoteric practices uncover a past invisible to the naïve observer. Absent approval from the deity (which, of course, many have claimed) *colonialists may not be able to find a better means to justify actions illegal by any people's customary standards*. (my emphasis, Kuklick 1991: 165)

From the 1960s onwards there had been a 'the new revisionism' (Pikirayi 2001: 23) which saw new versions of the exotic/foreign origins theories (Bruwer 1965; Gayre 1972; Hromnik 1981; Mallows 1986; Parfit 1992) emerge in direct confrontation to African nationalist use of Great Zimbabwe as an example of past African achievement. This 'new revisionism' led to direct government censorship of guidebooks at Great Zimbabwe during 1970s, which in turn prompted the archaeologists Garlake and Summers to leave Rhodesia (Frederikse 1982: 10–11). Thus the 'Zimbabwe Controversy' was revived and even more overtly politicised than it had previously been. As a result, today Great Zimbabwe is as frequently cited as an example of the politics of the past, and the 'genre of colonial archaeology' (Kuklick 1991: 162), as it ever was in the early part of twentieth century as an example of ancient 'white' civilisation, or in the 1960s and 1970s as an example of past African achievement.

Much emphasis has been laid upon the extremes of absurdity achieved by the early Rhodesian 'myths' concerning Great Zimbabwe, exemplified perhaps by Carl Mauch's early comparison of the site with Solomon's temple in 1872, and Bent and Hall's references to Arab or Semitic builders. One consequence of this emphasis upon such excesses of the imagination, is that the 'Zimbabwe Controversy' has often been represented in terms of a polarisation between the myths of 'fantasy' of Rhodesian apologists seeking historical and moral legitimacy for colonisation, and a 'neutral', 'scientific' quest for 'truth' about the past, represented by modern archaeology. Indeed it is startling how often comparisons of 'amateurs' and 'professionals', 'fiction' and 'fact' feature in accounts of the 'Zimbabwe Controversy'. This is clearly apparent in Paver's description of the controversy as 'a cavalcade of fact and fantasy'. Another good example is Summers's attempt (1963b: 37–40) to produce a neutral assessment of one of the most heated confrontations of the 'Zimbabwe Controversy' – that which occurred between Richard Hall and David Randal MacIver after the latter's excavations in 1905. Given that there was still a great deal of hostility towards archaeology from the Rhodesian public at that time (Garlake 1983: 4), it is not surprising that Summers sought to stress the 'clash of personalities' and 'personal antipathies' through which 'the controversy was nourished' (Summers 1963b: 37). But in attempting to restore Hall's reputation, he also managed to succinctly reinforce the fact/fiction distinction. Summers described what was found when he excavated in the Great Enclosure, deliberately including trenches dug by Hall and Randal-MacIver, which had featured strongly in the controversy.

> When we laid our trench system in the 'Temple' in 1958, it was deliberately sited to pass through the presumed positions of both Hall's 1903 and MacIver's 1905 trenches. When the trench was completed, it was possible to see the line of MacIver's small trench quite clearly and to follow Hall's much larger one. It was then absolutely clear that at the south-west end was the 'cement mass' with its underlying 'sand and ashes' just as MacIver described them. At the north-east end quite as clearly, there stood out the succession of daga and cement floors very nearly as Hall had drawn them in his section – they were not quite the same, for Hall had removed some of the upper layers entirely.
>
> MacIver and Hall were both right, but they had been looking at different places standing on the same spot but back to back!
>
> Elsewhere, Hall was proved to be equally reliable where *facts* were concerned and although his excavations technique was clumsy and untidy it was quite unfair of MacIver to have treated him with the disdain that he did. It is therefore only just that we should accord Hall his rightful place as an *observer* even if we disagree fundamentally with him as *interpreter*. (Summers 1963b: 39–40)

Two paragraphs below he concluded his chapter by stating,

> Let us, from now on, turn our backs on nineteenth-century antiquarianism and look forward to twentieth-century science to assist in learning what we can about Zimbabwe, whose 'mystery' now becomes knowledge to be revealed to those who will take the trouble to try to understand. (Summers 1963a: 40)

Summers therefore reinforced the distinction between 'observable fact' and its interpretation, and despite attempting to restore Hall's reputation as regards the former, he clearly fell on the side of the 'professionals', validating the claim of 'science' as a means of 'uncovering' the past. And these common references to 'science' and 'professionalism' are the means by which archaeologists at that time, and ever since, have authorised their own narratives over Great Zimbabwe's past, and marginalised those of others. Thus, as Garlake (1983: 4) has argued, during Summers's period as Curator of Archaeology of the National Museum from 1947, 'the alienation of archaeology in Zimbabwe from the rest of society, both white and black, was ... deliberately and considerably increased'. Through creating an opposition between 'professionals' and 'amateurs', 'fact' and 'fiction', the debate about the 'Zimbabwe Controversy' has led to the reification of the academic discipline of archaeology (and in a looser sense, that of 'history' too) as the only 'authority' capable of constructing 'valid' or 'true' historical narratives. If Rhodesian 'foreign origin' theories for Great Zimbabwe represent the *overt* political use of the past 'to justify actions illegal by any peoples' customary standards' (Kuklick 1991: 165), then this may be a much more *subtle* form of the politics of the past.

Elsewhere (Fontein 2000) I have discussed the politics of the past in relation to UNESCO and the World Heritage 'system'. I used the work of Kevin Walsh (1992), and Giddens (1990) to suggest that as a result of the European Enlightenment, a linear and progressive perspective of time and the past became embodied in the development of the modern disciplines of archaeology and history. As 'disembedding mechanisms' (Giddens 1990) these disciplines appropriate knowledge of the past, through claims to 'professionalism' and 'objectivity', and in effect, marginalise other ways of perceiving the past, such as through a sense of place, the body and memory. Using a concept developed by the French theorist Pierre Bourdieu (1991), the claim to 'professional status' is therefore a form of 'symbolic violence' through which different ways of constructing, and dealing with the past and landscape are marginalised by these academic disciplines.

In a colonial and racist context, where the very existence of an African past was being denied, it seems very justified that trained archaeologists should use claims to 'science' and 'objectivity' to debase the extremely biased constructions of the past of Rhodesian apologists. This is thrown into sharp relief however, if we consider how in a postcolonial context, such claims to 'objectivity' and 'professionalism' may be part of a process by which different ways of conceiving of the past, place, landscape, and importantly what to do with them, continue to be sidelined and ignored. This is especially the case when it involves a heritage site for which there are many different meanings and attachments, and competing claims of ownership.

This then, is the focus of this book. Based on recent fieldwork in Zimbabwe I explore how both knowledge of the past at Great Zimbabwe, and the management of the site itself, has continued to be dominated by these 'disembedding mechanisms' in postcolonial Zimbabwe. The continued alienation of Great Zimbabwe from local communities has resulted in both a *silence of unheard voices and untold stories* – the un-represented pasts of local communities – and the *silence of anger* – the alienation, and indeed *desecration* of Great Zimbabwe. Whilst the latter is the focus of the second half of the book, it is to the former that I now turn.

HISTORICAL AND ETHNOGRAPHIC REFUSAL AT GREAT ZIMBABWE

There are surprisingly few ethnographic and historical accounts of the local communities that surround Great Zimbabwe, and their attachments to it. Perhaps the most comprehensive and original ethnographic and historical description remains an unpublished but frequently cited PhD thesis by Mtetwa (1976). Apart from Mtetwa's work, Sister Mary Aquina's paper 'The Tribes in the Victoria Reserve' (1965) and the 'Delineation Report' (1965) by the Native Commissioner for Victoria District (now Masvingo District) maybe the only other comprehensive pieces of research into the ethnography and history of the communities that surround Great Zimbabwe.

It is also interesting to note that in terms of ethnographic representation, the writings, scribblings and journals of early explorers of the late nineteenth century, like Carl Mauch (Burke 1969) and William Posselt (1924), remain the main sources for archaeologists and historians writing on what Great Zimbabwe might have meant for locals in the nineteenth century. A further irony is that whilst professional archaeologists like Randal-MacIver (1906) and Caton-Thompson (1931) are often seen as having been the early champions for the African cause during the years of the Zimbabwe Controversy, the African voice is almost completely silent in their works on Great Zimbabwe based as they were on archaeological excavations. Instead, a much larger ethnographic presence is felt in the journalistic, antiquarian writings of the foreign-origin theorists, such as Bent (1896), and Hall (1905). Their books contained chapters outlining 'Camp life and work at Zimbabwe' (Bent 1896: 30), 'A day at Havilah Camp, Zimbabwe' and 'Zimbabwe Natives – Natives and Ruins – Natives (general)' (Hall 1905: 31,80), which give a detailed, if obviously biased, sense of the ethnographic context within which their excavations and explorations occurred. Compared to the dry archaeological excavation reports written by the 'professionals', these 'amateur' accounts are extremely informative in terms of the historical ethnography of the surrounding communities. And they firmly placed Great Zimbabwe in local historical landscapes.

Many of these early accounts (Bent 1896; Hall 1905: 84–5; Posselt 1924; Mauch 1972 (Burke1969)) describe the tensions and conflict that existed between the Mugabe, Nemanwa and Charumbira clans over control of Great Zimbabwe and surrounding land; one traveller even witnessed a battle in 1891.

> I was out at Zimbaye Ruins ten days ago. I had intended going there for a pleasure trip, but went there on duty. The party consisted of the Hon. Maurice Gifford, Captain Turner, Lieut. Chaplin, Dr. Brett Brabant, Nobbie Clark, Peter Forest and myself. There was a row between 500 natives attacking and 70 natives holding the little stone kopjes opposite Zimbabwe, on the southside. We were supposed to stop the natives

fighting, but our skipper and Gifford thought it would be rather good fun to see them scrapping. We enjoyed the fight immensely and watched it sitting on the walls of the Ruins. The attacking 500 lost 7 men and the fight lasted 2 hours. The Mashonaland natives are always fighting and quarrelling amongst themselves. I don't think we shall ever have much to fear from the Mashonas, they are too afraid of the white man. (Letter from R.C. Smith to his father, July 24 1891, National Archives of Zimbabwe, SA12/1/15)

In the early days of European exploration of Great Zimbabwe, 'local knowledge' of Great Zimbabwe was to some extent, being actively sought. Hall gathered his knowledge through a

series of conferences of the oldest natives authorities held at Zimbabwe during 1902 and 1903, at which Mr Alfred Drew, Native Commissioner, the Rev, A.A. Louw, Dutch Reformed Mission near Zimbabwe, and Dr.Helm, Medical Missionary, and other admitted authorities on native language and customs, have taken part, [which] will explain the local occupations for almost if not more than one hundred and fifty years. (Hall 1905: 81)

The recent occupation by different local clans and groups of Great Zimbabwe was not being denied by these theorists despite their determination to attribute its construction to foreign builders. The location of Mugabe's homestead on the hill (figures 2.3 and 2.4) is described by Bent, Posselt and Mauch, though by Hall's time this had moved to the north west, on a nearby 'low granite knoll called Pasosa' (Hall 1905: 10). Both Bent and Hall also dug and excavated recent graves found on the Hill.

In our work at Zimbabwe we unwittingly opened several of their graves amongst the old ruins. The corpse had been laid out on a reed mat – the mat probably, on which he slept during life. His bowl and his calabash were placed beside him. One of these graves had been made in a narrow passage in the ancient walls on the fortress. We were rather horrified at what we had done, especially as a man came to complain, and said it was the grave of his brother, who had died a year before; so we filled up the aperture and resisted the temptation to proceed with our excavations at that spot. After that the old chief Ikomo, [Chief Mugabe's brother] whenever we started a fresh place, came and told us a relation of his was buried there. This occurred so often, we began to suspect, and eventually proved, a fraud. So we set sentiment aside and took scientific research as our motto for the future. (Bent 1896: 79)

Richard Hall wrote of 'about fifty Makalanga graves' that he found on the 'Acropolis' in 1902–03, 'the remains in a score of instances were removed' (Hall 1905: 95). Apart from giving us a clear picture of the insensitive abandon with which early explorers conducted their digging, these accounts also show how in their efforts to remove 'kaffir debris' in order to get to the occupation levels of the 'ancients', these explorers both gained knowledge of Great Zimbabwe's position in local 'history-scapes', and simultaneously destroyed crucial features of those very 'history-scapes'. There seems to have been very little value placed on this knowledge, except to show that these Africans had only arrived recently, and knew nothing of Great Zimbabwe's actual construction. The fact that after these early accounts, and with the 'triumph of professionalism' (Kuklick 1991: 150), such ethnographic descriptions are no longer forthcoming, reveals a deep continuity between the different, opposed sides of the Zimbabwe Controversy. Whether one thought Great Zimbabwe was built by some foreign civilisation or by Africans, was irrelevant. All agreed that its significance as part of local historical and religious landscapes was not important because it did not date back to Great Zimbabwe's construction and 'original' occupation, whether it was placed in 'ancient' times or a medieval period.

Despite the 'triumph of professionalism', the change in focus from origins to economics, religion and symbolism in studies on Great Zimbabwe, not much has really changed in this respect. The bias against Great Zimbabwe's position in local 'history-scapes' remains, the academic lens still focuses on a time when Great Zimbabwe was built and occupied 'originally'. Even in the 1960s – when 'oral history' and the possibility of its use in the construction of 'objective' history was the new fashionable idea, and began to be accepted by archaeologists as a 'complimentary approach' (Summers 1963b) to the archaeology of Great Zimbabwe – this bias against the validity of 'local' oral history remained. For example, Abraham's work (1966) was largely based on oral traditions, but he used Rozvi informants, located in Mt Darwin (1966: 37), not the oral traditions of the communities surrounding Great Zimbabwe. Where local ethnography and history was used it was not 'freshly' collected so much as referenced to the writings of the early explorers, such as Carl Mauch (Burke 1969: 215) and Hall (1905: 93–94), both of whom described ceremonies at Great Zimbabwe that were interpreted by Abraham (1966), Summers (1963a) and others (Summers and Blake Thompson 1956; Robinson, Summers and Whitty 1961) in support of the idea that Great Zimbabwe used to be the centre of a major cult system either to *Mwari*, Chaminuka or both.[6] Local informants and their oral traditions were seen as inapplicable.

This idea of Great Zimbabwe as an *Mwari* or Chaminuka cult centre did not last long in academic archaeological and historical discourse, and by the 1970s two influential papers (Beach 1973b; Mtetwa 1976) from the History Department at the University of Rhodesia refuted the idea suggesting that informants' oral information had been accepted far too liberally. Abraham's use of Rozvi oral traditions was particularly criticised; the idea that Great Zimbabwe was once an *Mwari* Cult centre is now largely rejected in academic circles. After the initial enthusiasm for oral history, the continued quest for 'objectivity' led to a massive rethink of, and retreat from, its usefulness for 'professionals' investigating the past at Great Zimbabwe. Oral traditions are still considered problematic sources for archaeologists and historians trying to re-construct what Great Zimbabwe's builders and 'original' occupiers were up to, mainly because of the time span involved (Beach 1998: 48–9). The current local communities are out because they were not there then. The bias for a particular time period in history remains, and for historians and archaeologists Great Zimbabwe remains 'a primarily archaeological problem' (Garlake 1973: 76; Beach 1998: 49).

Even much more recent efforts to focus on how Great Zimbabwe was positioned as a sacred site in the local landscape during the nineteenth century (Matenga 1998), and its contemporary position today (Ndoro 2001), still rely very heavily on the accounts of late nineteenth century explorers, and early twentieth century antiquarians. In his comprehensive and widely accessible book on the Zimbabwe Birds, the former NMMZ (National Museums and Monuments of Zimbabwe) Regional Director at Great Zimbabwe, Edward Matenga (1998) makes a considerable effort to move beyond a mere account of the how the surrounding clans are later migrants who have nothing to add to the history of Great Zimbabwe. He shows that Great Zimbabwe did and does feature as an important sacred site in local 'history-scapes'. But there is very little evidence of fresh historical or ethnographic research into what these local 'history-scapes' actually involve.

This lack of representation of local histories is not just confined to the literature. Despite claims made at the re-opening of the site museum in 2000 that the new exhibition had been 'born out of a process of extensive consultation between staff at Great Zimbabwe and the local communities'[7], the site museum contains very little about local perspectives on Great Zimbabwe's past, or their continuing local attachments and claims to it. The new display is still based on an archaeological view of Great Zimbabwe's past. The monument is presented, as Webber Ndoro put it, as 'a bygone era' (2001: 112). But even Ndoro's own book (2001), which outlines some of the local concerns about the management of the site, contains disappointingly little fresh ethnographic or historical material from these local communities. Therefore, both in the recent literature and new exhibitions in the site museum there remains a problem of 'ethnographic refusal' (Ortner 1995) which results in repeated references to what early explorers like Carl Mauch saw and heard, and a heavy reliance on archaeological views on the past, rather than a contemporary ethnographic and historical presence of local voices. It is these local voices, pronouncing their variety of constructions of Great Zimbabwe's place in their historical landscapes, that I amplify now to fill the apparent silence.

Notes

1. As he wrote these lines Mauch was residing with another German, Adam Render, an elephant hunter and trader who had taken up residence in the area some years ago. After having been robbed, Mauch was 'rescued' by Render, a few days before the above account, from the kindly but enforced hospitality of a local 'chief', named as Mapansule in the journal (C. Mauch, 29 August 1871; Burke 1969: 136). It is likely that Render was, by this stage, already aware of the ruins, though it is not clear from Mauch's journal.

2. As Burke explained (1969: 4) 'it is not clear how Merensky developed these ideas, possibly from reading Portuguese sources which borrowed them from the Arabs, but he was assiduous in cultivating them. In 1862 he, in company with Rev. Nachtigal, tried to reach the ruins, of which he had heard from Sekukuni, chief of the Baedi, but his expedition was defeated by an outbreak of smallpox'.

3. The idea that what became 'Rhodesia', was once the mythical land of Ophir from where Solomon's gold originated was widely debated in the press in London and South Africa at the turn of the century. A unique collection of press cuttings of this debate was gathered by the British South Africa Company (BSACo), and exists today at the National Archives of Zimbabwe (S142/13/5), which illustrates how the Company was a very interested party in this public discourse.

4. In Carl Mauch's logic, this 'cedarwood' demonstrated that Great Zimbabwe had been ruled by the Queen of Sheba who had imported the wood from Lebanon. In fact, it is likely that the wood was *Tambootie*, or African sandalwood (*Spirotachys Africana*), an indigenous hardwood which fits Mauch's description very well, and has subsequently been found elsewhere in the walls of the Great Enclosure (Burke 1969: 190; Garlake 1973: 64; Matenga 1998: 22).

5. This Mr. R.M.W. Swan had accompanied Theodore Bent on his expedition in 1891, and carried out surveys of the ruins. He built up a belief that major sections

of the ruins had been built along sophisticated mathematical and astronomical concepts, and oriented along sighting lines for certain stars and solstices. According to him, this meant that a date could be set for Great Zimbabwe, as the bearings of the stars and solstices vary along with the Earth's ecliptic. Thus he came up with a date for the ruins on the Lundi river, of 2000 BC. A geologist called Henry Schlicter applied the dating method to Great Zimbabwe and arrived at a date of 1100 BC. Swan's theoretical meanderings were severely undermined when his basic measurements were challenged as wrong or taken from arbitrary points (see Garlake 1973: 68–69).

6. *'Mwari'* refers to the high Shona divinity who is above the level of even the most senior ancestors, and is sometimes equated with the Christian God. *Mwari* is also known by other names such *Musikavanhu*, 'creator of people', and is the centre of a regional cult system for which several major shrines exist in the Matopos hills (Daneel 1970, 1998; Werbner 1989; Ranger 1999; Nyathi 2003). Chaminuka is the name of a well known Shona ancestor who is popularly credited with having predicted the coming of the whites and the downfall of Lobengula. Ranger has argued this spirit's profile was greatly raised during the liberation struggle through the work of three writers (Gelfand 1959; Abraham 1966; Berliner 1978) whose main informant was the medium of Chaminuka, Muchetera.

7. Mr. Makonese, Chairman of the Local Committee of the Board of Trustees, NMMZ. Speech at the opening of the Great Zimbabwe Site Museum, 20 September 2000, NMMZ file J20.

CHAPTER 2

GREAT ZIMBABWE IN LOCAL 'HISTORY-SCAPES'

It is not my aim here to simply fill the silence of historical and ethnographic knowledge of Great Zimbabwe or its place in the historical landscapes of local communities. While this chapter is a step in that direction, my larger purpose is to explore how the past, place and landscape are sites of contestation between the clans of Nemanwa, Charumbira and Mugabe, and among different actors within them, as they seek authority and legitimacy in the present. Apart from looking at the differences in the pasts constructed by these clans, this chapter also investigates how these versions of the past and place are overlapping, interacting and borrowing from each other. They are located within a shared discursive as well as physical landscape – hence the term 'history-scape'. Furthermore, these 'history-scapes' are not isolated from wider historical discourses, sometimes individual members of these clans demonstrate their agency by relating their own clan's claims over Great Zimbabwe to the ideas of people outside of these communities, including historians, archaeologists and 'nationalists' of all sorts. This illustrates McGregor's point (2004: 1) about 'the entanglement of modern and traditional ideas' which defies any simple opposition between 'modern'/ 'European'/ 'scientific' and 'traditional'/'African'/'memory-based' ways of dealing with the past, which has been implied in the works of Kevin Walsh (1992: 11) and Pierre Nora (1989) among others, and was, as I have argued, central to the very 'disembedding' process by which archaeologists have authorised their own narratives. The way in which popular 'Rozvi myths' (Beach 1980) of Great Zimbabwe's origins have been put to use by local individuals and clans, which is discussed in the latter part of the chapter, demonstrates that oral traditions and memories are not so much isolated fragments of a past waiting to be reconstructed by outside researchers, but may be better viewed as cultural resources that are constantly being borrowed, utilised and engaged with politically in a constantly changing field of wider discourses.

THE NEMANWA/MUGABE WARS OF NINETEENTH CENTURY

The shared nature of the discursive and physical landscape within which local clans construct their versions of the past is most convincingly captured in the *mutupo* and *chidawo* (totem and praise name) combinations of the Nemanwa, Charumbira and Mugabe clans, who are the immediate groups surrounding Great Zimbabwe, and most intricately involved with the site. These *mitupo/zvidawo* (plural of *mutupo* and *chidawo*) are widely accepted, and used in formal greetings both within and between clans, and in this sense they are fairly concrete markers to the past that are almost beyond contestation. The *mutupo/chidawo* of the Nemanwa clan is *Shumba/Muguriri*; the totem being 'lion', and the praise-name meaning 'one who cuts'. The Mugabe clan's *mutupo/chidawo* is *Moyo/Matake*, being 'heart', and meaning 'one who crushes'; they are in fact a 'subclan' of the Duma group, all of whom share the common *mutupo* – *Moyo*. The Charumbira clan's *mutupo/chidawo* is *Shumba/Sipambi*, being 'lion' and meaning 'people who did not acquire land by force but received what they deserved' (Mtetwa 1976: 186). The origin of these *mitupo/zvidawo* relates to a story about the Nemanwa/Mugabe wars of the nineteenth century which was repeated to me in various versions by people from all of these three clans. As Chief Mugabe explained,

> *Chief Mugabe: The Manwa people and the Duma people started to fight again. As they were fighting, the Duma people started moving looking for the Manwas within the Manwa territory. The Manwa people also moving from their own territory coming into the territory of the Duma, So they were moving in opposite directions, some going one side, and others going where the others were coming from. The Duma people had their own policy that whenever we got hold of any Manwa people, we would crush their testicles.*
>
> *J. F.: So 'Matake'?*
>
> *Chief Mugabe: Yes, yes, 'Matake'. That is where the name that we are called now, 'Matake', comes from. The Manwa people also got to the Duma territory and found the chief seated at home, with just a few elders. And the soldiers that were there, were defeated by the Manwa soldiers. Then all the soldiers ran away and just the chief remained. When this chief remained alone, they got hold of him and cut off his hands. That is where the name 'Muguriri' comes from. So they are called the Muguriri people. So that is the story of the Zimbabwe Ruins.* (Interview with Chief Mugabe, 22/11/2000)

This is the most widely accepted version, which was included in the Delineation Report for Victoria District in 1965 (Delineation Reports, Victoria District 1965, National Archives of Zimbabwe). But other versions that I heard turned it around. A good example is the version offered by a respected elder of the Charumbira clan, which also describes how Charumbira fits into the scheme, and where their *chidawo*, 'Sipambi', comes from:

> *Nemanwa was just a small clan, but they had a strong relationship with Charumbira. When Nemanwa fought against Mugabe, Nemanwa came to Charumbira and said 'Charumbira I'm now being killed by Mugabe. Charumbira went there and fought against Mugabe and killed Mugabe's people. They actually [laughing] … using a stone … , to crush their testicles!!. So they crushed the testicles of the Mugabe's, and the Mugabes cut the hands of the Nemanwas. And then now today Nemanwa has the chidawo 'Muguriri' meaning he was cut, his hands. Mugabe is called 'Matake', because his testicles were crushed. So that is when the relationship began* [between Nemanwa and Charumbira], *when he said, 'you assisted me in the fight, and here is the country'. So from that day they started saying … we now have a relationship, you managed to help me during my war so we are going to stay together, and we won't separate … .* (Interview with VaMututuvari, 4/11/2000)

This reversal – where it is claimed Nemanwa and Charumbira did the crushing together, rather than Nemanwa having received it from Mugabe – illustrates how despite relatively 'concrete' pointers into the past (that is, the respective *zvidawo* of *Muguriri* and *Matake* for Nemanwa and Mugabe) there is still plenty of room for variations within a single framework.[1]

This story about the origins of each clan's praise name illustrates how the Nemanwa/Mugabe wars of the nineteenth century are the backdrop to the history-scapes of all three clans, and their continuing disputes over land, titles and the custodianship of Great Zimbabwe today.[2] These wars continued for a brief period after Rhodesian occupation of the country in 1890 until they were finally brought to an enforced end by an explorer called Willoughby in 1892 (Willoughby 1893: 6;

Mtetwa 1976: 195). Not surprisingly European involvement is often remembered in contradictory ways by members of the Nemanwa and Mugabe clans. Despite common accusations by both Nemanwa and Mugabe that it was the other who attempted to utilise the whites for their purposes, the most detailed analysis that exists of the Mugabe/Nemanwa wars of the nineteenth century (Mtetwa 1976: 192–8) very clearly shows that both Nemanwa and Mugabe attempted to use Europeans as military allies.

The wars came to an end when Willoughby rather cynically used it to his advantage to obtain labour for his excavations (Willoughby 1893: 6). He organised a 'peace conference', with the Native Commissioner Brabant present, and drew boundaries between the Nemanwa and Mugabe territories. The long-term consequences of this imposed peace was the alienation of Great Zimbabwe and surrounding lands from both the Mugabe and the Nemanwa clans. But as Mtetwa (1976: 196) has made clear, Nemanwa was left worst off, losing not only Great Zimbabwe, but all its territory, and even the title of Chief. While the Mugabe clan did lose control of Great Zimbabwe, and the *mapa* (grave sites) of their ancestors, the fact that they had been in possession of the site when the Europeans arrived, meant that their claims to it were taken much more seriously (Aquina 1965: 9, 11; Mtetwa 1976: 196). They lost some of their lands to European and mission farms, but retained a territory of their own, and the title of Chief. In contrast to both Mugabe and Nemanwa, Charumbira 'steadly enhanced both his political and military power and eventually emerged the victor' (Mtetwa 1976: 196). What remnants of the Nemanwa clan that remained in the area were living under the authority, and on the land, of Chief Charumbira, who they themselves (according to not only their own history-scapes, but also those of Charumbira and Mugabe) welcomed into the area in the first place.

Since the 1960s conditions have improved greatly for the Nemanwa clan, partly, it could be argued, through the efforts of writers such as Aquina and Mtetwa who questioned the uncritical acceptance of Mugabe's claims over Great Zimbabwe and the surrounding land. While they still remain with the title of Headman under Chief Charumbira, they have regained some of their own territory. In the 1970s Morgenster mission gave back a piece of land for Nemanwa resettlement below Bingura 'mountain', where the Growth Point stands now. After independence, the Growth Point was built and, significantly, named Nemanwa Growth Point. Mzero Farm was likewise returned by the mission, partly for Mugabe settlement, but mainly for Nemanwa people displaced by the building of the Growth Point. Two schools in the area, Nemanwa Primary School and Chirichoga Secondary School, illustrate how the Nemanwa clan's claims to the land have been re-embodied by the landscape – Chirichoga is the name of one of Nemanwa's founding ancestors. The Nemanwa claim to have been first in the area is now widely accepted, even if their claims to have 'germinated' at Great Zimbabwe are treated with scepticism, and they receive far more recognition as a claimant to the custodianship of Great Zimbabwe than they used to.[3]

The early 1980s saw the dispute between Mugabe and Nemanwa people re-emerge and reach new heights, as the former, especially Mugabe people of the Haruzvivishe house, contested the return of Mzero Farm to Nemanwa. They claimed it for themselves and occupied sections of Mzero near Great Zimbabwe not actually earmarked for resettlement. Eventually they were evicted in the later 1980s, but returned in 2000, taking advantage of the beleaguered ZANU PF government's land reform policies (Fontein 2005).

As far as Great Zimbabwe is concerned, with independence in 1980 and the adoption of the name Zimbabwe for the new state (as well as Great Zimbabwe's elevation to the status of World Heritage Site in 1986), the contest for Great Zimbabwe has been rekindled, despite the control of the site by National Museums and Monuments of Zimbabwe (NMMZ). If the Nemanwa/Mugabe wars of the nineteenth century were not specifically focused on Great Zimbabwe, but rather concerned the possession and exploitation of the wider landscape, as Mtetwa (1976: 188) has argued, then the opposite could not be more true today. The issue of the custodianship of the ruins is a particularly sensitive and emotive one, among a plethora of wider disputes over land and authority. Both the Nemanwa and Mugabe clans continue to stake their claims to its custodianship. Even the current Chief Charumbira now stakes a claim to Great Zimbabwe on the basis of his superiority as Chief over Nemanwa. Central to all these claims over Great Zimbabwe, as well as the wider disputes and contestations that occur both within and between these clans, lie different versions of the past. Despite the fact that these 'history-scapes' are 'shared', and hang on particular markers that are hard to avoid, they still allow room for contestation and manipulation. What is obvious then is that no clear historical narratives are produced, rather there is a multiplicity of historical discourse which is specifically concerned with justifying, or refuting, competing claims in the present. It is to these claims that we should now turn.

NEMANWA – *VAMERI*

While the value of local claims to Great Zimbabwe and its past may have been overlooked or dismissed by archaeologists on the basis that all the local communities are 'latecomers' to the site – that is they arrived in the area after Great Zimbabwe's demise – it is significant to note that from the perspective of the local communities, these archaeologists, and specifically NMMZ who actually manage the site, are themselves the 'latecomers' to Great Zimbabwe. Furthermore, in the disputes between different local clans, the claim to being the first people in the area is hotly contested. It is a very powerful way of asserting ownership over Great Zimbabwe, and the wider landscape. And it is a claim most frequently made by Nemanwa elders (see figure 2.1), who say that they are the *Vameri*; that is, they *'germinated'* in the area, as opposed to having arrived there from somewhere else.[4] As Chief Nemanwa put it,

Nemanwa does not know where he came from. We cannot say where we came from. We have been here for a very long time. Long back, the mountain of Great Zimbabwe used to speak, but now it is Matopos that speaks. If we go to Matopos, you'll find that they refer to the Nemanwa's as the 'Mumeri weZimbabwe', meaning to say the original people of Great Zimbabwe. All these other chiefs came here after Nemanwa, but the first one to come here was Chief Murinye, followed by Mugabe his younger brother, followed by Shumba, followed by Mapanzure and Charumbira. All these people came here hunting and looking for places to stay and as they did so they married Nemanwa's daughters, and that intermarriage led to them staying around here. (Interview with Chief Nemanwa, 15/1/2001)

This account firmly places the Nemanwa as the original occupants of the area, with other surrounding clans as later arrivals, who received their land from Nemanwa. A similar story about germination was repeated to me by Aiden Nemanwa, who is the most prominent Nemanwa spokesman, in which he actually located a particular spot on the landscape, a well or spring, within the area of Great Zimbabwe, from where an original ancestor, 'Chisikana' (meaning 'little girl') emerged.

J.F.: How long have Nemanwa people been here in this area?

Aiden Nemanwa: I have been trying to trace the history myself. No one has told me where they came from. The spirits of this area told the people that we germinated here, on this land, therefore we are custodians of this land.

We were told that there was an *Ambuya* [literally 'grandmother', but used as a word of respect for women past a certain age, of higher status, and for all female spirit mediums] *and that woman came from a well* [or spring] *and she gave birth to the forefathers of the Manwa people. That is the same well that was opened recently in July at Great Zimbabwe.*

That that Ambuya emerged from that well, shows that there is no history of Nemanwa people coming from elsewhere.

This Ambuya was just a small girl who came from the well when the Nemanwa people found her and she told them she came from the well. The name of the Ambuya was 'Chisikana' ['little girl']. *The name of the pool* [or spring or well] *is also 'Chisikana', as well as the river that comes from it.* (Notes of interview with Aiden 'Teacher' Nemanwa, 21/10/2000)

That the spring and the river are known as 'Chisikana' illustrates how place can embody the past and vice versa. As Chief Nemanwa described:

Take for example if a person climbs up a tree, then he falls down, then this tree will get the name of the person who has fallen from that tree, that's exactly what happened to Chisikana. There was a girl who was taken by an Njuzu [a form of water spirit often associated with healing]. *She had been taken by an Njuzu, and then later on she came out from the well, and that well obtained the name Chisikana from that girl who came out of that well.* (Group Discussion with Nemanwa elders, 18/7/2001)

The story of this little girl Chisikana, who emerged from the spring, is a central feature of a Nemanwa history-scape that closely links the past to the land; it allows the possibility of locating physically, a site of origin or 'germination' within Great Zimbabwe itself. However, the relationship between Chisikana and other known Nemanwa ancestors was often unclear. As others have also found, producing a clear genealogy of Nemanwa ancestors is very difficult.[5] This is, perhaps, not surprising given that forgetting may often be as relevant to memory and the politics of the past, as re-telling and re-inventing usually is (cf. Forty and Kuchler 1999).[6] Whilst most Nemanwa elders I spoke to traced their past back to a founding ancestor called Chirichoga who was the ruling chief when the Europeans arrived (Aquina 1965: 8), considerable confusion often arose about that ancestor's relationship with other well known ancestors including the girl Chisikana.[7]

The Chisikana story may be a recent addition or re-negotiation of the past which emerged from a 'traditional' ceremony held at the spring by NMMZ in July 2000, to which all the surrounding clans were invited.[8] That some members of the Mugabe clan also referred to this girl, Chisikana, in their 'history-scapes', illustrates how these clans share both a physical and discursive landscape, and how shared situations in the 'present' can become part of the positioned 'history-scapes' of these clans in the future.

Figure 2.1: The Nemanwa elders at Chirichoga Secondary School, after our group discussion (18/7/2001) (Author 2001). Chief Nemanwa is standing on the extreme right. Crouching in front of him is Aiden Nemanwa. VaChokoto is crouching second from the left.

This was further demonstrated by Ambuya VaZvitii when she described how she was the first *svikiro* (spirit medium) possessed at the Chisikana ceremony in July 2000, which according to her proved 'that the owners of this land are of Nemanwa' (Interview with Ambuya VaZvitii, 7 November 2000). Just as the Chisikana spring has been invested with new meaning through the way that ceremony has been remembered, so another very common Nemanwa story, about a person called Chinodziya Maranga who climbed the Conical Tower at Great Zimbabwe to 'prove' Nemanwa ownership of the site (Interview with VaChokoto, 6 November 2000), may refer to an occasion that actually happened, through which that particular point on the landscape then later gained special significance for the Nemanwa clan (see figure 2.2).

For the Nemanwa clan, the claim to have 'germinated' at Great Zimbabwe is the basis of their attempts to be recognised as its rightful custodians. For them, Great Zimbabwe is, or was, a sacred place were ceremonies used to be held that were intricately linked to the agricultural year ensuring rainfall and the fertility of the soil.

> *Nemanwa people used to gather once a month, once a season and once a year. The gatherings occurred to ask about specific things. About planting trees, cutting trees, the planting of crops, harvesting and so on. It was not just Nemanwa people who attended, but also people from surrounding areas. The Manwa people were the organisers. Their main purpose was for guidance for living their lives.* (Notes of interview with Aiden Nemanwa, 21/10/2000)

They acknowledge, therefore, that Great Zimbabwe was sacred not only for them but also for other people in the surrounding landscape. However, they see themselves as its

Figure 2.2: The Conical Tower in the Great Enclosure at Great Zimbabwe. (Author 2001)

Then these white people, they started to investigate, and try to find out who really was the owner of this place. (...) So they called the Manwa people and the VaDuma, and they gathered in the Great Enclosure. Then they said 'Whoever knows that his ancestors can allow him, must climb to the top of the Conical Tower'. There came a boy from the Manwa tribe, who was called Chinodziya Marange, and he said, 'Elders what is the problem, what is troubling you here?'. They told him the story ... "The VaDuma are now saying it is ours, and the Nemanwa people are saying it is ours and the white person said whoever knows that he can climb up that tower, then he is the owner of this place". This boy called Marange clapped and said 'Let me try elders'. And he just started to grab onto the tower, holding on, going upwards, upwards, upwards, until he reached the top of the tower. And people started ululating and playing drums. Then we wanted to see how he is going to come down. He went down, and got down properly, and then the people said, right then the story is over. But because these people who claim to know so much; the VaDuma are still saying its theirs. (Interview with VaChokoto, 6/11/2000)

original owners and custodians.[9] This role was granted to them by the *Voice* that used to speak there. As Chief Nemanwa's brother, Makwari Matambo put it, 'long ago that mountain used to speak, and that mountain ... said that that place belongs to Nemanwa' (Makwari Matambo, Group Discussion with Nemanwa elders, 18 July 2001).

The Nemanwa claim to have 'germinated' at Great Zimbabwe – to have been the original owners of the land – does not only relate to their claim to the custodianship of

Great Zimbabwe, it also relates to their concern about regaining the title of chief, rather than continue, as they do, as Headman under Charumbira (cf. Sinamai 2003).

> *At this moment, to make matters worse, those chiefs to whom Nemanwa gave land, who came after Nemanwa, they have been given a bigger title of chief, while I have the lower title of Headman. That's why our ancestors are much worried, and are angry, because of the truth that I am holding at this moment, its like saying Nemanwa doesn't have his land, the Government doesn't recognise Nemanwa's proper title.*
> (Chief Nemanwa, Group Discussion with Nemanwa elders, 18/7/2001)

The issue of titles is a particularly sore one for the Nemanwa elders. They resent being under Chief Charumbira, and feel they should be recognised as a chiefdom in their own right. Aiden Nemanwa blamed the colonial government.

> *Aiden Nemanwa: ... that Charumbira is above Nemanwa, that is the result of what the oppressive white government was doing. ... Nemanwa was given the title of headman, not considering the fact that Nemanwa are the original people living in this area, who gave the land to Charumbira and to the other chiefs, so Nemanwa should be given the title of chief.*
>
> *Our title is still there, we have to be given it, but what stops us being given it is that we are smaller in number than other surrounding chiefs, and they, the whites, took our land for their farms, but we gave the land to all the rest of the surrounding chiefs. Charumbira is mukwasha [son-in-law], Shumba is muzukuru [nephew, or sister's son], Mugabe is mukwasha, and Mapanzure ... It means we are great, because all these chiefs are under us, and we treated them well, gave them land. So at this moment we are crying to the Government, we are crying for our proper title and we are crying for our place Great Zimbabwe.*
>
> *J.F.: You said something about farms, this area was Mzero Farm?*
>
> *Chief Nemanwa: Yes, it was taken by the missionaries.*
>
> *Aiden Nemanwa: Yes, from the Dam [Mutirikwe Dam] upto Mucheke [where Masvingo town is today, about 27 miles from Great Zimbabwe], that was all Nemanwa's land.* (Group Discussion with Nemanwa elders, 18/7/2001)

Mtetwa has argued that the vast expanse of land that Nemanwa claim to have once ruled, is ridiculous (1976: 186), and indeed he may be right, but I would suggest that the reason such a vast area is claimed has to nothing to do with possible attempts by Nemanwa to reclaim that area – they are not actively involved in claiming back more land. Rather it has to do with buttressing their claim to have been the very first people in the area, to have 'germinated[10] from the landscape itself. This status of being the original owners of the land who gave wives and land to incoming clans is strongly coveted by Nemanwa because of their desire to be recognised as the rightful custodians of Great Zimbabwe, and to regain the title of chief, rather than continue as they do, as headman under Charumbira.

CHARUMBIRA – *SIPAMBI*

Despite Nemanwa grievances about their subordinate status under Charumbira, the history-scapes of Charumbira informants do emphasise a strong sense of a shared past of co-operation and affinal kinship relations between the people of the Nemanwa and Charumbira clans.

> *With Nemanwa we have never been in conflict. We started staying together and we started to marry from each other, and the people you see in Nemanwa are our cousins, to those in Charumbira, and the people in Charumbira are cousins to the people in Nemanwa. So we have never had any conflict, between us and Nemanwa.* (Interview with VaMututuvari, 4/11/2000)

Charumbira's *chidawo*, *Sipambi*, refers to the fact that they did not take the land by force, they were given it by Nemanwa, in reward for killing an elephant troubling Nemanwa people, and for helping Nemanwa defeat Mugabe in battle (Aquina 1965: 12; Mtetwa 1976: 186). Charumbira history-scapes therefore accept that Nemanwa arrived there first. Speaking about the ancestor Chainda, his medium (who takes his ancestor's name) related the story of how Charumbira entered the land and found Nemanwa there.

> *Chainda was a hunter, who came here from the Mbire-svozve area. Nemanwa was the owner of this land. Just like a hunter Chainda used to kill big game like elephants and one day he took a tusk to Chief Nemanwa. That's how the friendship began.*

> *Nemanwa was having a war with Mugabe whose chief was Chipfunhu. Then Chief Nemanwa asked 'are you going to help us?' So Chainda said, 'OK', so they planned and organised the war. They stayed in the bush with all their men, giving them mushonga [medicine/magic] that would prevent harm or injury if they were stabbed by spears or pierced by arrows. The Mugabes under Chipfunhu were defeated by the joint forces of Nemanwa and Chainda, where the present day Morgenster hospital waiting rooms are. The army ran away leaving Chief Chipfunhu hiding in a cave. He was captured, and they took him, and showed him to Chief Nemanwa … .* (Interview with VaChainda, 30/12/2000)

Apart from being granted land by Nemanwa, Chainda was also given a wife by Nemanwa creating the affinal kinship relationship that continues today. As Simon Charumbira, the current chief's uncle put it,

> *We call them fathers in law [tezvara], but there is a lot of intermarriage going on between the Nemanwa people and the Charumbira people. It is true with the Nemanwa people, some are marrying here, some are marrying here, others there. But originally Charumbira was the son-in-law [mukwasha] because he was given a daughter by Nemanwa.* (Interview with Simon Charumbira, 25/10/2000)

Because Chainda received a wife from Nemanwa, he was *mukwasha* (son-in-law) to Nemanwa who was *tezvara* (father-in-law), and Chainda's descendants are therefore *muzukuru* (sisters's children) to Nemanwa who are their *sekuru* (mother's brother). These kinship terms imply that particular types of relationship should exist between these clans.

The relationship between a woman's husband (the *mukwasha*) and the woman's father (the *tezvara*) is one of great formality and respect. The *tezvara* receives bride-wealth, *lobolla*, from the family of the husband and the *mukwasha* also is expected to perform certain tasks at family gatherings such as funerals and *bira* ceremonies to the ancestors thereby showing their subordinate status to their wives' families. But if the relationship between a man and his in-laws is characterised by formality, distance, and respect, then the *muzukuru/sekuru* relationship, between a man and his daughter's, or sister's, children[11] is the opposite. This relationship is very relaxed and informal, and a certain amount of cheekiness is tolerated by both sides, though there is still a sense of respect that is implied in the specific references to age that the terms *muzukuru/sekuru* embody.

If we relate this to the shared past constructed by Nemanwa and Charumbira informants, we can see that there is some value in considering whether the relationship between these groups is considered as a *muzukuru/sekuru* or a *mukwasha/tezvara* relationship. This is important given Nemanwa's title of Headman under Charumbira, which is the cause of great dissatisfaction among Nemanwa elders. While some Charumbira informants accepted that Chainda was *mukwasha* to Nemanwa, the relationship between their descendants is most frequently referred to today using the terms *muzukuru* and *sekuru* for Charumbira and Nemanwa respectively. In fact, some people I spoke to ridiculed Nemanwa for being 'ruled by their *muzukurus*'.

There is, therefore, an historical anomaly over the fact that the Charumbira clan holds the title of chief over Nemanwa. During our interview Chief Charumbira referred to his clan's nineteenth century military power, as demonstrated by their defence of Nemanwa in the war against Mugabe, to assert and justify their present supremacy over their *sekurus*, the Nemanwa clan (Interview with Chief Charumbira 13/8/01). But like Aiden Nemanwa, as well as other Charumbira and Nemanwa informants, he also explained Charumbira dominance over Nemanwa in terms of the interference of the colonial government. The chief's uncle, Simon Charumbira, put it as follows:

During that time they were fighting wars and who grabbed more land was more powerful at that time, so when the white man came, they could not say that Nemanwa was a more powerful chief than Charumbira because they found that Charumbira had a vast land whereas Nemanwa had just a small piece of land. (Interview with Simon Charumbira, 25/10/2000)

Such accounts recognise the role that Europeans played in ranking Charumbira as chief above headman Nemanwa.[12] Similarly, some people explained Charumbira superiority in terms of their 'better education' and greater sophistication than the Nemanwa chiefs, whilst others, most often members of other clans, not surprisingly, pointed towards collusion with the Rhodesian authorities.[13] In a similar vein, I was, of course, never told by any Charumbira person (and, interestingly, very rarely by any one else either) that for a brief period between 1950 and 1964 the Charumbira chieftainship had itself been reduced to a headmanship under Chief Shumba Chekai, a fact that is clear from the Delineation Reports (1965: 75).[14]

Colonial interference is also often used to explain the cause of an internal contest which, above all else, dominates Charumbira history-scapes today – the long running and continuing dispute over the right to the chieftainship itself. Chieftainship succession is often very contested, and can cause deep divisions within clans, but in Masvingo District the succession disputes that have plagued the Charumbira clan are without comparison. This is evidenced both by the historical depth of this dispute – which pre-dates the appointment of the current chief's grandfather, Mazha, by the Rhodesian Government in 1964 and the frequent articles on the subject that appeared in the local press during my period of fieldwork in the area.[15] Unfortunately I do not have space here to do any justice to the intricacies and complexities of this fascinating dispute. What I will stress however, is that this debate is grounded in competing versions of both 'tradition' – particularly the rules of succession – and the past – specifically the genealogy of the ancestors.[16] Both are extremely malleable and open to manipulation by different claimants to the chieftainship. For those 'on the losing side', so to speak, there is a sense that history itself is being cheated. This was very apparent in the words of one angry claimant, who referred back to the 1960s, suggesting that that was when things became distorted.

The thing should have been done properly, but it was a gradually planned thing to try to divert the clan into something else, that is not history.

If you go back and try and find out the history, you find that it seems as if it starts from 1959 or 1960. What genealogy is that? Any reasonable human being can see that this is false. ... And yet the history of the clan started long back, a hundred years back! (Interview with B.B. Charumbira, 19/7/2001)

The intense, and long running, internal contest over the chieftainship continues to dominate Charumbira history-scapes, overshadowing its disputes with other clans. They acknowledge Nemanwa's claim to have been in the area first even if they may question the idea that Nemanwa 'germinated' at Great Zimbabwe. They do not claim to have occupied Great Zimbabwe at any time and, therefore, have no direct claim to its custodianship for themselves. But that does not mean they are neutral bystanders. Chief Charumbira himself stated firmly that 'really Mugabe doesn't have a claim on the monuments' (Interview with Chief Charumbira, 13 August 2001). He also went on to suggest that through their dominance as chief over Nemanwa, the Charumbira clan is able to claim custodianship of Great Zimbabwe for themselves.

Most Charumbira people I spoke to, however, do not claim custodianship over Great Zimbabwe. Some readily acknowledged Nemanwa's claims, like the *svikiro* Ambuya Chibira who suggested that they should be the custodians because of their knowledge of the sacred places there.

I can just say that the Nemanwa people are the ones who are claiming the custodianship because Nemanwa was the one who used to go and fetch water from the sacred place, and coming with the water for us, when we were brewing beer, for us to bath. ... He is the one who knows the sacred places and the one who used to go in. (Interview with Ambuya Chibira, 4/12/2000)

Others suggested that both Mugabe and Nemanwa have valid claims. The current medium of Chainda, emphasised the need for the *masvikiro* (spirit mediums) and ancestral spirits of all three clans to decide, and not the chiefs.

Something has to be done about these masvikiro makuru [big spirits/spirit mediums] *with the names Chainda, Nemanwa and Chipfunhu. It is these masvikiro that can only solve this problem.* (Interview with VaChainda, 30/12/2000)

In common with other spirit mediums and 'traditionalists' who are eager to promote the role of the *Mhondoro dzeNyika* ('the lions of the land' referring to senior ancestors) and *chikaranga* ('tradition'), VaChainda stressed that tensions between chiefs is one of the problems that often prevents the role of ancestral spirits from being recognised.

One other thing is the disagreement between Nemanwa and Mugabe, because each are claiming the area for themselves. Each claim that they have built the walls, that's why I say the big Mhondoro dzeNyika should have been consulted to find out who actually is responsible for the making of the walls, even Inzwi [the Voice] *would also answer because the proper procedure would have been followed, and the truth would have been found.* (Interview with VaChainda, 30/12/2000)

As a school teacher not fully resident in his area, a senior member of AZTREC Trust (a conservation group committed to promoting the role of *chikaranga* and the ancestors – see Daneel 1998), and a Charumbira *svikiro* not recognised as such by his own chief, he is perhaps much more able to hold such a detached view on Great Zimbabwe then other similarly active *masvikiro*. The most obvious example here is Ambuya VaZarira, a senior VaDuma *svikiro* who is also involved in wide efforts to gather *masvikiro* and other 'traditionalists' together to promote the role of *chikaranga*. Unlike VaChainda, she is very determined that Great Zimbabwe belongs to the VaDuma clans, and that she herself is its custodian.

MUGABE – *MATAKE* AND THE VaDUMA CLANS

The Mugabe clan is one of the VaDuma clans in what is Masvingo District today, which historically formed part of a wider 'Duma Confederacy' (Mtetwa 1976: 12) until its 'decline' and 'fall' at the end of nineteenth century. The VaDuma trace their ancestry back to a Pfupajena, whose descendant Chief Masungunye is still often recognised as the most senior of all the VaDuma clans (see figure 2.4). Of the four VaDuma clans in Masvingo District, Murinye arrived first, followed by his sons (or in some versions his brothers) Mugabe, Shumba Chekai and Chikwanda. All the VaDuma clans share the *mutupo/chidawo* combination *Moyo/Gonyohori* (Aquina 1965: 9), but only the Mugabe clan has the *chidawo* '*matake*', reflecting the fact that only they were involved in the nineteenth- century wars with Nemanwa and Charumbira over Great Zimbabwe and the surrounding land. Today the Mugabe clan continues to maintain a strong claim on Great Zimbabwe on the basis of both having been the last pre-colonial residents of the site, and because some of their most senior ancestors, including Chipfunhu, are buried there. Samuel Haruzvivishe described how his grandfather was the last member of the Mugabe clan to reside within Great Zimbabwe.

> When Chipfunhu died, my grandfather took his elder brother to bury him in Masvingo [Great Zimbabwe]. At Huhuri, near the 'danger sign' [on the road to Chikarudzo behind Morgenster Mission] there is a grave of Chipfunhu there, but it is a fake. He was not buried there, he was buried in Masvingo. That's when my grandfather Haruzvivishe went to Fort Victoria saying he wanted the chieftainship of his father, and he was given it, Haruzvivishe from his elder brother.
>
> Chipfunhu was killed around 1894 and my grandfather got the chieftainship in 1895. So he then lived as chief for a long time until 1928 when he died, the year I was born, 33 years as chief.
>
> He used to live in there [Great Zimbabwe], there in the ruins. When you are in the ruins there is a range of hills that arise from west to east, that hill is called 'Mutuzu'.[17]
>
> [...]
>
> Soon after the death of my grandfather, no one lived there because it was declared a national park, and they put up a fence, saying it was the property of the Government. (Interview with Samuel Haruzvivishe, 4/3/2001)

The reports of early explorers (for example, Bent 1892, Hall 1905, Posselt 1924) confirm that Mugabe people of the Haruzvivishe house did inhabit the Hill Complex at Great Zimbabwe until the early part of the twentieth century,[18] and that they used to bury their dead amongst its walls, crevices and caves. Several photographs held at the National Archives of Zimbabwe show Mugabe homesteads on the Hill Complex at Great Zimbabwe (See figures 2.3 and 2.4).

Figure 2.3: Photograph of Mugabe homestead on the Hill at Great Zimbabwe (taken in 1890) (National Archives of Zimbabwe, G 5560).

The Mugabe clan's main claim over Great Zimbabwe is that the graves or *mapa* of their ancestors are located on the hill there. It forms the basis of their claim to the site's custodianship. Just as Nemanwa's claim to have 'germinated' at Great Zimbabwe is unique to them, so is the claim that the *mapa* of their ancestors are located within the site, unique to the Mugabe clan. It marks perhaps the most fundamental difference between the claims of the two clans. This difference is highlighted because the sacred features ascribed to Great Zimbabwe by people from all the surrounding clans, including the *Voice* and other mysterious sounds that used to be heard there (which I describe in greater detail in Chapter 5) are remarkably similar. The Chisikana spring is the obvious example which again illustrates how these clans share a discursive landscape. According to Matenga (2000: 15) 'Mugabe says the spring belongs to their great aunt, *Vatete* Chisina, who was abducted by a mermaid'. While I never heard the name 'Chisina' mentioned, one Mugabe person did mention the now familiar name 'Chisikana'.

> *I was told that Great Zimbabwe, that's where our forefathers were buried. I was told this when I was very young. We were told it was a sacred place. There is a stream which is called 'Chisikana'. It was said there used to be a mermaid that was seen there, njuzu in Shona. I am told it was the daughter of one of our forefather's wives. She was carried by the njuzu, that's why it was called 'Chisikana'.* (Interview with Radison Haruzvivishe, 2/11/2000)

Not surprisingly, Mugabe people dispute Nemanwa's claim to have 'germinated' at Great Zimbabwe. They often argued that Nemanwa used to be a 'policeman' or

Figure 2.4: Photograph of people at the Mugabe homestead on the Hill at Great Zimbabwe (taken in 1894 by Miss Alice Balfour) (National Archives of Zimbabwe,12118(6)).

advisor under Chief Mugabe, and that he lived far away from the ruins. They also claim that much of what makes up the territories of Nemanwa and Charumbira today used to belong to the Mugabe clan. As Samuel Haruzvivishe put it,

> *Manwa was under Mugabe. Charumbira was invited to this area by Nemanwa, but Nemanwa was under Mugabe. Nemanwa was the 'policeman' for Mugabe, he stayed at Mupata Chidziwe, near Mount Nyanda. He was there to protect the area, and stop people from crossing the Tokwe River. Charumbira was staying in Chibi, not as a chief but as a resident. He came here for hunting, and crossed the Tokwe river, and was received by Manwa. Those two are now saying they are the owners of Great Zimbabwe. They were helped by the white people and the missionaries.* (Discussion with Samuel Haruzvivishe, in Ambuya VaZarira's presence, 28/10/2000)

Apart from these contests with Nemanwa over land and Great Zimbabwe, there is also an internal dispute over the Mugabe chieftaincy that has emerged since the appointment of the current chief in 2000. As with the dispute over the Charumbira chieftaincy, this dispute revolves around genealogies and the rights of different houses to claim the chieftainship. Again various people made accusations of political interference in the succession process. I also heard rumours of a simmering boundary dispute between the Mugabe clan and their 'father', Murinye. In fact, both the Shumba and Murinye clans claim some sort of superiority over the Mugabe clan; the former as an elder brother and the latter as father of them both.[19]

Both the Shumba and Murinye elders that I spoke to also made claims over Great Zimbabwe, though these do not appear in the work of Aquina (1965) or Mtetwa (1976).

Figure 2.5: Ambuya vaZarira. (Author 2001)

During our group discussion, the Shumba elders claimed that some of their ancestors too were buried at Great Zimbabwe, and therefore Great Zimbabwe belonged to the VaDuma clans in general (Group Discussion with Shumba elders, 26 June 2001). For their part, the Murinye elders I spoke to emphasised that although Murinye was never directly involved in the disputes between Mugabe and Nemanwa, he was the overseer of Great Zimbabwe and rather acted as elder to them both (Interview with Chief Murinye and his '*Dare*', 10/01/2001).

Despite the obvious tensions between the claims of these other VaDuma clans, and those of Mugabe, I never sensed that there was an active dispute between these clans over Great Zimbabwe. While Mugabe people were keen to emphasise the fact that their ancestors were buried at Great Zimbabwe, their claims were also often framed in more general terms as those of the VaDuma people. Partly this reflects the close links between the history-scapes of these VaDuma clans partly, but I also suspect it is related to influence of one particular VaDuma *svikiro* who is recognised as the *Vatete* (paternal aunt) of all these clans – Ambuya VaZarira (see figure 2.3).

Ambuya VaZarira is an extremely active *svikiro* who has extensive links across all the VaDuma clans (see figure 2.4), and not just those of Murinye, Mugabe and Shumba. Her authority is based on her mediumship of the ancestral spirits, VaZarira and Murinye. Apart from being a senior VaDuma *svikiro* she is also involved in variety of wider efforts to lobby for the role of 'tradition' or *chikaranga* in Zimbabwe today. As a former member of AZTREC (see Daneel 1998), and its more recent splinter group AZTREC Trust, she has also developed strong links with some of the *Mwari* shrines at Matonjeni in the Matopos. As I will elaborate in the next chapter, these multiple interests and allegiances require very careful positioning on her part. Her view on Great Zimbabwe is mainly focused on restoring its previous sacredness, through the co-operation of chiefs and *masvikiro* from across Masvingo Province, and indeed the country as a whole. As she made clear during one of our interviews, the co-operation of *masvikiro* is vital for the revitalisation of Great Zimbabwe.

> *People are conflicting, but they are not respecting the Mountain. It is important that instead of conflicting, the masvikiro should be able to communicate. A svikiro from the Charumbira people should be free to approach a svikiro from other people, like myself, Ambuya VaZarira and say 'how best can we, masvikiro, make this mountain important?' Rather then for them to be afraid, and conflict as masvikiro, you see.*

[...]

> *So that is my request for masvikiro, let all the masvikiro work together and then we will make our mountain sacred, the mountain of Great Zimbabwe.* (Interview with Ambuya VaZarira, 19/11/2000)

In this sense her perspective is similar to that of the Charumbira *svikiro* VaChainda, however, she is also adamant that as a senior VaDuma *svikiro*, she herself should be recognised as its custodian. She often stressed that Great Zimbabwe belongs to the VaDuma clans and that her role, as the '*Zitete* of Great Zimbabwe' ('the great, great aunt of Great Zimbabwe'), is to 'sweep the *mapa*' of VaDuma ancestors buried there (Interview with Ambuya VaZarira, 16 August 2001).

Given that other historical and ethnographic studies of the VaDuma clans (Aquina 1965; Mtetwa 1976) emphasise that it is the Mugabe clan specifically who made claims over Great Zimbabwe and not Murinye or Shumba, it seems that Ambuya VaZarira may herself have had a part to play in elevating the Mugabe claim to include all the VaDuma clans. That this is not rejected by members of the Mugabe clan reflects her status among them. Samuel Haruzvivishe himself is a close associate of Ambuya VaZarira, and he has involved her in his efforts to re-occupy parts of the 'game-park' next to Great Zimbabwe and, less successfully, into the dispute over the Mugabe chieftiancy. With her wide influence among 'traditionalists' across the province, and even at Matonjeni in the Matopos, it is clearly in the Mugabe interest to accept her claim over Great Zimbabwe, and by extension that of the VaDuma in general.

Mugabe people may also have another interest at stake in emphasising the wider VaDuma claim to Great Zimbabwe. This relates to the idea, popular across Zimbabwe today, that it was the Rozvi who built Great Zimbabwe. Some Mugabe people were able to harness this idea for their cause by stressing that the Rozvi and the VaDuma were closely related, as evidenced by the fact that they share the totem *Moyo*. This illustrates how the history-scapes of the clans surrounding Great Zimbabwe are not isolated from wider historical discourses. And it is not just Mugabe people who made use of the Rozvi idea, rather people from all three clans were able to adopt and manipulate this popular idea to suit their particular claims.

THE 'MYTH' OF THE ROZVI AND GREAT ZIMBABWE

As Beach (1980: 220, 1994a: 191–208) argued at length, the Rozvi have been greatly mythologised and 'inflated beyond their real stature' by 'professional' historians and archaeologists, as well as by various 'amateur' African, and specifically Rozvi historians. Beach argued that much of the early basis of the Rozvi myths were the result of uncritical and un-referenced historical research and writings by members of the Rhodesian Department of Native Affairs in the 1920s, like F.W.T. Posselt, and his contemporary Charles Bullock (Beach 1980: 222). Beach also blamed later researchers for failing to question their generalisations. Following Beach, Pikirayi has argued that the extensive archaeological projects carried out in the 1950s by Summers (1963), Robinson (1966) and Whitty (1961) built upon 'a mistaken though seemingly conventional understanding of the more recent periods in the history of the Zimbabwe plateau' (2001: 29). In particular, they associated Great Zimbabwe with the Rozvi 'empire'.

The work of these archaeologists was closely related to that of various historians of the period (including Gelfand 1959; Abraham 1966; Ranger 1967) who were fired up with the possibilities that oral traditions offered as a way of uncovering a hitherto denied African past. Great Zimbabwe was identified by Abraham as the site that 'housed the Rozvi monarchy and the Mwari/Chaminuka cultural nexus' (Abraham 1966: 36). He was not alone in making the direct connection between the Rozvi, the *Mwari* cult and Great Zimbabwe; others included Aquina (1969–70), Daneel (1970) and Summers (1963a). It was also a period of African nationalism, and as Ranger (1982) later noted, the works of Gelfand (1959) and Abraham (1966) – and we could add Ranger's own book (1967) *Revolt in Southern Rhodesia* to this list – fed directly into a African nationalist discourse wherein the Rozvi 'empire' held special position (Beach 1980: 223–4).

But the 'inflated stature' of the Rozvi was not just the result of feedback from the writings of academic historians; various Rozvi activists and revival movements, as well as the wider nationalist movement, were also a very active part of this process. In chapter 5 of his *A Zimbabwean Past* (1994a: 191–211), Beach provides a fascinating account of Rozvi revival movements and various attempts by different Rozvi historians to collect Rozvi histories. He shows how these attempts were intricately involved in both local disputes over land and authority between different Rozvi rivals, and the much grander goal of recreating the title of *Mambo* (paramount chief/king) over all the Rozvi groups.

Since the euphoric days of the 1960s, archaeologists and historians have retreated from both the idea of the Rozvi as Great Zimbabwe's builders/occupiers (Garlake 1973: 180), and the idea that Great Zimbabwe was a previous centre for either the *Mwari* cult shrines now in the Matopos (see Ranger 1999) or a *Mhondoro* cult . Beach, in particular, queried the uncritical use of oral history of the 1960s which precipitated the spread of the 'Rozvi myth'. As he put it 'the picture given by historians of the 1960s has been cut down to size by further research, and it has been shown … that Zimbabwe ended as a culture almost two centuries before the Rozvi became powerful' (Beach 1980: 224).

But despite the fact that 'professional' historians and archaeologists have rejected the idea that the Rozvi built Great Zimbabwe or ever resided there,[20] these ideas have remained prominent in popular discourse. In 1990, various articles appeared in national newspapers about a young oral historian, Zebediah Ntini, who claimed to have uncovered evidence to suggest that 'the ruins were built by the Vasiri people of the Rozvi Dynasty under Chief Kasiri'. As a journalist for the magazine *Moto* reported,

> Other experts on the subject have also dismissed Ntini's story as unfounded and mere myth.
>
> [...]
>
> At a recent meeting of the Oral Tradition Association of Zimbabwe … the … Regional Director of the National Museums and Monuments of Zimbabwe, George Mvenge, totally rejected Ntini's story as 'misdirected and unfounded'. He said every nation or society had its own myths about its origin and it was up to the experts to interpret the myths professionally in order to establish the truth. (Johanis 1990: 10)

Apart from revealing how historians and archaeologists, and especially those of NMMZ, have continued to maintain their position of authority over constructions of the past through claims to professionalism, expertise and the truth, this episode also shows how despite the objections of these 'professionals', the idea of the Rozvi as the builders of Great Zimbabwe remains a potent idea in popular discourse.

THE ROZVI IN LOCAL HISTORY-SCAPES

The Rozvi do feature very strongly in local discourses of Great Zimbabwe's past. Many people I spoke to, though not all, accepted the widely held view that the Rozvi built Great Zimbabwe. Often the Rozvi are presented as a mysterious and mystical people with a special talent for building, and somehow closer to *Mwari/Musikavanhu* (God/ Creator).

> *The idea of everyone in the country is that the Rozvi people are the ones who have built the Monuments, because they are talented at building. And they were said to have to be closer to God, because there is no one else who can do what they were doing.* (Interview with Ambuya Sophia Marisa, 8/3/2001)

This obviously supports Beach's argument that the Rozvi have been 'inflated beyond their real stature by mythmaking' (Beach 1980: 220). However, Beach's implication and emphasis is that this reification of the Rozvi has occurred through feedback from the work of 'professional' historians and archaeologists, as well as Rozvi historians, and I imagine he would have argued that the above excerpt is an example of the 'contamination' (Beach 1998: 49) of oral traditions by the Rozvi myth. A similar perspective is offered by Matenga (1998: 13), where he used the example of a Nemanwa elder crediting Great Zimbabwe's construction to the Rozvi to illustrate how 'confused' oral traditions have become. I would argue that the quest for 'objective' narratives of the past prevents historians from exploring how an idea like the 'Rozvi myth' is a resource that can be utilised and manipulated or denied and rejected, in discursive constructions of the past that are always politically situated in some way.[21] This is reflected in how particular individuals made very specific use of the Rozvi myth to support their clans' claims, whilst members of competing clans either made different use of the Rozvi idea or rejected it completely. Perhaps most interesting of all is how different people within a single clan may make different uses of the Rozvi idea, while still supporting the same claims.

A good example here is a comparison between what different members of Duma clans, and specifically the Mugabe clan, had to say about the Rozvi. Several Mugabe people suggested that the VaDuma and the VaRozvi are closely related, drawing on the fact that they both share the same totem *Moyo* (heart). As VaHaruzvivishe put it,

> *We are not claiming that we built the ruins, that is us the Mugabe people, but we are just the guardians/custodians. We look after that place, because it happens to be in our area ... we used to look after that place from long ago. We have Ambuya VaZarira, it was her who used to work together with VaMugabe to look after that place ...*
>
> [*pause*]
>
> *But if we want to give it more thought, in a way we can say it was us that built it, because we are VaDuma, VaRozvi, VaDuma-VaRozvi.*
>
> *People say that it was the VaRozvi who built that hill, isn't it? So how wouldn't they have helped each other? MuDuma and MuRozvi are brothers, so why wouldn't they have helped each other? Because even today, we are the same people, of the same group, VaDuma-VaRozvi, that are still living here, so we are the right custodians of that Mountain.*
>
> *Even now we have a proverb, which says 'Duma harina muganhu'* [literally 'Duma has no end'] *meaning that every place in Zimbabwe belongs to the VaDuma.* (Interview with Samuel Haruzvivishe, 4/3/2001)

In such accounts, popular ideas about the mythological/spiritual status of the Rozvi, and their connection to the *Mwari* shrines in the Matopos, are being utilised to strengthen the VaDuma position by claiming close kinship ties between the VaDuma and the Rozvi.[22] But other Mugabe informants, including Chief Mugabe himself, denied that the Rozvi and the Duma were related at all, stating rather that the Rozvi were driven out of Great Zimbabwe by Mugabe. Of course both these versions maintain the central theme in the Mugabe claim on Great Zimbabwe that their ancestors lived and were buried at Great Zimbabwe.

It is important to note that VaHaruzvivishe and the current the chief are not close associates, as the former does not accept the other as entitled to the chieftainship. Also Samuel Haruzvivishe is a very close associate of Ambuya VaZarira, for whom, as a senior *svikiro* for all the Duma clans, it is of greater interest to emphasise the VaDuma link with the more widely known and respected VaRozvi, rather than become too involved in the parochial chieftaincy quarrel of one particular Duma clan. In Ambuya VaZarira's presence, Samuel Haruzvivishe tended to put more emphasis on the VaDuma generally, and especially Ambuya's particular role at Great Zimbabwe. Yet when I spoke to him at his homestead without Ambuya VaZarira, he focused more on his own forefathers who used to live at Great Zimbabwe. This reminds us that we are not just dealing with the perspectives of groups, but of individuals within groups who are simultaneously involved in different 'political' struggles at different levels at the same time. Samuel Haruzvivishe, for example, is involved in struggles at a national level to lobby for the role of 'traditional leaders' and for Great Zimbabwe's sacred role to be recognised; at a regional level to ensure the VaDuma and Ambuya VaZarira are its recognised custodians; and that among the VaDuma it is the Mugabe clan specifically that has a historical claim to Great Zimbabwe and disputed parts of the wider landscape. Furthermore, within the Mugabe clan it was the Haruzvivishe house that once lived at Great Zimbabwe, and moreover, are

Figure 2.6: Duma elders preparing to slaughter a goat at a *bira* held at Ambuya vaZarira's homestead in January 2001. Chief Masungunye is standing third from the left; Chief Murinye is third from the right; Samuel Haruzvivishe is second from the right; and Ambuya vaZarira's son Peter Manyuki is holding the goat on the extreme right (Author 2001).

now in line for the chieftainship. Of course, it is not just Samuel Haruzvivishe who is involved in such a kaleidoscope of disputes, to different extents all people are constantly involved in a variety of webs of interest.

Some Mugabe people suggested that the Nemanwa people were based some distance away from Great Zimbabwe as servants, lookouts or 'policemen', for the VaRozvi who were at Great Zimbabwe. This claim is, of course, strongly disputed by some Nemanwa actors, because it throws doubt on the central Nemanwa claim to have 'germinated' at Great Zimbabwe, and perhaps it belittles them to be described as servants for someone else. I also suspect that an articulate person like Aiden Nemanwa is conscious of some VaDuma claims to be related to the Rozvi. Therefore it is not surprising that he firmly stated that the Rozvi came when the Nemanwa were already there, and that 'it is a lie that the Rozvi built Great Zimbabwe'. Rather, he attributed it instead to 'the spirits that were talking to the Manwa people'. But this is not an opinion shared by all Nemanwa people. I was intrigued to watch a difference of opinion on this issue emerge as I talked with Nemanwa elders who had gathered for a group discussion. As he had done in a previous one-to-one interview, Chief Nemanwa put forward his idea about how it came about that the Rozvi built Great Zimbabwe. He was then contradicted by another Nemanwa elder.

J.F.: So was it the Nemanwa people who built the ruins?

Chief Nemanwa: No it was the VaRozvi, they came here, running away from a war far away, they came and hid in nhare [underground hole/cave/passage] *they were taking it as a place of refuge and they found Nemanwa already here.*

[…]

VaChinorumba: Masvingo [literally 'the stones', refering to Great Zimbabwe] *were not built by VaRozvi.*

Chief Nemanwa: When the Rozvi people came to this place, they were people with no fixed place, they used to move around a lot from this place to another, then they would build their camp there, and when the war occurred in that area, they would move to another place, until they came to Great Zimbabwe. And that's when they started building those kopjes and those stones were being collected from various places, within the country and some from South Africa.

J.F.: But you Sekuru [turning to VaChinorumba] *say that the Rozvi did not build that place.*

VaChinorumba: They did not build that place, how could they do that? It was built by Mwari. Those structures were built by Mwari, because it is impossible for a man to build something on those steep rocks of that mountain. (Group Discussion with Nemanwa Elders, 18/7/2001)

What is interesting is how, while attributing Great Zimbabwe to the Rozvi, Chief Nemanwa was still able to 'spin' it in way that emphasised that the Rozvi came to area after the Nemanwa, thereby maintaining the central claim that Nemanwa 'germinated', or at least got there first.

Chief Nemanwa was not the only Nemanwa person to suggest the Rozvi built Great Zimbabwe. I was told by another person that Great Zimbabwe was built by the Rozvi and Nemanwa together. This account also maintained the central Nemanwa claim to have 'germinated' at Great Zimbabwe by emphasising that when the Rozvi came, running away from wars with the Ndebele, they found the Nemanwa already there. Another way in which some Nemanwa informants were able to integrate the idea of the Rozvi with the Nemanwa claim to have germinated at Great Zimbabwe involved the common story of the spring at Great Zimbabwe from where the girl Chisikana emerged. Several informants suggested that that she had been accidentally left there by the Rozvi when they left Great Zimbabwe, and was later found by Chirichoga Nemanwa. She was then brought up by Nemanwa and eventually married them. This implies that the Nemanwa are closely related to the VaRozvi, which is striking because of the similarity with some VaDuma/Mugabe claims that the VaDuma and the Rozvi are like brothers. Clearly there is some 'cultural capital' (Bourdieu 1991) to be gained from close association with the Rozvi because of the wider popularity of the idea that the Rozvi built Great Zimbabwe. It is a way in which a local claim to the custodianship of Great Zimbabwe, can be merged with, situated within, and legitimised by a much wider view about the origin of Great Zimbabwe.[23]

While local actors and groups do borrow ideas and 'cultural resources' like the Rozvi 'myth' from the wider landscape of discourses within which they are situated, they are, of course, concerned to maintain their own claims on Great Zimbabwe. Thus the central claims of each clan over the site – that their ancestors germinated or lived or were buried there – are sustained in nearly all the different versions that people put forward about the relationship between the Rozvi and Great Zimbabwe. Furthermore, whilst there is an understandable interest for local groups claiming ownership or custodianship of Great Zimbabwe, to engage with the wider field of discourses in order to legitimise their claims, there is also a danger that through doing so they may undermine their own claim. It is after all not uncommon for spirit mediums claiming to be possessed by Rozvi, or even 'national' ancestors such as Chaminuka and Nehanda, to turn up at Great Zimbabwe out of the blue claiming it as their own.[24] I suspect that it is for this reason that some people were particularly keen to emphasise that Great Zimbabwe had not been built by any people at all, but rather by *Mwari/Musikavanhu*.

Both Aiden Nemanwa and Ambuya VaZarira are particularly active and articulate local players who often emphasised that Great Zimbabwe's sacredness goes beyond just a local or provincial level – that it is a, if not the, national sacred site. Yet both are also keen to emphasise that their clans, or even themselves individually, are the rightful persons to act as guardian/custodian for the site. It is for this reason that they deny strongly that the Rozvi built Great Zimbabwe. Whilst Ambuya VaZarira was keen to emphasise the closeness between the Rozvi and the Duma, and acknowledged that the former did once occupy the ruins, she was in the same interview very emphatic that it was *Musikavanhu* that built Great Zimbabwe.

> *It was not built by the Rozvi, but it is a thing that came from the word of Musikavanhu. We can see that is the brain of Musikavanhu, it is not of a person. How could a human being make those stones to be formed like that? What would that person be?* (Interview with Ambuya VaZarira, 19/11/2000)

This is remarkably similar to Aiden Nemanwa's claim that Great Zimbabwe could only have been built by the spirits or *Mwari* himself, except that Aiden was, of course, keen to emphasise that Nemanwa was named as Great Zimbabwe's keeper by those very spirits/*Mwari*. Given that there is a sense of a real threat of Rozvi elders and *masvikiro* claiming Great Zimbabwe as their own, it is understandably a preferred option to suggest that *Mwari/Musikavanhu* was responsible for building Great Zimbabwe. This perceived threat – posed by the possibility of Rozvi spirit mediums claiming Great Zimbabwe for themselves – was revealed to me when Aiden Nemanwa told me (his version of) why an attempt to hold a very large *bira* ceremony at Great Zimbabwe in the early 1980s, involving the slaughter of several hundred cattle, failed.

> *Some Rozvi went to Harare, five of them, so they said they wanted to get cows from the chiefs, each chief was supposed to provide cattle. They gathered 250 cattle, so that they could slaughtered here. They asked the minister of Lands to be given a farm. On which to keep these cattle. They went to all the districts, and the last district they went to was this district Masvingo. They sent a message before they came here, saying we want to come and collect animals from all the chiefs in Masvingo to slaughter at Great Zimbabwe. When they came there was a big meeting with all the local chiefs. We asked them why they wanted these cattle, and the Rozvi group answered us saying 'we are the ones who have given the chieftainships to all the chiefs in this country'. And that's when I disputed with them. I said 'I don't want to tell lies about the land. If you say that you have given all the chiefs their land, and titles then you are lying, because it was Nemanwa that gave land to all the chiefs here, Murinye and his sons, Mugabe, Shumba, Chikwanda, and also Mapanzure, Nyajena, Charumbira and Zimuto. All were given their chieftainships and land by Nemanwa. Yet the Rozvi people, they did come here long ago, and they passed through. So I told them they were lying.*

> *So those words were carried to minister Shamiyarira, then on to Muzenda [Vice-president], and VaMuzenda told them to the president, and when the president accepted that information, he said, 'If that*

is the case all the animals have to be brought back to the chiefs'. That is why the bira failed, and never happened. (Group Discussion with Nemanwa Elders, 18/7/2001)

From a local perspective there are, therefore, both advantages and disadvantages in agreeing with a wider view that the Rozvi built Great Zimbabwe. On the one hand, working the idea of the Rozvi into their own claims on Great Zimbabwe can have a legitimising effect. On the other, acknowledging the Rozvi as Great Zimbabwe's builders can undermine the claims of these local clans, especially when other people turn up at Great Zimbabwe (as they regularly do) claiming to be possessed by very senior Rozvi ancestors. This reveals that there is clearly a tension between Great Zimbabwe as a sacred site for local clans, and Great Zimbabwe as a shrine of national significance. Local actors are keen to acknowledge Great Zimbabwe's national importance, as long as their clan, or even they themselves, are acknowledged as the rightful custodian of it.

CONCLUSION

What I have been trying to show in this chapter is that the discourses of local clans about Great Zimbabwe's past, and the surrounding landscape, are not isolated from either each other, or from much wider discourses, both 'popular' and 'academic'. The 'history-scapes' of the Nemanwa, Charumbira and Mugabe clans are situated within a shared discursive and physical landscape, and constantly interact and negotiate with wider discourses in a politically engaged manner. These 'history-scapes' are not necessarily unified, static or internally consistent; they are constantly changing, moving, and dependant, in part at least, upon the agencies of different actors within, as well as between, these disputing groups.

While some may argue that the Rozvi example shows how oral traditions have been 'doctored' and 'distorted' in the twentieth century, I would suggest that it exemplifies the nature of oral traditions and memories as cultural resources that are **not** isolated fragments of a past waiting to be reconstructed by a neutral, outside researcher; rather they are constantly being borrowed, negotiated, manipulated and engaged with politically in a continuously changing field or landscape of wider discourses. I also find it hard to accept that this is an innovation of the last century, that before 1890 oral traditions were somehow 'more neutral', and 'more true' than they are now, as is implied by descriptions such as 'contaminated', 'distorted' and 'doctored'. Indeed, it is the continued dominance of a 'professional' archaeological view on Great Zimbabwe, and its 'quest for objectivity', which holds back the representation of other perspectives of the past and landscape, and thereby maintains *the silence of untold stories and unheard voices* around Great Zimbabwe. But if oral traditions and memories are better seen as cultural resources put to use by groups and individuals in the construction of fluid, changing and politically situated 'history-scapes', then the legitimacy of any version of the past must, in part, be dependant on the authority of the person or persons articulating it. The authority of individuals within their own communities, clans and beyond depends upon, among other factors, kinship, age, status and reputation. Perhaps most of all, however, it depends upon the ability of actors to perform their narratives and social roles convincingly. This I shall call the performance of the past.

Notes

1. It should be noted that only a few of my informants made this reversal, and these may simply have been 'mistakes' – but those few informants were all from the Charumbira clan. The few academic historians who have collected oral traditions from this particular area (Aquina 1965; Mtetwa 1976) do not mention this story as having been told to them reversed as above, but then one of them does not actually mention the '*Matake*' part of the story at all (Aquina 1965).

2. Many of the accounts of early explorers (Mauch 1872; Bent 1896; Hall 1905: 84–85 (Burke 1969); Posselt 1924) describe the tensions and conflict that existed between the Mugabe, Nemanwa and Charumbira clans over control of Great Zimbabwe and surrounding land; one traveller even witnessed a battle in 1891 (Letter from R.C. Smith to his father, 24 July 1891, National Archives of Zimbabwe, SA12/1/15).

3. See for example Hove, C. and Trojanov 1996 *Guardians of the Soil: Meeting Zimbabwe's Elders*, Munich: Frederking and Raler Verlong.

4. Apart from the claim to have 'germinated' at Great Zimbabwe, Nemanwa history-scapes also referred to having lived at the site, on a small hill called Mutuzu (Group discussion with Nemanwa elders, 18 July 2001 and interview with Ambuya VaZvitii, 7 November 2000). This is also reflected in the name of one Nemanwa ancestor that Aquina (1965: 8) referred to – Garabwe (literally 'lives amongst the stones'). In the Delineation Report (1965: 101) this name is mentioned not as that of an ancestor but rather as a previous Nemanwa *chidawo*, before *Muguriri* was adopted.

5. Both Aquina (1965: 8), and the Native Affairs Delineating Officer (Delineation Report 1965) describe the problems they faced trying to establish a clear genealogy of the Nemanwa clan. Aquina attributed this to Nemanwa secrecy, which is obviously a more attractive option than the explanation of senility which was offered by the Delineation Officer in his report (Delineation Reports 1965: 101).

6. The lack of clarity surrounding Nemanwa's past stands in marked contrast to the detailed accounts presented by Charumbira and VaDuma informants about their genealogies. Of course, their pasts and genealogies are not by any means consolidated or mutually coherent. In fact, they are highly contested largely because they are directly relevant to fierce chieftainship wrangles, and boundary disputes that continue today. Indeed this may explain the lack of knowledge among Nemanwa elders of their own genealogies. As Nemanwa is a very small clan numerically and territorially, its conflicts and disputes with other clans have taken on far greater significance than internal disputes. As a result, debate over exact genealogies may have been minimal in comparison, and there is, therefore, a lack of knowledge precisely because it has not been an issue of the same proportion.

7. Aquina (1965: 8) traced the genealogy of Nemanwa ancestors back to a Goko Vameri – a name which in itself refers again to germination. I heard no mention of this name during any of my interviews with Nemanwa people. In July 2001 I held a lengthy discussion with a group of Nemanwa elders at Chirichoga Secondary School. During this discussion it was agreed that Chirichoga was 'the person that gave birth to all the Nemanwa people' (Interview with Nemanwa

elders, 18 July 2001), but his relationship with other known ancestors was very unclear. Apart from Chisikana, other ancestors that were mentioned included Muntunguzuma, MaChibwa, Kuwangepi, Mahuni, Madingura, Ziyavizwa, but each time they were cited in various different orders and sequences – dates were rarely, if ever given. According to Aquina's version (1965: 8) some of the names mentioned in my discussion with Nemanwa elders in 2001 came between Goko Vameri and Chirichoga, whilst in my discussion they all came after.

8. This 'Chisikana' ceremony was unique, and epitomises recent efforts by museum authorities (NMMZ) to respond to local concerns about the management of the site. See also Chapters 5 and 9 and Matenga (2000).

9. During another discussion, Aiden Nemanwa refined his view of the relationship of other clans with Great Zimbabwe, vis-à-vis Nemanwa's role as its original custodian. 'That place is occupied by two types of spirits. The first is that *Voice* that used to speak at Zimbabwe, and then the other type are the *vadzimu* of other people, of people who died there; of Nemanwa, of Charumbira, of Mapanzure, of Mugabe, of Murinye. All the people that surround the area, they have their *vadzimu*, so when others show that they were in Zimbabwe, they can go and appease their *vadzimu* there, but they cannot appease our *vadzimu*, or the One that used to speak at Zimbabwe. Only Nemanwa can do that' (Aiden Nemanwa, Group Discussion with Nemanwa elders, 18 July 2001).

10. Of course, the claim that Nemanwa 'germinated' in the area is disputed by both academics and members of other local clans. The few academics that have worked in this area (Aquina 1965; Mtetwa 1976) find the claim hard to accept 'objectively', and have therefore searched for a place where the Nemanwa clan must have migrated from. Mtetwa (1976: 185) suggested that 'the Manwa broke away from the old Mtoko-Budya shumba-Nyamuzihwa/hora dynasty in the late seventeenth or early eighteenth century and found Great Zimbabwe uninhabited …'. Mtetwa (1976: 185) and Aquina (1965: 12) both suggest that when Nemanwa arrived they found *Rombo* and *Gwadzi* people living in the area, with whom there was an 'uneasy co-existence' (Mtetwa ibid.) until later with the help of the recently arrived Charumbira, the *Rombo* and *Gwadzi* people were forced to leave. It is perhaps not surprising that I never heard any mention of *Rombo* and *Gwadzi* people from Nemanwa informants during my extended fieldwork in the area. This may indeed be an example of a deliberate and convenient *loss* of the past. Putting aside the case of the *Rombo* and *Gwadzi* people, the basic historical account (e.g. Matenga 1998 or Pikirayi 2001) of the past that academics have accepted does now see Nemanwa as the first of the current local clans to have arrived at Great Zimbabwe and occupied the area, followed by the VaDuma clans of Murinye, and his sons (or brothers) Mugabe and Shumba, and then by Charumbira.

11. The terms *muzukuru* and *sekuru* also refer to 'grandchildren' and 'grandfather' respectively. This is significant because it relates directly to the type of relationship that exists between these kin.

12. This apparent from the Delineation Report which described Nemanwa's lands as 'going over to the European area and his people scattered', and the Nemanwa 'that remained loyal live in the eastern portion of Charumbira's country' (Delineation Report 1965: 101).

13. One informant implied all three (Interview with A. Govo, 27 June 2001).

14. The Mugabe clan was also placed, for a while, under Chief Shumba, and again this was rarely mentioned by Mugabe informants, though on a few occasions it was referred to as an example of the kind of 'historical mistake' Rhodesian authorities often made.

15. Both Aquina (1965: 14–15), and the Delineation Report (1965) described the controversy and disputes that embroiled the appointment (by the Rhodesian Government) of Mahza, the current chief's grandfather, in 1964 after 'the wrangle [had] continued for years, with neither side giving in' (Delineation Report 1965: 75). Since then the dispute has deepened because the chieftainship has remained within the Mazha house, going first to his son Zephaniah, and then to his grandson, the current Chief, rather than switching, collaterally, to the next 'house'. There were many local newspaper reports about this continuing dispute during my fieldwork, e.g. *Masvingo Star* 14–20 July 2000, 5–11 Jan. 2001, 14–20 Sept. 2001, 16–22 Nov. 2001; *The Mirror* 16–22 March 2001, 16–22 Nov. 2001.

16. Succession is supposed to occur collaterally – that is from eldest brother to youngest – according to the genealogy of the ancestors. It is supposed to switch at the death of a chief to the next house in line for the chieftainship. Each house represents an 'original son' of the 'original' ancestor. Arguments and disputes therefore circulate around both the 'correct' names and sequence of the sons/ houses, and as well as the identity of the central 'original' ancestor. In the Charumbira case the system of 'collateral succession' has itself also been challenged, as the chieftainship has been passed from father to son, and hence remained in one house, since Mazha's appointment in 1964. This is often explained locally in terms of the political interference of the state. Zepheniah was a senator in the Rhodesian Senate, while Fortune's appointment as chief in 2000, (after many years acting as 'interim' chief following the death of his father) as well as his coinciding appointment to the Zimbabwean Parliament, and, more recently, to the post of deputy minister of local government are linked to his close ties with the upper echelons of the ruling ZANU PF party.

17. Nemanwa people also often told me that they used to occupy a hill within Great Zimbabwe called Mutuzu. This again illustrates how the history-scapes of these clans are located within a shared discursive and physical landscape.

18. It is not clear when people were removed from the Hill Complex itself. It is possible that the Mutuzu hill referred to by Samuel Haruzvivishe is in fact the Hill Complex, but most people identified Mutuzu as a smaller hill behind the Great Enclosure. It is likely that Haruzvivishe Mugabe was removed from the hill when the first fence was erected between 1911 and 1913, and that he then took up residence on Mutuzu, until the whole estate became declared a National Monument in the 1936, or when the entire area was fenced in the 1950s.

19. While Mugabe people often agreed that Shumba and Mugabe were both sons of Murinye, it was never framed in terms of superiority. Some accounts describe Matorefu Mugabe as Murinye's brother (Aquina 1965: 11). Similarly, I was very rarely told that for a brief period in the 1950s, Mugabe was demoted to headman under Chief Shumba (Aquina 1965: 11). When it was mentioned, it was used to exemplify the kind of 'historical mistake' often caused by Rhodesian interference.

20.   The Rozvi are not the only Shona dynasty that has been connected to Great
      Zimbabwe. After the nationalist period of the 1960s, and the rebuke that the
      Rozvi idea received from Beach and others in the 1970s, writers in the 1980s
      (Mufuka 1983; Mudenge 1988) began to focus on the oral traditions of the
      Munhumutapa state, which suggested that its founding ancestor Mutota was
      the last ruler at Great Zimbabwe, who moved north looking for salt and founded
      the Mutapa state. This does not mean that the importance of the Rozvi had
      disappeared from popular discourse. Indeed as Beach has shown (1994a:
      191–211), the histories produced by Rozvi revivalists in the 1960s, often
      incorporated Great Zimbabwe, the Rozvi and the Mutapa Dynasty into one
      coherent whole.

21.   Matenga does mention this in his suggestion that 'as the groups vied for political
      supremacy they could have doctored oral testimonies to support territorial claims'
      (Matenga 1998: 13). But I would question Matenga's implication that this
      'doctoring' is something that has only occurred since 1890.

22.   It is interesting to note that Aquina (1965: 9) suggested that there did used to be
      a connection between the Rozvi and the Duma, which Mtetwa (1976) has denied
      for lack of supporting evidence. But Aquina also noted that 'there exists great
      hatred between the Duma and the Rozvi, which is expressed outwardly by the
      Duma purposely abstaining from wearing the Ndoro, the royal emblem, and by
      changing the meaning of their Mutupo Moyo' (Aquina 1965: 9–10). This hatred
      is not apparent in the claims that the Rozvi and the VaDuma are like brothers.

23.   VaChinomwe, a Charumbira *svikiro*, suggested, very impartially, that both
      Nemanwa and Mugabe are descendants of the Rozvi; 'they are the Rozvi who
      remained in the Monument' (Interview with VaChinomwe, 9 November 2000).

24.   This happened on at least one occasion during my period of fieldwork in the area,
      when a group of spirit mediums claiming to be possessed by a whole host of widely
      recognised ancestors including Chaminuka, Nehanda and Kaguvi turned up at
      Great Zimbabwe, much to the ire of members of all three competing clans. A file
      held at the Conservation Centre at Great Zimbabwe (NMMZ ref: G1(I)), which
      contains twenty years of correspondence between NMMZ and local chiefs, indicates
      that this has been a very frequent occurrence at Great Zimbabwe. The most recent
      occasion that I am aware of was in May 2003 when Dick Marufu, a spirit medium
      from Zaka, turned at Great Zimbabwe intending to conduct ceremonies there
      (*The Herald* 10 May 2003; 14 May 2003 & *Daily News* 13 May 2003).

CHAPTER 3

'TRADITIONAL CONNOISSEURS' OF THE PAST[1]

If they come and say, 'you know nothing', when we know our soil, myself, Zarira, I will tell them, those in office, and say 'You are learned and educated in your universities, Zarira has the university of the soil. You have got your own lawyers, but Zarira's lawyer is the soil' He is a big lawyer, this one! [Ambuya VaZarira laughs as she indicates the soil]

(Interview with Ambuya VaZarira, 10/11/2000)

It is clear that among communities that live around Great Zimbabwe National Monument, and among different individuals within them, the past is a site of great contestation, as they jostle to justify their claims to the site and the surrounding landscape, as well their own individual positions, influence and authority. Within the limits of certain fixed markers – like the names of certain people and places, as well as ancestors, totems and praise names, and well known stories about past events – individuals are sometimes able to exert a high level of agency to renegotiate and manipulate stories according to their collective, and individual interests. Their authority within their own communities, clans and beyond, depends on their kinship and descent ties, their status and age, their political allegiances, and their reputation as knowledgeable of the tradition or *chikaranga*. Perhaps most of all, it depends on their ability to perform adequately within their roles.

The *performance of the past* – as narratives performed at formal gatherings or informal visits; or the procedural 'correctness' of a ritual performance; or the credibility of the possession of a spirit medium (*svikiro*) and even of the spirits themselves – all become a crucial factors to be assessed, considered and discussed by the recipients and participants, on whom any 'traditional connoisseur' depends for his or her authority and legitimacy. The authority of a 'traditional connoisseur', and the narratives they construct, are therefore dependent to some degree upon wider society, and not just the legitimacy they receive from 'above', be it the spirit world, or the government.

'TRADITIONAL CONNOISSEURS'

I call 'traditional connoisseurs'[2] those people who are considered in rural communities in Zimbabwe to be knowledgeable about a clan's past and the 'tradition' – *chikaranga* or *chivanhu*. This includes chiefs (*vaishe*) and their aides, village heads (*sabhuku*), and other respected elders, as well as spirit mediums (*masvikiro*) and healers (*n'anga*). Their status and authority on the past is often based on kinship and age. Old people – especially elderly men who are members of the chief's clan, and even more so his immediate family or his 'court' (*dare*) – are generally regarded as knowledgeable about the past. In particular, members of a chief's clan are able to claim their authority, as Alexander has put it, 'by right of their position as autochthonous owners of the land' and their 'access to a body of knowledge which derives from their position as living representatives of the ancestors' (1990: 2 cited in Maxwell 1999: 175).

Apart from the importance of age and kinship for status as a 'connoisseur' of tradition and the past, there are also significant gender restrictions. Rural societies in Zimbabwe are very patriarchal and it was rare that I was told that I should talk to a woman, even an old woman, about the past. There is a very significant exception to this rule in that there are far more female than male *masvikiro* and *n'anga*. It is interesting to note that in a cultural and social context in which women are very silent and marginalised, in these roles they are able sometimes to exert extraordinary influence.[3] A lot of the women I spoke to were, therefore, *masvikiro* or *n'anga*, and Ambuya VaZarira, a senior *svikiro* of the VaDuma clans, shines out as one very good example of a woman who through her status as a *svikiro* commands considerable respect beyond just her own clan or a collection of related clans. Of course, this is not to say that women who are not influential *masvikiro* are not knowledgeable about the past, only that they are not considered to be so, and therefore would not be consulted on 'traditional' matters. Similarly while young and middle aged people are rarely considered to be authorities on the past and 'tradition', this does depend upon status; younger or middle-aged men who are accepted *masvikiro* can have influence, as do active sons or close associates of chiefs and *masvikiro*.

The 'traditional leaders', particularly a chief and his closest associates, of an area are able to exercise some control over narratives of the past by virtue of their position, which often enables them to act as 'gate keepers' of knowledge. Very frequently, I would be required to get a chief's permission before others, who are maybe considered more knowledgeable than the chief himself, were willing to speak to me as an outsider. A good example here is Aiden Nemanwa, who insisted on several occasions that I must go and get permission from Chief Nemanwa before he would speak to me, which I endlessly did, only to be referred back to Aiden Nemanwa by the chief. In other instances I sensed I was being deliberately dissuaded from visiting certain elders or *masvikiro*, often by not being told of their existence, because of personal rivalries, or the 'political' differences that exist within and between clans.[4]

Apart from exercising some control over access to people, it was also clear that in group situations some people were much more vocal than others. This became particularly apparent when I conducted a group interview with elders of the Shumba clan that I had previously arranged with Chief Shumba. Having arrived at the prescribed place and time, and then led into a derelict room where our meeting was to be held, my research assistant and I were aghast when more than 25 people, all men of various ages, gradually gathered in a large circle on the floor. I was told to sit on one of two chairs in the centre of the room, next to a close relative of the chief who sat on the other chair. Whilst I was slightly preoccupied with the logistics of carrying out an interview with so many people, it soon became very clear that the discussion was dominated by only a few of the older men, in particular the elder beside me who was representing the chief. Apart from occasional outbursts of laughter, most of the people there, especially the younger men, limited their contributions to muffled replies of '*ndizvozvo*' and '*ndizvo chaizvo*' ('it is so' or 'it is true') when prompted into agreement by the dominant speakers. In this way, a kind of group consensus of a common past was being constructed around a discussion conducted by a only small number of those present, whose authority derived from age and proximity to the chief. In my field notes

I also reflected on my own role (because I organised the meeting for the purposes of my research), both as a catalyst in the consolidation and construction of a group's past, dominated by the perspectives of a few privileged people, and in the reinforcement of existing power relations based on age and proximity to the incumbent chief.

> It soon becomes clear to me that for some of the younger people present, this session may in fact be a 'history lesson', and for all involved, it is a kind of 'consolidation-of-agreed-group-history' session. Reflecting on this, it occurs to me that my role in initiating this group session, as an academic pursuing my own research, is quite heavily implicated in the consolidation of these people's notion of their history. It also becomes clear to me that most of the answers to my questions are given to me by the man seated next to me, who occasionally turns around looking at certain other elders for their comments or disagreement. One or two elders take up the cue, adding details here and there, rephrasing what has already been said. One man seated to my right, wearing a hat, with a slightly mischievous look on his face occasionally breaks in to question something or to add a humorous comment, but he is an exception, most people agree with what is being said by the man seated next to me on the chair, with a nod of their heads, or with an out break of laughter when a joke is made. (Fieldnotes of Interview with Shumba elders, 20 July 2001)

Of course, researchers are always implicated through their very presence, as well as the situations they create and the questions they ask, in the very power relations that they are studying. So it is reasonable to argue that because I approached a community via the chief and his elders, I was reinforcing established power relations, even though I would seek out informants individually once I received the blessing of the chief and his elders. Furthermore, it is also clear that through arranging a group interview, the chief and his associates themselves were using me to reinforce their version of their clan's past. But to overstate this would ignore a very important dynamic which is apparent in the section of my field notes quoted above. There is a sense in which group consensus is important, illustrated on that day by the way in which the dominant speaker frequently turned to face other people in the group, requesting their approval and comments. Some elders took up this cue and made their comments; one in particular made sarcastic and humorous comments to the amusement of those present. His additions were tolerated among lame protests to remain focused on the point of the gathering coming from the 'authority' sitting next to me. Thus participants were offered a chance to participate, to offer additions, and occasionally, to differ.[5]

This emphasis on participation, allowing people to air their views, and an attempt to consolidate a consensus, was a common feature not only of meetings that I organised, but also gatherings and ceremonies to which I was invited. Sometimes small *bira* ceremonies (which usually involve spirit possession) and parties, for which beer had been brewed, turned into very raucous affairs, as people, fired up with alcohol, asserted their rights to be heard, sometimes over stepping marks of respect and dignity to the amusement or annoyance of others. Of course, the emphasis on participation, group knowledge and consensus does not necessarily negate power relations; chiefs, male elders and *masvikiro* are more likely to contribute and be heard than young men, and women, many of whom might not be present in the first place. But people are able to use such opportunities to challenge the authority of the narratives of others, or to strengthen their own authority and legitimacy. Therefore, there is a sense in which authority and power are not just based on structural positioning in society, but also on the agency of actors themselves and their ability to perform their narratives convincingly. Obviously the 'performances' of 'traditional connoisseurs' involve many different aspects in relation to what their position/roles are – an elder, a chief and a *svikiro* all have different roles so will 'perform' differently – and depending on the situation, whether a family meeting, a clan ceremony, or a council meeting. But if we accept that traditional leaders are 'traditionalists', because to some degree at any rate their authority depends upon 'tradition' (therefore the term 'traditional connoisseur'), then references to it become central to the authorisation of their narratives of the past.

REFERENCES TO 'TRADITION' AND THE AUTHORISATION OF NARRATIVES, CLAIMS AND POSITIONS

In their personal and clan rivalries these 'traditionalists' often accuse each other of not knowing or following 'tradition' (*chikaranga*) correctly, or in the case of a *svikiro*, for not being genuinely possessed. People frequently utilised a strong notion of 'truth' in terms of *chikaranga* and the knowledge of the spirits, or as Ambuya VaZarira put it, the 'university of the soil'. Often this 'knowledge' or 'wisdom' of the spirits was contrasted to the 'new wisdom' (Aiden Nemanwa in Hove and Trojano V 1996: 79) or 'education' of Europeans, which an important Charumbira *svikiro*, VaChainda,[6] described as the *uchenjera hwebhuku* (lit. 'knowledge/wisdom of books').

The 'truth'/'knowledge' or 'wisdom' of the spirits and *chikaranga* was, however, based on very ambiguous and variable notions of what these very 'traditions' actually consist of, and what the knowledge of the spirits actually is. As is acutely revealed by the succession disputes of the Charumbira chieftainship where the genealogies of ancestors and chiefs, as well as the system of succession itself, have become the focus of much contestation, it is clear that 'tradition' is very malleable indeed. This is corroborated by Maxwell in his work, *Christians and Chiefs in Zimbabwe* (1999),

> traditional leaders used 'tradition' as a set of strategies which enabled them to enhance their own social, economic and political influence. But theirs was not a fixed model of the tradition. (Maxwell 1999: 174)

Most frequently, people I spoke to authorised their narratives by stating firmly that it was knowledge that had been passed to them from their fathers, who themselves were told by their fathers, and so on. Chief Nemanwa did so in relation to the story of the girl Chisikana, who emerged from a spring at Great Zimbabwe.

> *This story has been passed down from one mouth to another, even ourselves we were told this information, and even our ancestors were told the same information.* (Chief Nemanwa, Group discussion with Nemanwa elders, 18/7/2001)

Alternatively, a story would be authorised with direct reference to the spirits. Councillor B.B. Charumbira claimed his information had come from the ancestor Chainda himself, speaking through a possessed medium.

I am able to tell you a lot. Me, as some who was born here, can tell you the history that I was told by the ancestral spirits that used to possess masvikiro like Masikati Zvitambo and Jack Mataruse.

I called them to this place in 1955, I wanted to learn the history of my people and our chieftainship. Masikati Zvitambo was possessed by Chainda. I entered into the house to ask that mudzimu about our history. I said, where did you come from? He said we originated from Tanganyika. (Interview with B.B. Charumbira, 19/7/2001)

Apart from 'truth claims' based on references to 'tradition' and the words of the ancestors themselves, in practice, of course, people also frequently referred to what they may have read in books, learnt at school or gleaned from other 'non-traditional' sources. A brilliant example of this occurred when Radison Haruzvivishe showed me a photocopy of a map of Great Zimbabwe and the surrounding chiefdoms and their boundaries that he said came from National Archives, and been drawn in the 1960s during the Smith Regime. On this map Mugabe's territory was much greater than today's boundaries, including Great Zimbabwe itself, and much of what is now under Chief Nemanwa. According to Radison Haruzvivishe it showed each chiefdom's territory – 'how it should be', and it is clear from the following quote that he intended to use the map to substantiate both Mugabe claims over Great Zimbabwe itself, and the efforts of his family to re-occupy parts of what had previously been Mzero farm.[7]

I think that this thing here [the map] that we managed to grasp, satisfies anyone who has any questions. Because now we have the map, this map was compiled during the Smith Regime, its different from other current maps which are there. Some of the current maps which are there were not indicating Great Zimbabwe to be under Chief Mugabe. They indicate Great Zimbabwe to be under Chief Murinye or maybe Chief Charumbira. So Nemanwa for the time being is under Charumbira, it is no longer under Mugabe. That is why Nemanwa also has a thing about the monuments. That's why it is important we have this map, so we can see who was who, and who is who. (Interview with Radison Haruzvivishe, 2/11/2000)

This example illustrates very clearly that these actors are able to draw on a wide range of sources in their efforts to legitimise their own agendas. It also illustrates that the opposition between 'traditional knowledge' or 'wisdoms' and the knowledge of 'books' or 'education' is but a conceptual construct through which people try to authorise their own positions and narratives, and through which they make sense of what they see as their own marginalisation in society and their alienation from a place like Great Zimbabwe. As a construct which can be used for self-empowerment, it does not necessarily limit the pool of resources available for 'traditional connoisseurs' to use to legitimise their claims, nor does it prevent them from drawing on a wider plethora of tools for their narratives.

It is quite startling how many of the most influential 'traditionalists' have relatively well-educated backgrounds. Aiden Nemanwa and the *svikiro* VaChainda, for example – both of whom strongly espoused the importance of *chikaranga*, and the ancestors, and blamed Great Zimbabwe's 'desecration' on their marginalisation – are both teachers; the former is retired and the latter is a deputy headmaster. I confronted VaChainda about this apparent contradiction between his role as a school master, and his role as a *svikiro*, and his answer shows how he is able to separate and keep distinct these different roles.

J.F.: Can I ask you a slightly flippant question? You are a teacher yourself, so how do you overcome this contradiction? You are a svikiro and a teacher; you work with books and you work with the spirits of the past, how do you see the relationship between the two?

VaChainda: Ahh, they are very different, when I am at school, I forget about being a svikiro, and I do the job as professionally as I can. I have to agree with 'pachokwadi pavo' [their truth], like what they really want. Because 'kuchenjera' [education] and reading that is very different. What I do as a svikiro, when I am possessed is not that. He talks saying 'Us, we have to do this, this and this' that is what happens, its different. (Interview with VaChainda, 30/12/2000)

It is obvious that these 'traditional connoisseurs' are much more complex than such a label implies. It should be clear that they are pragmatic agents not restricted by apparent contradictions between their narratives, and their daily lives. Therefore, Ambuya VaZarira, who spoke of the 'university of the soil' in the quote at the beginning of this chapter, on other occasions lamented rising school fees which many rural people were unable to afford. Indeed, only to 'outsiders' does it appear like a contradiction. While they frequently lamented the loss of respect for 'tradition' and the ancestors, it was not suggested that the clock should be turned back, so to speak. Ambuya VaZarira made it clear that it is 'respect' for the ancestors that is seen as crucial.

Wearing clothes and other Western things, we can't leave them, we can't blame that, but what we want is the actual manners, respect for the elders, because sometimes you find just a young kid saying 'pfutseki' [swear word from English 'for God's sake'] *to an elder, to VaZarira. Is that good manners?*

That is bad, swearing doesn't make the vadzimu angry, but scolding them does, very angry. (Interview with Ambuya VaZarira, 17/02/2001)

In a similar vein, it was obvious that despite the rhetoric in support of 'traditional knowledge' against the 'knowledge of the book', many people saw me and my academic research, as a means by which they could attempt to further their personal and group interests in relation to Great Zimbabwe and the wider landscape. Thus, Ambuya VaZarira agreed to carry out a series of interviews with me on the condition that I collected the material into a single document which I would then give her as her 'life history'. Similarly, members of the Nemanwa clan became very keen that I held a group meeting with them when they became aware that I had been carrying out a number of interviews with representatives of the Duma clans, including Ambuya VaZarira. They were concerned that I should get the 'correct' information. I became a resource which could be used to further individual or clan interests via a means not always available.

While I have chosen to use the notion of 'traditional connoisseur' to emphasise that most of these people strongly espouse the importance of 'tradition', it is clear that they are pragmatic rather than dogmatic in their use of means with which to justify their arguments and further their interests. We are not really talking about clear cut definitions here; as complex characters, with multiple interests, some people are 'more traditionalist' than others; some are more concerned about the 'silence of Great Zimbabwe', or the role of 'traditional leaders' like the *masvikiro*, than others. As Ortner has put it, there is a 'the multiplicity of projects in which social beings are always engaged, and the multiplicity of ways in which those projects feed on as well as collide with one another' (Ortner 1995: 191). Because of this 'multiplicity of projects', people who do not identify themselves, and are not identified by others as 'traditionalists', are sometimes forced to appeal to 'tradition' to legitimise their own positions as 'traditional leaders'.

Chief Charumbira is a good example in this respect because while support for him is often expressed in terms of his education, wealth, and political connections, he also faces considerable opposition from within his own clan. People dispute his historical right to the chieftaincy, and question his knowledge of 'tradition'. Thus, he is forced to engage in debates about the genealogy of the Charumbira ancestors, and he even attends and sponsors 'traditional' *bira* ceremonies, because it is 'expected' of him. Furthermore, when I interviewed him,[8] he seemed to suggest that he was an authority on 'tradition' because of his membership of the Chief's Council in the National Parliament, rather than on the basis of what he had been told by his forefathers or by possessed mediums.

The authority of 'traditional leaders' is clearly based on an interplay between, on the one hand, 'structural' factors – such as their specific role or status in society as a village head (*sabhuku*), or a chief (*Vaishe*), or a spirit medium (*svikiro*) derived from the political authority of the 'state', or the spiritual authority of the ancestors – and on the other hand, the support they receive from society. The latter is, of course, dependent on an individual's personality and charisma, and their ability to make alliances and to perform in their roles. Authority comes both from 'above' and 'below', and depends on the ability of individuals as agents to negotiate the two. This applies to both chiefs and *masvikiro*, because both claim the authority of the ancestors – the former as their living representatives and descendants, and the latter as their mediums through which they make their wishes known – and both in a sense mediate between the spiritual owners of the land, and society. Yet whilst now chiefs also derive their authority from the recognition they receive in local and national state structures,[9] and therefore they also mediate between the state and society, *masvikiro* receive no formal recognition or government allowances. This has sometimes caused friction within the 'traditional leadership', and for *masvikiro* it accentuates their reliance upon recognition from society for their authority. This authority is particularly dependent on their ability to perform convincingly, as liminal, ambiguous characters situated between the world of people, and the world of the spirits.

THE AMBIGUITY OF *MASVIKIRO*

Spirit mediums, or *masvikiro*, occupy a liminal, ambiguous position between the world of people and the world of the spirits. They are both distinct from the ancestors and yet in ceremonies and rituals they become embodied by them, as 'vessels' or 'vehicles' through which the ancestors speak to their descendants, and make their wishes known. Therefore they represent a paradox, as Lan noted:

> When possessed, the medium is thought to lose all control of body and mind. He may be referred to as *Homwe* which means pocket or little bag. The medium is simply a receptacle, the vessel of the spirit. He has no specialised knowledge or unusual qualities

of his own. But this attitude to the mediums contains a paradox. Although the medium is thought of as an ordinary person, when a particular woman or man is selected from all others, they are marked out as extraordinary, as unique. The medium combines in one body two contradictory aspects; he has no specialised qualities and he is as close as anyone can come to divinity. He has no influence on the will of the ancestor, yet the ancestor cannot act without him. He is a person of no special powers and he is a source of the most significant powers on earth. (Lan 1985: 49)

While people often point out the distinction between the person of the medium and the spirit that possesses them, in practice this boundary is never clear cut. *Masvikiro* often take on the name of the spirit that possesses them and are treated with a great deal of respect even when they are not possessed, as if they are the ancestors themselves. Furthermore through dress, and through the rituals, taboos and prohibitions which feature in *bira* ceremonies, but are also frequently adhered to outside of specific ritual situations, their liminal, ambiguous position between the worlds of people and the spirits is marked for society.

Lan presented the 'ritual prohibitions' of mediums systematically and argued that they are 'shaped by a conception of what the chiefs of the past were like and how they were treated by their followers' (Lan 1985: 69). Thus he suggested that the clothing of mediums related to the dress of Shona chiefs since the sixteenth century, and the removal of shoes and hats in the presence of *masvikiro* is more than an unusual show of respect, but rather relates to how certain Mutapas were addressed in the past. According to him, mediums were also not allowed to see rifles, because certain chiefs of the past forbade people to appear armed before them out of concerns for their security. He developed his argument to suggest that, because the ritual prohibitions of mediums related to what the chiefs of the past did, or were thought to do, and the fact that for mediums these prohibitions apply throughout both their daily lives and their 'ritual lives', they are creating an 'illusion' that they themselves are those very chiefs of the past.

> By adhering to these prohibitions the mediums present the illusion that they are not simply the mediums of the chiefs of the past but that they actually are those very chiefs returned physically to earth. (Lan 1985: 68)

Lan called this 'the great spectacle of the past',

> the dead ancestors of present chiefs returned to earth, the history of the land displayed and acted out, the heroes of the past available once more not to rule but to give the benefit of their wisdom, to tell the truth only they, being dead understand. (Lan 1985: 68)

Like much of Lan's narrative, the reasoning is coherent and elegant, but there is a sense in which it is perhaps too elegant. In Masvingo, I did not encounter either the same severity or consistency of taboos and ritual prohibitions that Lan described for *mhondoro* mediums of the Dande region of Northern Zimbabwe (Lan 1985: 68–70). While there are some ritual taboos that mediums do carry into their daily lives, outside of the ritual sphere they are by no means as expansive as Lan describes. The extent and particularities of taboos seemed to vary from medium to medium. Most wear 'western style' clothes on a daily basis, though with a predominance of black, rather than the ritual garments of black, white and red cloths and animal skins they put on once possessed. They are sometimes distinguishable from other people by more subtle paraphernalia such as black and white beads, and copper bangles that are worn under their clothes, and the staffs they often carry. Apart from these subtle indicators, and barring one or two exceptions, *masvikiro* in Masvingo are, for the most part, fairly indistinguishable from the rest of society, if one does not know them, or observe them too closely. Perhaps the difference relates to regional variation and the fact that most of the *masvikiro* I met were not *mhondoro* mediums as such, but mediums for clan ancestors related to specific chiefdoms and lineages. However, the key point is that there is no sense that *masvikiro* are collectively part of some conscious attempt to create the illusion that they themselves are the great chiefs of the past, manifest in the present. Rather, as they clearly state, their authority is based on their roles as *mediums* of the dead; as 'vehicles' through which the dead speak to the living, the world of spirits meets the world of people, but not as the dead themselves.

Where blurring and ambiguity do occur is on the point of agency and power. The 'daily' respect that is shown to mediums is not because they themselves are the ancestors. Rather, as mediums chosen by the ancestors, they are protected and nurtured, and watched over by them. In their dreams and visions they are warned by the ancestors of coming events, and dangers to avoid. One gets a sense of the ancestor hovering around them, and this is what induces the respect. The prohibitions that *masvikiro* carry out in their daily lives are explained in terms of the instructions and peculiarities of the ancestors themselves; they are observed out of respect for them and because of the threat that the ancestor might make them sick or leave them altogether, if they are not observed. They are not carried out to convince people that the medium is the very person of the ancestor. If there is illusion involved in the 'ritual prohibitions' that mediums variably enforce, then it more likely relates to their efforts to convince others that they are authentic spirit mediums, rather than authentic ancestors.

The narratives of mediums about their own pasts often illustrate this blurring between the agencies of the ancestor and themselves, whilst simultaneously keeping the person-hood of the medium and the ancestor separate. In her long accounts of how she became a medium, Ambuya VaZarira described miraculous, and some very unpleasant, events in her life. These show how from the moment she was born, until she finally became possessed by the spirit Murinye in 1966, she was shadowed by the spirits of VaZarira and Murinye, which affected her behaviour and gave her abilities that she could not explain. She began with an account of her own birth, when she described how when her mother was pregnant with her, her fourth child, she became very sick and the previous medium of VaZarira turned up unexpectedly at the house, to claim the child.

So then when my mother was pregnant with me, this one [her elder sister who is sitting next to us] *was born easily, my mother did not have any problems with her. The other one that followed was also born easily, the next one again was born with no problems. Now when it was my turn, I caused a lot of trouble for my mother. My mother was ill/sick for three days. People long ago didn't like hospitals. VaZarira had gone to Uzeze, in Murinye; every year she used to go to see the Sabhuku* [kraal heads] *and the Vaishe* [chiefs] ....

Then when she was there, when my mother got ill, she saw it when she was in Uzeze. And so she came, and found that my mother had been sick for three days, and her teeth had blackened, her face had changed due to illness. And she came into the homestead, and asked 'where is the woman who is ill?' She was shown where the woman was, and she entered the house wearing the leopard skin and her walking stick. As she put one foot inside the house, while my mother was seated near by, … as she put one foot inside, my mother gave birth, and I came out. My mother never had any more trouble.

So VaZarira was very excited and she untied her leopard skin and hit me with it, whilst I was just a small baby.

Poooo! She was saying 'you!' – my father was a teacher, he used to go to church, he was a teacher and at the same time a priest – and then she hit me, and I started crying and crying. And she said to my father 'I know you are a Christian, you are a 'dutch' [ie member of the Dutch Reformed Church who set up and ran Morgenster mission] *but let me rest in your daughter, I am no longer going to be away from her, I will not be removed again. She will be the svikiro raVaDuma* [spirit medium of the Duma people]. *VaMurinye, VaMugabe, VaShumba, if they refuse this child then they will have brought death. If they don't receive this daughter, they will bring death into their own homesteads'.*

Then my father sighed and said 'If the VaDuma people argue and refuse to receive this child, they will be poorer and poorer, I can even see it now'.

So there was actually an agreement. Even now, if anybody does not agree that I am the one, then I will just say 'VaZarira has made an agreement, I don't have anything to do with this' (Interview with Ambuya VaZarira, 10/11/2000)

This illustrates the way in which *masvikiro* can narrate the events of their lives in terms of the spirits who later possessed them. In a sense, their narratives of their own pasts justify their authority as *masvikiro* in the present, by deflecting responsibility for the events of their lives onto the ancestors. The entanglement of Ambuya VaZarira's own agency with that of the spirits that possess her is even starker in her descriptions of her childhood, and the abilities she claimed she had as a child, but was, at that time, unable to explain.

When I went to school, my father had already died, and my mother was left alive, she sent me to school, but I was very intelligent compared to the others, and this was making me wonder. I could hear my brain in my head going click, click, click. I used to have a lot of brains. During those years there used to be a choir. Then that time, it showed me something, because I could sing, but I could not understand how or what I was singing. My sisters they know it, I could not be surpassed in singing.

[…]

In school, the games we used to play, when I was then at the mission, ahhh! [laughing] *my behaviour used to show that there was something there.*

[…]

In singing there, it started again, the same way. It came again, the real thing, when I was at the Sutu school even at football I was excelling, not knowing why or how I had been born, or that it was the same old woman from so long ago [ie the spirit VaZarira].

Then one day when I was back at school, in the classroom I saw that I could no longer see properly. If I left the classroom, going towards our dormitories, I could see again, and then I'd go back to class and could no longer see. So I wrote a letter to my mother. I did not know what was really behind me, because I was not really worried about the name that I had … I was even confused about it , I didn't understand it. So I wrote a letter from Gutu to Zaka where my mother was, and told my mother that I could no longer see. Then my mother took a bucket of finger millet and went to a n'anga to investigate, because my mother knew. Then the n'anga said 'This child has a big name, she has a zita renyika [name of the country]. *The spirit that is on your child is one of the chiefs' And she said yes, and was told 'go and buy a white cloth to tie around her head, that is where they are, they are the ones that are preventing her from seeing in class'*

And so my mother bought that white cloth, and came with it …

So when I came from school, playing with others in the yard I had to put on my head dress, I put it on for two days. The third day when I went back to school, I discovered that I could see again, and I was seeing what was on the black board, and my brains came back again. Then my mother said 'Don't loose that head dress. Every time you come from school, you should tie it' She never told me what was the secret behind it. (Interview with Ambuya VaZarira, 10/11/2000)

This passage illustrates how *masvikiro* can envisage the ancestors who so drastically affect their lives, and ultimately possess them, as a force that is both separate to them, watching over them, but also inside of them, giving them unusual capabilities and talents, and very capable of causing sickness and misfortune if they fail to follow their instructions. This suggests that rather than creating the illusion that they are the ancestors themselves, *masvikiro* often emphasise the ambiguity of their agency, as an entanglement of that of the ancestor and their own.

In relation to mediums among the Karo of Sumatra, Mary Steedley has described this as the 'edgy double agency of spirit/medium'(Steedley 1993: 15), and argued that

> stories of spirit encounters raise with particular urgency questions of belief, agency, and authenticity – questions that lie at the heart of ethnographic experience and its representation. What to believe? Do others believe? Do I? Is this experience (of mine, of others) 'real'? How to evaluate the evidence of unseen experience, or of experience shared yet incomprehensible? (Steedley 1993: 14)

Importantly she stressed that

> This is not simply a matter of Western scientific rationality at its epistemic limits; Karo too are troubled , for reasons of their own, by the edgy double agency of spirit/medium.
>
> … spirits provoke a way of reading narrative experience against the grain of credibility; as uncertain, duplicitous, always open to revision, *bottomless.* (Steedley 1993: 15)

The cultural and geographical context of her work may be very different, but what I want to draw out here are these questions of belief, agency and authenticity that do not just confront the 'outside' ethnographer but all those participating. If the authority of the ancestors is beyond question, then the authenticity of the medium and his/her narrative is consistently questionable, and ultimately dependent upon the belief of others.

As both Bourdillon (1987) and Lan (1985) have described in some length, the process of becoming an accepted spirit medium is a long one that can take many years, and often begins with prolonged and unexplainable illness, and diagnosis of its spiritual causes by a *n'anga*. This may then be followed by a long series of *bira* ceremonies during which the spirit is encouraged to possess the aspirant medium, and then to gradually reveal its identity so that it can be established who the ancestor is, and therefore what type of medium the 'patient' will become. Eventually, in the case of very 'big' spirits, there will some kind of 'testing procedure' (Lan 1985: 52) at a large public ceremony, which may involve the recital of the 'correct' genealogy of the ancestor, or the selection from a pile of carved staffs, the one previously used by the last medium of that ancestor.

Obviously the procedures may vary greatly across different regions and different clans, but the key point is that individuals do not just become mediums by themselves. They need the support of their close kin, other members of their clan, and, importantly, the mediums of other ancestors of that clan in order to become accepted. Throughout the process of becoming a medium, a person relies on others; first the *n'anga* that makes the initial diagnosis; then close kin who help organise and pay for the *bira* ceremonies in which the spirit is encourage to possess the patient; and ultimately members of the whole clan, and other established mediums who have to accept the medium as genuine.

Even after the medium has become accepted, it is still possible that the spirit can leave the medium if he/she does not follow the correct taboos, or doubt may later be cast on authenticity of the spirit itself, and it may be suggested that the spirit is actually a mischievous lesser spirit like a *mashave* pretending to be the ancestor of great chief of the past. At any point later on, depending upon the performance of the medium, it may be decided that the medium is a fraud, or even when the medium is considered genuine, it is acknowledged that he/she may occasionally fake possession (Bourdillon 1987: 241). Furthermore, they may have to contend with rival claims to the mediumship of the same ancestor. What all this suggests is that there is always room for doubt at any particular possession ceremony, and for any spirit medium. Thus *masvikiro* must constantly perform convincingly as *masvikiro* and of course, maintain their allegiances with elders within their clan, and outside, to ensure their popular support upon which their authority, in practise, depends.

THE PERFORMANCE OF THE PAST AND THE DENIAL OF AGENCY

The most important moments when *masvikiro* must perform convincingly occur during possession ceremonies. Contrary to the entanglement of the agencies of the medium and their ancestor that often pervades in their narratives of their own pasts, at these ceremonies, the distinction between the ancestor and the medium becomes crucial. In these carefully prepared rituals and ceremonies, they *are* seen to become the ancestors themselves, and at these moments there can be no ambiguity. Here they do represent the 'great spectacle of the past' that Lan described, and it becomes very important for the credibility of the medium that he/she 'speaks and acts the part of the spirit and behaves in a way that is markedly in contrast to his normal behaviour' (Bourdillon 1987: 237). After possession a medium is not supposed to know what the spirit has said, and he/she is usually informed of this later by his/her attendants. So it is during possession ceremonies that the agency of the medium and the spirit are perhaps most clearly defined and contrasted, particularly at the very moments of possession itself, and the authority of a medium derives from being able to convincingly deny his/her own agency.

At *bira* ceremonies home-brewed 'traditional' millet beer is drunk, and snuff taken, and the spirit is encouraged to possess the medium through the combined and overlapping rhythms of drumming, dancing and singing, sometimes accompanied by *mbira* (thumb piano), and maracas. On those occasions all the 'ritual prohibitions' are observed: shoes and hats are removed; beer is poured from traditional beer pots and passed around in kalabash gourds; no metal cups or plates are used. Once possessed the mediums put on their 'ritual attire' of black, white and red cloth, and animal skins. Clasping their staffs or their 'ritual axes', their behaviour becomes tense, their movements awkward, and they speak with a different voice, clearly demonstrating that the ancestor has 'come out'. People in the 'audience' become very respectful to the possessed medium, ululating and clapping, as they are addressed by the spirit. As the following description of a possession that I witnessed shows, the medium becomes the ancestor through visible signs that differentiate him/her from his/her normal self.

> The clouds have become darker now, and the air is warm and moist as we walk back to the *bira* which sounds like it is in full swing. The sound of drumming, singing, ululating and whistles spread through the air. As we approach we hear some loud screams and Madende says 'Some one is getting possessed' – just what I was thinking.

> Taking off our shoes, we enter the area. The men are standing up, dancing and clapping, along with some of the women. There is a man, middle-aged, wearing no shirt or anything on top. I recognise him from this morning but now he is different. His face is contorted, and he trembles and jumps, and stretches his arms and muscles, gasping, occasionally screaming. Quickly and suddenly the drumming stops, people sit down, and the man in the trance is beckoned to go and sit down under the tree, on a rock. He struts around, stumbling, looking around him with quick piercing glances, making jerky movements. Referring to him as Sekuru, Ambuya again beckons him to sit down. This he does, and VaHaruzvivishe and Ambuya VaZarira both approach him, and sit down on the ground in front of him, as does a man in a smart suit. The possessed man continues to flex his muscles, and beats his chest. He begins to speak in a squeaky whisper, and the women ululate, VaHaruzvivishe claps. The spirit says '*Makadii*', indicating to everyone, and everyone claps in time, slowly. The spirit addresses VaHaruzvivishe, who replies politely, clapping his hands constantly. Ambuya VaZarira also speaks to the spirit. Extreme politeness is used all round. The possessed man fiddles with his staff and asks for snuff, which he is given. He fidgets and takes snuff almost constantly. He is offered a pot of beer, that is put in front of him. The talking continues, and after talking mainly to VaHaruzvivishe for a long time, the discussion opens up as other older people address the spirit.

> After an hour or so it is clear the discussion is over, people ululate and clap to thank the spirit. The possessed man gets up slowly, takes a final heavy dose of snuff and slowly walks out, towards the bushes. He stops and shudders, and beats his chest before disappearing behind some bushes. After a few minutes, a large woman with a particularly large voice, who was seated in front of me, follows him taking with her, his shirt and jumper. The man returns alone a short while later, wearing his clothes. People ignore him now just as moments before they were showing him so much respect. He looks confused, dizzy and slightly dazed as he enters the group again, and quietly sits down. (Fieldnotes of *bira* ceremony of the Haruzvivishe Family, 29 October 2000)

So while in the narrative descriptions of their own lives, the agency of the medium and that of the ancestors may become blurred, in possession rituals the agency of the medium and the ancestor become sharply contrasted and clearly defined. It is important that for a medium to be taken seriously, there is a marked difference between his/her normal behaviour and when he/she is possessed. Thus the performance of a medium when he/she is becoming possessed is vital for him/her to be considered authentic. This is true especially for young aspirant mediums still trying establish their authority as it is believed that it is easier for more experienced and established mediums to become possessed.

Later on, during the same *bira* ceremony described above, another much younger medium arrived as we were eating. Without any drumming or dancing he was spontaneously possessed, and the manner in which people treated him suggested that they doubted the authenticity of his possession. So as to convince people of his authenticity, the possessed medium/spirit then described in a long monologue, everything that had been discussed and said by the previous spirit, before that medium had arrived.

> As we begin eating some more people arrive, including two young men. One wears a hat and carries a staff with some leopard skin on it. A *svikiro*? I think to myself.

> People are evidently hungry and the food is quickly consumed by the old men. Just as they are finishing eating, and the women are being served their food, the young man in the hat, sitting in front of us, lets out a gasp, and again loudly. He holds his head, as his hat falls to the ground. His eyes are closed, as he burps loudly and deep gurgling sounds come from his throat. He trembles and before long he is grabbing inside a plastic bag that a young woman has just brought him, pulling out a hat made of leopard skin and a whole leopard skin. He puts these on, fumbling with his eyes still firmly shut, then stands up and puts a black, white and red cloth around his waist. Before long he is standing in the middle of the circle, and begins a speech in a clear, if occasionally faulting, and then deliberate voice. Muffled ululation and clapping begins, but this time, no one approaches, and the tranced man is not offered a place to sit, and remains standing. The possessed man begins by saying that he can prove he is possessed by telling them what was said before the medium arrived.

> 'The first thing I am going to do is I am going to tell you what was said by the first spirit that came out, when I was not here'

> He talks about the issue of the beer that is supposed to be brewed by the Mugabe people. He also mentions issues of corruption and killing. It is almost as if he really knew what had been discussed in his absence. (Fieldnotes of *bira* ceremony of the Haruzvivishe Family, 29 October 2000)

This episode illustrates both how an unconvincing performance by a medium can cast doubts about the authenticity of his possession, and how people in that situation may react in quite subtle ways by not showing the respect they would otherwise do if an established medium had become possessed. It is also interesting how the possessed medium attempted to substantiate the truth of his possession by referring to knowledge that as a medium he could not have known because he was not there, but which as a spirit, he would, of course, know. Still later on during the same ceremony, after he had come out of his trance, the same medium again referred to the distinction between the knowledge of the medium and that of the spirit, when he claimed that he knew nothing of what had happened when he had been possessed. This occurred after a very dramatic twist in the turn of events, when the possessed medium began to treat Ambuya VaZarira with a great deal of disrespect, which resulted in Ambuya VaZarira furiously storming off, and then directly challenging the authenticity of the medium. The argument began, interestingly, over Great Zimbabwe, and the efforts that were being made to 're-traditionalise' it.

The possessed man goes on to say that there should be a 're-traditionalisation' of Great Zimbabwe, and outlines the long procedure that this will involve. He stresses that Chief Mugabe, VaHaruzvivishe, the medium himself and other Mugabe elders should all be recognised by the whole country as the custodians of Great Zimbabwe. Once this is done, there should be a big *bira rokuchenesa* [cleansing ceremony] involving all the chiefs of the country, to cleanse the land and settle the spirits of the those who were killed during the war of liberation. This should be organised by the Mugabe people. Each chief must bring a sheep and a sample of soil from their lands. The sheep will be burnt and mixed with the soil, and each chief will bring the soil back to their territories and put it into the rivers. This is the only way that the people of Mugabe can become recognised in whole of society.

Ambuya VaZarira is now speaking to the possessed man. She agrees that beer should be brewed and such a ceremony be organised, but questions the procedure that the spirit put forward.

Referring Ambuya VaZarira as '*muzukuru*' [grandchild/or niece or nephew] the possessed man asserts his authority and says 'I am not trying to argue with you'.

Ambuya VaZarira seems irritated by the manner in which the possessed man is confronting her, and asks 'Do you think it is so easy that we can just do it? Can we just do it in the way that you are saying it?'

The possessed man replies 'Yes, it is possible, the only problem is that most of you young people are just afraid; you are afraid to approach the Museum people. You are afraid to approach even the Government!'.

He constantly refers to Ambuya VaZarira in disrespectful terms as *muzukuru*, or iwe [you!] and seems to be challenging her authority as the senior medium of the Mugabe people, as he describes the land as 'my land' and even calls Ambuya VaZarira '*murandakadzi*' [woman slave/servant], as if she were younger than him, and his spirit superior even to her spirit.

Ambuya VaZarira is furious now, and stands up to confront the possessed man face to face. Though she is not possessed she looks very fierce, but the possessed man stands there in his leopard skin, clasping his staff, and insists that she is just a medium, and should respect him. Ambuya VaZarira grabs her own leopard skin and staff, and stomps away into the surrounding bush, followed shortly by several other women. The possessed man sits down, and with a short gasp, a burp and a sneeze, the spirit is gone. He takes off his leopard skins, and the black and white cloth and puts them in his bag.

Ambuya VaZarira returns, still furious and directly challenges the medium as to his authenticity. The man shrugs his shoulders, and says she is talking to the wrong person, because the spirit has gone.

Ambuya VaZarira addresses the rest of the Mugabe people present, and referring to earlier discussions when they requested her to come back to live close by, she says: 'You see, Mugabe sons, you know I can even leave you around. You were talking about me coming back, but you see what is happening. You are not giving me any respect. You know I can forget about you, I am a big spirit, I don't only care for you, the Mugabe people. I have a lot of chiefs who are concerned about other issues, that are more important than the issues I am addressing for you now. I sympathise with your position that is why I am helping you . I am the spirit of the rest of the Duma people, I am not only meant for you the Mugabe people. How can you disrespect me, like you are doing right now? When you know that I am the founder of this land, I am the greatest spirit of this land then there is no way you can survive, you people, without me.'

Things begin to quieten down, and Sekuru Haruzvivishe announces that the beer is now finished, and the proceedings are coming to a close. The very last of the beer is passed around. There is a final session of drumming, and singing, before people begin to drift away. (Fieldnotes of *bira* ceremony of the Haruzvivishe Family, 29 October 2000)

The dramatic events described above illustrate how even an established medium such as Ambuya VaZarira can be challenged from within her own clan, by a medium who claimed his authenticity by denying his own agency. She, however, reasserted her own authority to the people present by emphasising the superiority of her spirit as an ancestor for all the VaDuma clans, and not just the Mugabe people. In a sense, she was drawing on her wider support base and allegiances to legitimise her authority among people of the Mugabe clan. This illustrates that apart from performance in rituals, allegiances and alliances within a clan, or a group of clans, are crucial for the authority of *masvikiro*. *Bira* ceremonies in which *masvikiro* become possessed are usually organised by themselves or by close supporters from within their clan, so challenges to a *svikiro*'s authority can be limited, though as the above example shows, this is by no means certain.

CLAN LOYALTIES

Masvikiro are often accused of being fraudulent by people outside of their clans, especially where there are long running rivalries between neighbouring clans over boundaries and territory, as exists around Great Zimbabwe. Thus I was often told by people from the Nemanwa and Charumbira clans that Ambuya VaZarira is not a genuine *svikiro* at all, and vice-versa. So to a certain extent, *masvikiro* are bounded within inter-clan disputes. Even if they dispute the legitimacy of the chief of their own clan, this does not mean that they support the *masvikiro* of other clans. But *bira* ceremonies tend to occur within clans, so disputes between *masvikiro* of different clans rarely emerge as they do not meet in these circumstances.

Of course there are exceptions to this. Ambuya VaZarira is particularly active outside of her own Duma clans, and organised various *bira* ceremonies to which she invited *masvikiro* and other 'traditionalists' from a much wider area to address much larger issues. Once she asked me to pass on an invitation to one of her *bira* ceremonies to several *masvikiro* from the Charumbira clan, but they all offered me excuses saying they were unable to attend. The *bira*, as a result, was still dominated by the Duma elders despite Ambuya VaZarira's own efforts to reach across clan divides, which illustrates how clan loyalties, and concern with one's support from one's own clan, can prevent *masvikiro* from making alliances with *masvikiro* from beyond their own clans. This is particularly the case in relation to the clans surrounding Great Zimbabwe, because of the intensity of the disputes that surround that site.

This was also very apparent in the comments that different *masvikiro* made about the events that occurred at a 'traditional' ceremony organised by the National Museums and Monuments of Zimbabwe (NMMZ) at Great Zimbabwe in July 2000, to reopen the 'Chisikana' spring. Most people who attended described how the representatives of the Nemanwa, Charumbira and Mugabe clans that had been invited formed three separate groups around their own *masvikiro* rather than integrating together as a whole. One *svikiro* from the Nemanwa clan suggested that the purpose of the *bira* was to show which of the local clans was the right custodian of Great Zimbabwe, and described how she became possessed first, implying that this meant her spirit was the most superior.

> *That is why you find now that even the Museums people know that this land that we have here is sacred land, and it has its own people who take care of it. That's why they called the three surrounding chiefs, when they were trying to see who is the real custodian of this land. And then they discovered that the owners of the land are of Nemanwa, since the first people to be possessed were from Nemanwa. And some even went into the spring, swimming in the water showing that the spirit that they had was an njuzu. I'm not trying to be proud in terms of spirits but the first spirit that came out was the spirit that I bear with me here. Then came some other spirits, and then came the spirit of VaMafodya.* (Interview with Ambuya VaZvitii, 7/11/2000)

She also claimed that the only spirit she heard talking was from another Nemanwa medium, who said that the fact that no Mugabe spirits 'came out' illustrated that Great Zimbabwe belonged to the Nemanwa clan, and following them the Charumbira's, but not the Mugabe clan.

> *Because of the dispute that is there, during the movement, when people were approaching the well, the Mugabe people started to make a separate group and immediately after that, the spirits just possessed people. After my spirit had gone, the only spirit that I heard talking was the spirit that had possessed Peter Muzvimwe. He is the one who said, 'I have come, there are others who came prior to me. Some of my friends. But I am going to talk, myself, Chinja' These are the words spoken by Chinja.*
>
> *'I have seen that some of my spirits have come' referring to the spirit of Nemanwa, 'and also I have seen the spirits of Charumbira people, but I have not seen the spirit of the Mugabe people.'* (Interview with Ambuya VaZvitii, 7/11/2000)

This shows how the performances and actions of possessed mediums at this ceremony can be narrated in terms of the disputes between different clans over Great Zimbabwe. Ambuya VaZarira, in turn, dismissed the Chisikana ceremony and ridiculed the performance of *masvikiro* of the Nemanwa and Charumbira clans.

> *On that day of the bira held at the spring, other masvikiro went into the muddy water that had just gushed out. They were in there like wild pigs just making a scene, and I told them 'Go in, I have nothing to do with it, because I do not know who made the ruins'* [laughing] *Nothing worthwhile was done there, beside masvikiro just going into the mud like wild pigs!* (Interview with Ambuya VaZarira, 27/12/2000)

While the Chisikana ceremony was unique because it was organised by NMMZ, and the three disputing clans who were invited to participate would not normally attend or organise *bira* ceremonies together, it does illustrate how disputes between clans intricately involve the *masvikiro* because their own authority and support is based on their role as mediums for the ancestors of specific clans. Therefore, while being the basis of their authority and support, clan loyalties may also limit and hold back *masvikiro*, where their agendas and interests go beyond those of specific clans.

JUGGLING A MULTIPLICITY OF PROJECTS AND ALLEGIANCES

Like all 'traditional connoisseurs', and indeed 'social beings' everywhere, *masvikiro* are, to varying extents, involved in a 'multiplicity of projects' which 'feed on as well as collide with one another' in many different ways (Ortner 1995: 191). Ambuya VaZarira is perhaps the exceptional example of this that emerges from my fieldwork in Masvingo. Her ambitions go beyond the local, that is, restoring Great Zimbabwe for the VaDuma clans, or regaining the *mapa* (grave site) of her ancestor VaZarira on Mount Beza. She is involved in much wider efforts, with other *masvikiro* and 'traditionalists' from places across the country, to lobby for the role of 'tradition', for the ancestors and traditional leaders to be recognised by the government. As she herself often emphasised in our discussions, she is not just concerned with the particular and specific concerns of the Duma clans, but also the state of the country as a whole.

> *I've got a big problem with this land. I'm not worried if it is the land of Masungunye or of Mugabe, or if it is the land of Shiku.*

> *I am worried about the whole land, the country itself, that is what we are looking at, these other things can come, but the concern of the spirits is in the winds of the land, how the country is run.* (Interview with Ambuya VaZarira, 10/11/2000)

In these efforts she has made allegiances far beyond the VaDuma clans themselves. Like many other *masvikiro* in Masvingo province (VaChainda, and Ambuya Chibira are examples from the Charumbira clan) she was very involved with both the first AZTREC set up under Daneel's ZIRRCON (Daneel 1998), and then later its splinter group AZTREC Trust.[10] Although she is no longer involved with either, she maintains contact with many of the *masvikiro*, *manyusa* (*Mwari* cult messengers) and other 'traditionalists' she met there. She also has close contacts with several different groups of war veterans in Masvingo province and beyond, and has organised several large *bira* ceremonies and meetings with war veterans, *masvikiro* and *manyusa* from other areas of the country as well as the usual entourage of VaDuma elders, and chiefs. The focus of these meetings and *bira* ceremonies was not just specific issues relating to the VaDuma clans, but also the wider concerns of 'traditional leaders' about the political and social turmoil, and indeed violence that steadily increased throughout the country during 2000–2001, in connection with the government's controversial land reform programme. In relation to these efforts, she has also organised various trips to the Dzilo shrine at Matonjeni in the Matopo hills.

While these allegiances and alliances depend upon both her credibility as the *svikiro* for the ancestors VaZarira and VaMurinye, and the continued support of her own Duma clans, at same time, this 'multiplicity of projects' can also buttress and empower her own authority among these clans. Her connections with the *Mwari* shrines at Matonjeni certainly seem to have empowered both her efforts to be recognised as the right custodian for Great Zimbabwe, and her legitimacy within the VaDuma clans. But it did seem to me that she was often treading a very fine line between all these different alliances and allegiances, particularly in relation to the meetings and *bira* ceremonies she held in 2001 with different groups of war veterans in order to discuss the highly politicised 'fast-track' resettlement programme being implemented by the ruling party. This was brought into sharp relief when her son told me that they had been visited by the CIO (Central Intelligence Organisation) shortly after a *bira* ceremony in July 2001. Given that Ambuya VaZarira is very outspoken in her criticism of the government's perceived marginalisation of traditional leaders, especially *masvikiro*, and the ancestors, and even 'tradition' itself, it is clear that she has to be careful not to undermine her own position by becoming a political threat to the ruling party.

Masvikiro can clearly court controversy in the 'multiplicity of projects' that they are involved in. While multiple alliances can empower their authority, they can also undermine them if, for example, the stakes get higher, more powerful forces such as the state are threatened, or the underling support of others is eroded. Ranger (1982) has described in detail how, during the liberation struggle, a particular medium in Makoni District, Muchetera, was eventually killed by ZANLA (Zimbabwe African National Liberation Army) guerrillas, despite having been instrumental in raising the profile of his spirit – Chaminuka – by co-operating, if not co-authoring a book by Michael Gelfand (1959) which 'fed straight into the rather different interests of the cultural nationalists' (Ranger 1982: 352). Muchetera had used Gelfand's research as a means of gaining white support, because the local support he claimed in Makoni district did not in fact exist. As a result of this need for 'white recognition and patronage' (Ranger 1982: 359), Muchetera courted the white administration and was ultimately used by them for propaganda opposing African nationalism and the war of liberation being fought by guerrillas in the bush. This is why he was killed by ZANLA guerrillas. This is perhaps an extreme example, but it does illustrate how alliances that *masvikiro* make to buttress their support and give themselves legitimacy can also undermine and threaten their authority, and in the context of civil war, even their lives.

When Ambuya VaZarira became possessed by the spirit of VaMurinye at a *bira* ceremony in January 2001, she demonstrated how she is able to manage her 'multiplicity of projects' and multiple allegiances. I quote my field-notes at length because they illustrate how she co-ordinates the tensions between her individual aspirations for the custodianship of Great Zimbabwe; her 'local' role as a senior Duma *svikiro*; and her wider concerns about the state of the country and the role of 'tradition' and the ancestors generally, through her performance as a great ancestor of the past. Note especially the way in which the spirit of VaMurinye that possessed her, both challenged and reconciled with different characters present, drawing them close, offering them beer, and respecting them, while at the same time gathering their respect through her authoritative and harsh performance. This episode shows clearly that the performance of *masvikiro* does not just involve adequately demonstrating the denial of their own agency, but also a highly delicate juggling of alliances and allegiances with others, whose support is depended upon.

Lunch is over, and I am outside having a fag. People are now all gathering in the kitchen hut, drumming and dancing is about to begin.

The drums start and soon a strong rhythm is pulsing around the room. Dancing is dominated by the daughters of Ambuya, and her 'aides'; various Ambuyas and daughters in law.

Ambuya VaZarira takes centre stage, slowly at first, but soon dancing fiercely, her feet stamping the dusty ground in rhythm to the drums. The kitchen is full now, even Chief Murinye has made his way into the hut, and sits on the side waiting, as are Chief Masungunye and VaHaruzvivishe. The dusty, pulsing air is filled with expectation. The tempo of the drums increase, matched by the dancing of the women, who dance together in a line, approaching and retreating, approaching and retreating.

Ambuya VaZarira jerks and makes odd movements as she dances. She stops, stands still clasping her chest, and slowly turns around, looking at everyone in the room, the others carry on dancing. She sticks out her left hand, pointing to a leopard skin by the women's side of the hut. One of her daughters brings it to her. It is draped around her shoulders. Still the drumming continues and again she dances before stopping. Her face carries an expression of deep anguish, pain, terror as she stands still before again slowly turning around looking around the room. She looks over to the right side, and slowly wanders through the dancers, coming down to sit on the left of the hut, in the darkest corner of the hut, not far from behind the open door. One of her aides accompanies her. As she sits down, with her leopard skin around her shoulders, her black head scarf off her head, the women ululate, the drums stop, the dancers sit down. The spirit has come out. Again the women ululate and the men clap, as the possessed medium coughs and splutters and gasps. Chief Murinye's son, and Ambuya's son Peter organise quietly with hand signals, indicating for Chief Murinye to sit on a chair that is brought, close to the medium, and on her other side Masungunye sits on a chair.

After some moments while this is organised, the medium still coughing and spluttering, starring at expectant faces in the room with wild piercing eyes, the spirit begins to speak.

'I am Murinye the fighter, the lion of the forest. No one should argue with me or deny what I say.'

The spirit medium's voice is changed, slightly deeper, and more uncontrolled, sometimes almost grunting , sometimes squeaking. The spirit Murinye announces that this bira has taken place to inform the Duma chiefs of the '*kukanya*' [destruction] of the Duma land. Traditional leaders are being sidelined in daily developments that concern them very much.

Referring to the ex-combatants as '*Vana veSango*' [children of the bush], the spirit Murinye encourages them to work under the guidance of the chiefs and the masvikiro in the fast track resettlement program that is now happening across the land.

'Why neglect me after we worked together during the war? I healed the wounded by removing the bullets. I helped the poisoned. I worked with the masvikiro from other regions, such as those in Chipinge, I even went to as far as Mocambique. I gave hope to the comrades, and asked them to take their guns and go and fight. But why is it that you are now forgetting me? Because you have money, because you are driving cars, you forget about yesterday, what service I offered during the war.'

The spirit continues, urging the chiefs to organise bira ceremonies to straighten things 'or else there will be bloodshed'. People listen with respect, occasionally breaking out into ululation or clapping in unison.

The spirit notices VaHaruzvivishe, and referring to him as VaMugabe, urges him to come closer which he does, sitting on the ground before the possessed medium. Turning to Chief Mazungunye, the spirit asks the chief about recent problems in Bikita, referring to the violence that surrounded the recent by-election. And then the spirit asks the chief, '*Wakabika doro rebirara ndakakuudza here?*' [did you cook beer for the ceremony that/as I told as you to do?] Chief Mazungunye looks a little stunned, taken aback and admits he did not arrange the bira as told previously, because all his brothers who should work with him, are Christians. The spirit then tells him that should not matter, he should still have cooked beer as told.

Chief Mazungunye is very quiet now, so much so that the spirit even demands to know why he is so quiet and is not talking.

The spirit of Murinye then turns to three ex-combatants who are present, urging them to come closer. The spirit asks for and offers them a pot of beer as a token of their part in the fast-track resettlement program, and as a promise that he, Murinye, would still accept their request for assistance in what they are doing. The spirit tells the ex-combatants that they should come to meet Ambuya VaZarira for advice and consultation at a later date as right now there are too many people present.

Through out the rest of the time that the spirit is there, he keeps reminding the ex-combatants, the three present, to urge the rest of them to desist from neglecting and looking down on the traditional leaders, because otherwise there will be chaos in the country.

The spirit has been there for at least an hour but still carries on. He encourages the chiefs to take their own initiative in organising traditional bira ceremonies and not to only follow the orders of the *masvikiro*.

Murinye also introduces his 'grandson' or *muzukuru*; Ambuya VaZarira's youngest son, referring to him as having carried his walking stick and bow and arrows. Again this refers to what Ambuya VaZarira herself said early about her youngest son, who was born just before she first got possessed by the spirit of Murinye. So the spirit requests now for beer from his *muzukuru*; beer that has been brewed and thanked for, to be given to the people present.

Finally   ɪre leaving, the spirit of Murinye states fiercely that he is unhappy with the situation at Great Zimbabwe, especially the fact that permission has to be sought to sweep the '*dare*' and '*mapa*' at Great Zimbabwe, from authorities who are not even related to the VaDuma people. Murinye remembers on one occasion when Ambuya VaZarira, his medium, had to seek permission from a rude officer, who stood there with his hands in his pockets, making her wait for a long time, before finally allowing her in, instructing her not to take long cleaning the place.

The medium has been possessed for a long time now, and the spirit is beginning to indicate that it wants to go now, but still it is there, adding several times that there are two winds, *Mhepo dzenyika*, at Great Zimbabwe. One cold, and one hot, and that it is the hot wind, *mhepo inopisa*, that could lead the whole country into trouble if tradition is not followed.

The people inside the hut have been very patient, listening politely, but I can see that they are becoming tired, and it is clear that the spirit is about to go, as he drops his voice, speaking quietly to those around him. Some of the younger people there (I was surprised to see some of the kids actually inside the hut, during the consultation with the spirit) have already left, and Dzingi and I go outside to discuss what has been said. Before long, when the women are ululating, and as I go back in, I can see Ambuya VaZarira leaning back against the wall, eyes closed. The attention is no longer on the medium, and with a few more gasps and accompanying ululations from the women, the spirit has gone. The spirit medium lies back and rests, while the ex-combatants share out the pot of beer they were given by the spirit. When they indicate to me, I approach, and I am offered a gourd of the strong beer. Ambuya VaZarira asks me if I have eaten *sadza*, and say yes, I have (surprised somewhat by this question, as we ate a while ago in the lounge & she was there), and asking me what with; I said chicken and liver, before going back to my seat with a cup of beer I was given. Ambuya is back, the spirit has gone. (Field notes of *bira* at Ambuya VaZarira's home, 26–27 January 2001)

One of the most striking things revealed by this event is the way in which Ambuya VaZarira's broad 'multiplicity of projects' are conceptualised and presented as a coherent whole through her possession by VaMurinye – 'the fighter, the lion of the forest'. Concerns about the 'destruction of the Duma land', the neglect of 'traditional leaders', the 'fast track' land resettlement programme, and the 'situation at Great Zimbabwe', are unified and given spiritual authority through the words spoken by this ancestor. And importantly, Ambuya's own authority is invigorated and reified through references to the role that she, and her spirit, played during the *chimurenga*, the war of liberation, supporting the freedom fighters.

In the backgrounds of many *masvikiro* lie their experiences of the war of liberation, and their narratives of the roles they and their spirits played are often used to justify their authority in the present. Ambuya VaZarira told me that it was during the war that she became widely accepted among the Duma clans as the *svikiro* for both the ancestors VaMurinye, and VaZarira, though as is clear from the excerpt above, VaMurinye in particular is identified as a 'fighter'. Other *masvikiro* I spoke to made similar claims.

There has been a great deal of literature on the role of spirit mediums and the ancestors during Zimbabwe's war of liberation (Lan 1985; Ranger 1985; Daneel 1995, 1998; Maxwell 1999). Ranger (1985: 187) has shown how with the rise of nationalism in Zimbabwe in the 1960s there came a cultural revival which manifest itself in a regeneration of 'traditional religion'. Lan (1985) highlighted how guerrillas in the Dande in northern Zimbabwe were provided with legitimacy by the spirit mediums of *mhondoro* ('royal ancestors') who drew them into a grand Korekore cosmological system. There has since been a retreat from an initial over-generalisation of the role of spirit mediums in mobilising peasant support for the war, as some writers have stressed guerrilla coercion of the rural peasantry (Kriger 1988, 1992), and others the role played by Christian churches and missions (Daneel 1995, 1998, also Maxwell 1999). Of course some of these differences reflect local and regional variations, but taken as a whole they represent the 'ethnographic thickness that Ortner (1995: 190) has called for, which by 'filling in the black hole would certainly deepen and enrich resistance studies … [and] … should reveal the ambivalences and ambiguities of resistance itself'. The 'ambivalences and ambiguities' that therefore characterised Zimbabwe's war of liberation are reflected in my descriptions of the 'multiplicity of projects', agendas, interests and alliances that *masvikiro*, and indeed all people, are always involved in. It seems obvious when one considers the complexity of individuals and their multiple interests and alliances within any society, that the study of any large scale political and military resistance movement will reveal the ambivalences and ambiguities that Ortner described.

There is not enough space here to do justice to the complex and important ways in which the legacy of the 2nd *chimurenga*, the war of liberation, has continued to be utilised and manipulated not only in the strategies of *masvikiro* in Masvingo province, but also by war veterans (Kriger 2003) and the ruling party (Ranger 2004).[11] It is a theme we will return to in later chapters. For now the most salient point is that the war of liberation, and the role of the ancestors and their *masvikiro* in it, have become a powerful means by which spirit mediums can assert their authority and legitimacy among rural communities in the present. But conversely, this legacy of the second *chimurenga* is also the basis of wide-ranging grievances about the marginalisation of the ancestors and *chikaranga* by the government since independence. Spirit mediums and other 'traditionalists' often frame this perceived marginalisation of 'tradition' in terms of a betrayal of the ancestors, and readily attribute to it droughts, floods, car accidents and any other 'troubles' that occasionally effect the country.[12]

Among rural communities around Great Zimbabwe in Masvingo district, such grievances often relate directly to local concerns about not only the representation of the site's past, but also its management. This was illustrated by the claim of the spirit VaMurinye, that 'it is the hot wind, *mhepo inopisa*' at Great Zimbabwe 'that could lead the whole country into trouble if tradition is not followed'. It is a view that was mirrored by VaChainda, who scorned the role of *uchenjera hwebhuku* at Great Zimbabwe, and lamented the demise of the *masvikiro*.

> *VaChainda: Because if we look back at the chimurenga, we can see there wasn't much need for education …* *what used to happen is that the **kuchenjera** [education/knowledge/learning] of people that* *have died, what we call the **svikiro**, was being relied upon. They relied a lot upon the masvikiro because* *masvikiro tell the truth … but today because power is in their hands, its over for the masvikiro, … now* *they are concerned with the **uchenjera hwebhuku** [knowledge of books]. That's why we are* *experiencing a lot of trouble in this country …*

> *J.F.: So in terms of Great Zimbabwe, do you think that they are relying on education, what you call* *'kuchenjera', and they are not relying on the masvikiro?*

> *VaChainda: That's the truth … they are using kuchenjera. kuchenjera is ruling … Its not from the* *masvikiro.* (Interview with VaChainda 30/12/2000)

Apart from demonstrating how concerns about Great Zimbabwe lie within a plethora of wider issues (which can stretch beyond the very localised desires of clans and individuals to be recognised as the owners or custodians of the site) what such use of the *chimurenga* legacy also suggests is that for *masvikiro* such as VaChainda or Ambuya VaZarira – whose ambitions go far beyond local clan concerns, but whose local support must be maintained – it is of vital importance to be seen to be actively engaged with broader social and political discourses. As Spierenburg (2003: 173) has pointed out, the extent of mediums' mobilising power is not always clear and I have argued in this chapter that *masvikiro* constantly rely upon, and perform for, others around them to substantiate their own authority and legitimacy. In this sense, Ambuya VaZarira, and others like her, may have no choice but to be involved in the wider 'multiplicity of projects' in which she engages, despite the risk involved as exemplified by the case of Muchetera which Ranger (1982) described. It remains to be seen how *masvikiro* across Zimbabwe have engaged with recent events such as the 'fast-track' land reform programme, and the widespread and increasing social, economic and political turmoil effecting the country.[13] It seems certain, however, that despite their marginalisation by central government, Ambuya VaZarira and others like her, will continue to engage with the broader issues affecting communities across rural Zimbabwe as well as their own particular concerns, and those of their clans.

CONCLUSION

This chapter has focused on the means by which 'traditional' leaders in rural Zimbabwe establish the legitimacy of their narratives, positions and authority. While references to 'tradition' and the spiritual authority of the ancestors are a key part of this process, of equal if not more importance is the ability of actors themselves to perform their narratives and roles convincingly. Authority and legitimacy may be derived from 'above', that is the spirit world or the government, but just as crucial is the support they receive from 'below', that is society. This is particularly true for spirit mediums, who, unlike chiefs and village heads, are not recognised in local and national state structures and receive no formal recognition or government allowances. This accentuates the need for spirit mediums to perform convincingly as liminal, ambiguous characters situated between the world of people, and the world of the spirits. On a day to day level this performance of spirit mediums is not so much about creating the illusion that they are the ancestors themselves, as Lan (1985: 68) argued, but rather involves an emphasis on the ambiguity of their agency, as an entanglement of that of the ancestor and their own. Only during possession ceremonies, when the distinction between the medium and the ancestor is sharply contrasted and clearly defined, are mediums seen to become the ancestors themselves, and it becomes crucial for them to be able to convincingly deny their own agency.

Apart from the importance of demonstrating the authenticity of their mediumship through convincing performances during possession ceremonies, the authority of spirit mediums is also based on their alliances and allegiances both within their own clans and beyond. While wider allegiances maybe based on the underlining support of their own clans, the support of their own clans can also be buttressed and empowered by the 'multiplicity of projects' with which spirit mediums are sometimes engaged. In the case of Ambuya VaZarira, her links with the *Mwari* shrines of the Matopos certainly seem to have strengthened her authority within the VaDuma clans. But wider projects and allegiances can also threaten the authority of spirit mediums, and therefore, as demonstrated by Ambuya VaZarira's possession by the ancestor Murinye in January 2001, the multiplicity of projects a spirit medium is involved in may also make demands on their performances during possession ceremonies.

Finally, the authority that spirit mediums can derive from the claims they make about their role, and that of their spirit, during the war of liberation suggest that it has perhaps become increasingly important for spirit mediums to engage actively with the wider issues effecting rural society today. The apparent paradox that we are left with then is that while spirit mediums may base their authority on their role as intermediaries between the past and the present – as vehicles through which the ancestors can communicate with their descendants – in practise it may be vital that they are at the very forefront of events as they take shape, promoting not the maintenance of the status quo, but rather change in the order of things.

Notes

1. An earlier version of this chapter was first published in 2004 as ' "Traditional Connoisseurs" of the Past: The ambiguity of spirit mediums and the performance of the past in southern Zimbabwe' *Occasional Paper* no 99 (University of Edinburgh).

2. An early anonymous reader of this chapter commented that 'the use of the term "connoisseurs" is … an unfortunate choice of phrase' and 'highly inappropriate in such a religious/political context'. Along with analysing spirit mediums' narratives in terms of 'performance', its use underplays, according to the same commentator, the importance of the content and social context of these narratives. I appreciate how, taken in isolation, this chapter could provoke such a response. However, placed in its proper context within a book that focuses precisely on both the content and context of these narratives, its purpose and importance becomes clear. The point is to consider how the status and legitimacy of any narrative of the past depends, to some degree at least, on the ability of its narrator, writer or performer to engage with an audience. In the case of spirit mediums, like archaeologists, historians and heritage professionals, this involves claiming authority as an expert or specialist, hence the term 'connoisseur'; defined in one dictionary I have to hand, as 'a person with special knowledge of the arts, food or drink' (Collins 1999: 166), and broadened deliberately in this case to include the 'past', however defined.

3. In a recent paper, Ranger (2003) has explored the role of women in what he calls 'ecological religion' from pre-colonial Zimbabwe to the present. He argues that 'despite the constraints of patriarchy, men and women play complementary roles in Zimbabwe's eco-religious ideology and practice' (2003: 86).

4. This was particularly the case among members of the Charumbira clan, where there is a very long-standing and bitter chieftaincy dispute.

5. This dynamic was also very apparent during the group interview with Nemanwa elders, which I described in the previous chapter, when a dispute emerged about the role of the Rozvi in relation to Great Zimbabwe's past, during which the chief himself was directly, though politely, contradicted by several of his elders.

6. Interview with VaChainda at his homestead (30/12/2000).

7. Unfortunately I was unable to get a copy of this map, or a reference for it from National Archives. I did examine it at the time and it looked genuine. That Mugabe's territory was much larger and included a large section of Nemanwa's current area corroborates with other sources which clearly show that, for a considerable time, Nemanwa people were removed from their land and lived under Headman Nemanwa on Chief Charumbira's land. For more detail on recent efforts by members of the Mugabe clan, in the context of recent 'fast-track' land reform, to reoccupy land that used to be Mzero Farm see Fontein (2005).

8. Interview with Chief Fortune Charumbira (13/8/2001).

9. While it was expected at independence that the role of chiefs would diminish because of their perceived complicity with the Rhodesian state (see Bratton 1978: 50, Lan 1985: 149, also Ranger (1982) cited in Alexander 1995: 185), in the event there was a re-emergence of their role in rural areas during the 1980s and 1990s as the government increasingly offered chiefs and headmen concessions in the from of salaries and the return of some of their responsibilities. The 1999 Traditional Leaders Act has been seen by many as part of the ruling party's continuing efforts to co-opt chiefs and 'extend its hegemony deeper into rural areas at a time of political discontent' (Chaumba et al., 2003b: 599). The most recent stage in this ongoing process was the announcement of their support of the government's land reform programme by the National Assembly of Chiefs meeting at Great

Zimbabwe in May 2004. At the same meeting chiefs were also promised higher allowances and access to a new car loan scheme (*The Herald* 6 May 2004, 8 May 2004 and *Zimbabwe Independent* 14 May 2004, see also *The Masvingo Star* 23–29 July 2004). Although, as Alexander (1995: 186) has correctly pointed out, 'the spirit medium is more accurately seen *as part* of a traditionalist faction', and they often work very closely with chiefs, they have never been offered any of the concessions that chiefs have received.

10. Daneel (1998) describes the origins and early years of the traditional environmental movement he helped launch under the auspices of ZIRRCON (Zimbabwean Institute of Religious Research and Ecological Conservation) and its sister organisations AZTREC (Association of Zimbabwean Traditional Ecologists) and AAEC (Association of African Earth Keeping Churches). Ambuya VaZarira played an important role in Daneel's AZTREC until she switched her allegiances to a splinter group called AZTREC Trust, formed by a disgruntled former senior member of AZTREC and ZIRRCON, who had been accused of financial mismanagement. According to Daneel (1998: 299), her 'defection' disappointed the Duma chiefs who kept their allegiance with the original AZTREC. By the time of my fieldwork, however, she was no longer involved with AZTREC Trust, and her support among the Duma chiefs was strong. Both AZTREC Trust and Daneel's AZTREC still continue their environmental projects, though there was a sense in which support of these two NGOs among different communities in Masvingo district was aligned according to clan divides.

11. Ranger (2004) describes the recent emergence of a new, ruling party-dominated, discourse of national history which he calls 'patriotic history'. Highly politicised, and monolithic in nature, one of 'patriotic history's' main characteristics is its constant reinforcement of the ruling party's liberation credentials. Great Zimbabwe's position in this 'patriotic history' is discussed in the last chapter.

12. See Spierenburg 2003 and Derman 2003 for recent accounts of how spirit mediums in northern Zimbabwe express their sense of their own marginalisation by the state by attributing drought to the anger of the ancestors. See also Fontein (2005).

13. In another paper (Fontein 2005) I have begun to examine the way in which a shared *chimurenga* legacy has been invoked on the ground by spirit mediums and war veterans involved in the occupation and redistribution of farmland in Masvingo Province (see also Chaumba et al., 2003b). The emerging literature on the role of spirit mediums, chiefs and others in recent land reform in Zimbabwe is still of a very preliminary nature (see Chaumba et al., 2003a, 2003b; also Alexander 2003), and it remains a very fertile area for further research.

CHAPTER 4

THE SILENCE OF GREAT ZIMBABWE: FROM MYSTERY AND UNTOLD STORIES TO A SILENCE OF ANGER

Zimbabwe (After the Ruins)

I want to worship Stone
because it is Silence
I want to worship Rock
so, hallowed be its silence.

for in the beginning there was silence
and we all were
and in the end there will be silence
and in the end, we all will be.

Silence speaks to fool and wise man
to slave and king
to deaf and dumb
to blind man
and to thunder even.

for in the beginning there was silence
and we all were
and in the end there will be silence
and we all will be.

The mind that dreamt this Dream
massively reading unto time and space
the voice that commanded
the talent that wove the architecture:
friezes of dentelle

herring bone
check patterns,
chevron
and all
the many hands that put all this silence
together,
the forgotten festivals at the end of the effort;

All speak Silence now – Silence.

And behold these stones
the visible end of silence
and when I lie in my grave
when the epitaph is forgotten
Stone and Bone will speak
reach out to you in no sound
so mysteries will weave in your mind
when I'm gone
Because silence cradles all –
the space and the universe –
and touches all.

M. Zimunya (1982: 99–100) – displayed in
the Great Zimbabwe National Monument
site museum.

The idea of silence at Great Zimbabwe is a powerful one that has frequently been used in many, very different, representations and writings on Great Zimbabwe. In the above poem, written by a well-known Zimbabwean poet since independence, there is a sense of nostalgic silence; silence as sacred and timeless, something carrying a universal wisdom from a mysterious past to an unknown future. Certainly the romantic idea of silence as 'mystery' and awe is one that has been frequently used in relation to Great Zimbabwe. This is not altogether very surprising given that at certain times of the day, or on days when there are few tourists and visitors, Great Zimbabwe is indeed a very quiet place. Wandering around the towering stone walls and huge boulders of the hill complex in the evening or on a quiet misty morning, I was often struck by a profound silence that seems to resonate almost from within the stone walls themselves. And this experience of silence is often reported in literature on Great Zimbabwe. Richard N. Hall, one of the Rhodesian antiquarians of 'foreign origins' persuasion, wrote of his first night at Great Zimbabwe in 1902, having been hired by the BSA Company to explore and open up the ruins.

> To describe this grand ruin in one article would be utterly impossible, and any statement of one's first impression, on walking about the Temple amid massive titanic walls must be altogether inadequate. At any rate, one experienced an overwhelming and oppressive sense of awe and reverence. One felt it impossible to speak loudly or to laugh. (Hall in *Rhodesia* 12 July 1902, National Archives file S142/13/5; see also Hall 1905: 4)

The experience of such profound silence at Great Zimbabwe must have influenced the early explorers and antiquarians as they developed their fantastically romantic theories of its past. A few months after he wrote the above account, Richard Hall described 'Sunday at Zimbabwe' for a South Africa newspaper.

> Wandering about the elliptical temple at Zimbabwe on a Sunday morning, one is faced at every turn with texts for innumerable 'Sermons in stones'. The houry age of these massive walls is grandly and silently eloquent of a dead religion – a religion which was but the blind stretching forth of the hand of faith groping for the deity. (Hall in *The Rhodesian Times* 27 September 1902, see National Archives S142/13/5; see also Hall 1905: 12)

Such romantic references to silence and mystery at Great Zimbabwe have not disappeared as the 'Zimbabwe Controversy' has been bypassed and debates on Great Zimbabwe have moved on from the early focus on the question of origins. An article by Ken Mufuka in the daily newspaper *The Herald* (24 July 1982) in 1982 carried the headline, 'SILENT WITNESS TO PAST MYSTERY. The enigmatic soul of Great Zimbabwe'. The 'mystery' Mufuka referred to was not about 'origins', as for him that question was easy to solve. As he put it, 'the more one looks at the monument, the more one appreciates the grandeur of the Shona – Karanga civilisation to which it is a silent witness' (Mufuka 1983, 1984: 9). Mufuka's 'mystery' concerned the 'nature of DZIMBAHWE', and his purpose was an explicitly nationalistic one of telling 'the story of Great Zimbabwe as the natives whose ancestors built it have told them'[1]. Mufuka's description of Great Zimbabwe as a 'silent witness' to a 'Great Shona – Karanga civilisation' illustrates how notions of silence and 'mystery' continued to be used after independence in 1980 to describe Great Zimbabwe.

Silence as 'mystery' has also been used to create romantic images of Great Zimbabwe for marketing and tourism purposes, both during the Rhodesian era and after Independence. As Kuklick notes (1991: 156), when Rhodesian stories of Great Zimbabwe's 'ancient origins' 'could no longer be proclaimed unabashedly, the ruins were instead promoted as "Rhodesia's Mystery"'. Despite continuing efforts by the National Museums and Monuments of Zimbabwe (NMMZ) to present the site as undoubtedly of 'African origin', the promotion of the site as a 'mystery' has continued after Independence. This is illustrated by the headline, 'THE MYSTERY AND MARVEL OF GREAT ZIMBABWE' that appeared in the 'Travel Supplement' to *The Herald* and *The Chronicle* on 24 June 1981, which outlined the 'slow recovery' of tourism after the drawn out war of liberation.

But if images of silence at Great Zimbabwe have emphasised a *romantic* idea of 'mystery' around Great Zimbabwe, it has also been used to represent a *lack of knowledge* or history of Great Zimbabwe; a past not yet 'uncovered' or even literally, the lack of a *knowable* past. Recalling a visit to 'King Solomon's Mines' around 1904, Mrs Archibald Colquhoun wrote in the imperial journal, *United Empire* in June 1914:

> The strangeness of these Rhodesian relics of a forgotten past lies in the utter silence in which they are enveloped, no inscription no voice from bygone centuries such as we get in Egypt or Asia Minor. (Colquhoun 1914: 486–7)

She continued

> They constitute a historical puzzle of the most controversial kind. The buildings and rock mines spread across Rhodesia have no known history. The natives who inhabit the country, and have done so for 700–900 years, know nothing of the buildings. (Colquhoun 1914: 486–7)

This idea of 'silence' as an unknown or unknowable past, and particularly an unknowable *African* past, was based on the lack of written history and a deliberate disregard for oral traditions as viable constructions of the past. Writing for the *Science Digest* (February 1968) L. Sprague De Camp put it as follows,

> If the unmarked stones of Zimbabwe could speak they could tell us who piled them up and when and why; but they cannot. In the absence of written records, the history of Negro Africa is inevitably vague and sketchy before the coming of the whites. (L. Sprague De Camp 1968: 18)

It was this refusal to accept oral traditions as legitimate sources for knowledge of the past that was one of the main inspirations for Ken Mufuka's *DZIMBAHWE*, as he clearly stated at the end of his introduction.

> And yet in returning the history of our people to themselves, we had to battle with intellectual imperialism as well. *Archaeologists* who could not speak any African languages insisted that there was no oral evidence worthwhile. Thus they arrogated to themselves the role of chief interpreters of a culture they knew miserably little about. We hope that we have delivered the first blow in that battle. (Mufuka 1983: 8)

Mufuka was by no means the first or only historian to see the value of oral traditions. While oral history has become an accepted source of historical enquiry since the 1960s, especially through the work of historians like Jan Vansina (1965, 1980, 1985), it is instructive to note the amount of criticism Mufuka received from widespread quarters for the uncritical, and romantic application of oral traditions to Great Zimbabwe. The work of Zimbabwean 'TV' historian and now Minister for Education, Chigwedere (1980,1982, 1985), has received similar criticism (Garlake 1983:15). Even the 'cognitive archaeologist' Thomas Huffman has been criticised for his 'inappropriate' application of both Venda ethnography, and recent Shona oral traditions, to a much earlier time period (Beach 1998). Academic historians – contrary perhaps to 'nationalist' historians like Mufuka and Chigwedere – are clearly very critical about how and when oral history should be used. Maintaining a grip on the need for 'objective' narratives of the past, means that 'controversial' narratives are easily shot down. This limited and restricted acceptance of oral traditions indicates that for historians and archaeologists, oral traditions do not have the same purpose or value that they carry for oral informants themselves. In this sense oral historians 'co-opt' oral traditions for their own purposes, and use them according to their own rules.

While the study of oral history has become much more common and acceptable among academic historians generally, in relation to Great Zimbabwe it is still seen as problematic mainly due to the length of time that is involved. In Beach's opinion, 'even when they exist oral traditions about Great Zimbabwe … have to be regarded with caution unless it can be shown how they could have survived over a span of some six or seven centuries' (Beach 1998: 49). Thus for many historians and archaeologists and other scholars of the past, archaeology remains the favoured route into Great Zimbabwe's past (Garlake 1973: 76; Beach 1998: 49). This has resulted in the *silence of unheard voices and untold stories* of local communities surrounding Great Zimbabwe that I have been describing. The 'history-scapes' of local communities are marginalised because they do not refer to the time period when Great Zimbabwe was built and occupied 'originally'. There has been, and continues to be, a bias in favour of the period when the 'Zimbabwe state' emerged, existed, and according to some abruptly ended. While the existence of these 'unheard voices' and 'untold stories' are sometimes acknowledged by the archaeologists at NMMZ who run Great Zimbabwe, they are not clearly represented in the site museum or in the published literature.

The 'history-scapes' of local communities are not methodologically or empirically 'gathered' or 'uncovered'. They are told, retold, and renegotiated according to the political interests of competing clans and individuals. They consist of what historians and anthropologists may call 'myths', but are also often set in opposition to the knowledge of 'books' or 'education' by these actors themselves. More importantly, they may involve a concept of time and the past that is not linear and progressive, but rather is based on the ancestors who are potentially contactable. Going one step further, it may be that these histories are as of yet incomplete because the ancestors themselves may not yet have had a chance to speak. As VaChainda in an interview put it,

> *J.F.: What do you think about the history that you were told at school, and the history that the Museums present of Great Zimbabwe? Do you think its true?*
>
> *VaChainda: What used to be experienced at Great Zimbabwe and what we have in the books today, are different things.*
>
> *J.F.: How?*
>
> *VaChainda: Because someone who does things and someone who reports, are different.*
>
> *J.F.: What would you like to be seen done at Great Zimbabwe? What history would you like to be told about Great Zimbabwe?*
>
> *VaChainda: I can say I would only agree with a history of Great Zimbabwe, if it was told by the mhondoro dzeZimbabwe* [senior/royal ancestors of Zimbabwe – ie the country], *about how they came to this place … how it was built and why. Then I would agree.* (Interview with VaChainda, 30/12/2000)

What is very clear from this statement is the idea that the ancestors themselves should be consulted directly to gain a 'true' history. As we have seen, the distinction between the 'truth' or knowledge of the ancestors, and that of 'the book' relates to a much wider range of grievances of 'traditionalists' across postcolonial Zimbabwe. These grievances are often directed against the government, the state, or the ruling party, and focus on the perceived sidelining of 'traditional leaders' and *chikaranga* generally. Chiefdomship wrangles often invoke these grievances, especially when the state is seen to be somehow involved, manipulating who gets chosen, and therefore opposing *chikaranga*. Great Zimbabwe fits amongst these grievances in various ways. For many 'local' actors, it has to do with their particular claims on the site. For others it takes a more national perspective involving a perceived necessity to hold a national ceremony or *bira* to thank the ancestors for their help during the liberation struggle. Sometimes these two coincide. What is clear is that in relation to Great Zimbabwe this perceived opposition between a knowledge of books/education, and the knowledge of 'tradition' – or specifically the ancestors and spirits – is not just relevant to the *representation* of Great Zimbabwe's past, it also concerns the physical *management* of the site.

In this context I can introduce another form of silence at Great Zimbabwe. This is silence experienced as a *loss* of something; silence representing desecration, or to put it in the words of Aiden Nemanwa, '*the silence of anger, not happiness*' (Hove and Trojanov 1996: 80). This *silence of anger* refers to something that many informants talked about in relation to Great Zimbabwe; that it used to have a *Voice*, or voices and sounds that could be heard early in the morning, or in the evening. Some described it as the *Voice of Mwari* that now speaks at the *Mwari* shrines in the Matopos (Daneel 1970; Ranger 1999; Nyathi 2003), while others described less specific sounds of people going about their daily business, milking cows, grinding corn, whistling, ululating and drumming. People also referred to sounds of cattle bellowing, cocks crowing, and goats bleating. All these sounds had no visible source and, importantly, are no longer heard. This silence represents Great Zimbabwe as a desecrated site; the voices and sounds have gone.

THE SILENCE OF ANGER

The construction of the past and the management of the physical remains of it (that is, heritage sites) are intricately interdependent. And it is for this reason that it becomes overtly political (with a small 'p') and an important issue for the 'traditional connoisseurs' whose voices are unheard. This is how Great Zimbabwe features most among a list of grievances that are centred on a distinction between, as my informant VaChainda articulated it, *Kuchenjera* and *Chikaranga*; the knowledge of books/education versus 'traditional knowledge' and the ancestors themselves.

Writing on the development of the representation of the past and heritage in museums in a European context, Walsh argued that pre-industrial, rural society's 'awareness of the past was an experience which was entirely more organic than that understanding which was to develop in the modern world' (Walsh 1992: 11). It was an 'awareness of the past' built on a sense of place, which Walsh considered to be

> a more organic form of history, one which recognises the crucial contingency of the past processes on present places. Places, natural and man-made features, acted as 'time-marks', physical phenomena which exists in the present but possess, for those who know them, a temporal depth which gives them special meaning. (Walsh 1992: 11)

Walsh suggested that the processes of industrialisation and urbanisation helped create a view of time as linear and progressive, and the past became viewed as increasingly distant, or 'foreign' as Lowenthal (1985) has put it.

> Towards the end of the nineteenth century, there does seem to have been a developing awareness of the importance of the past, but this importance was increasingly neutered by the developing perception of the past as something which was separate and had limited contingency for modern societies. (Walsh 1992: 12)

In a previous paper (Fontein 2000) I argued that this was reflected in the emergence of the academic disciplines of history and archaeology. Fundamental to the development of these disciplines was a combination of the idea of progress through linear time, an enchantment with a distant and separate past caused by a period of massive social change, and the search for objective truths (Fontein 2000: 8). While Walsh constructs an opposition between two notions of the past – a pre-modern idea of the past that is closely linked to the present through a sense of place, and a modern one based on linear, progressive time, and is distant from the present – I would suggest that this itself reveals a particular linear and 'progressive' perspective on time on his part.[2] Instead I would argue that there exist different ways of perceiving and constructing the past, and that they compete for authority but are not necessarily mutually exclusive. In relation to my fieldwork, people living around Great Zimbabwe do construct and perceive the past through a sense of place, but also through memory, oral traditions and importantly, through the ancestors themselves, in dreams and through possessed mediums. This involves a performance of the past, which is in a close relationship with the present. But as I have already discussed, these constructions of the past also often borrow from wider discourses of the past, including those of the 'professionals'.

If a sense of place does feature in the construction and perception of the past, then it also works in reverse. Place itself, and what should be done with it, is perceived, imagined and constructed through a sense of its past.[3] And this is in no way exclusive to 'traditional connoisseurs' from among surrounding clans. Rhodesian uses and abuses of Great Zimbabwe clearly demonstrate how ideas about the past can inform practices and approaches to places and landscape in the present. These ranged in degrees of audacity and absurdity from the burial of the remains of the 1893 Alan Wilson Patrol at the site (until their removal to assist Rhodes's post-humous appropriation of the Matopos in 1904 – see Ranger 1999: 31–32, 40) to the 'psychic' seances of H. Clarkson Fletcher (1941) in the Great Enclosure in the 1930s, when contact was made with several dead Rhodesian heroes, including Alan Wilson and Richard Hall, alongside more 'ancient' characters such as 'Abbukuk' the 'high priest', 'Ulali' the 'last of the Queens of Great Zimbabwe', and of course her lover 'Ra-set'. While the early onset of tourism further shows how Great Zimbabwe was captured in the Rhodesian imagination from the very start – Bent (1896: 64) mentioned how he was frequently visited by tourists during his stay at the ruins in 1892 – the existence of two competing hotels, and a golf course within, or just outside the boundaries of the reserve by the mid-1930s,[4] illustrates how impudently this Rhodesian imagination of the past in turn, informed approaches to the site's management. And archaeologists and 'heritage professionals' today, are no different either. Great Zimbabwe as a place or landscape, and approaches to its management, are similarly imagined or constructed through particular ways of constructing and imagining its past. This is most clearly evident in terms of heritage management.

Heritage management does seem to take its lead from archaeological and historical perspectives of the past – the development of the two are closely 'entwined' (Ucko 1994: 271). Of course it involves much more than archaeology and history, they are to some extent separated (see Lowenthal 1998) by the fact that 'heritage management' caters for a larger, wider audience – most obviously the tourist industry but also, for example, nationalist sentiment.[5] But they do share a concept of the past; in particular, a concept of the past as 'distant' and separate. This is reflected in how Great Zimbabwe fits into the contemporary surrounding landscape, distanced and separated through both time and space from local communities. The following passage from Bender's work on Stonehenge (1998) is illustrative here.

> In the context of the contemporary obsession with preserving and commodifying the past, it becomes particularly urgent that we take the measure of the landscape, both theoretically and in practice. More often than not, those involved in the conservation, preservation and mummification of the landscape create normative landscapes, as though there was only one way of telling and experiencing. They attempt to 'freeze' the landscape as a palimpsest of past activity. But, of course, the very act of freezing is itself a way of appropriating the land. For the Heritage people freezing time and space allows the landscape or monuments in it to be packaged, presented, and turned into museum exhibits. We need to recognise that this is just one way of handling the past. We need to work against this passive, nostalgic, heavy-with history notion of landscape. (Bender 1998: 26)

If this is true of Stonehenge, a similar situation has existed and continues to exist at Great Zimbabwe. Many local actors, 'connoisseurs' or not, complained bitterly about having to pay at the gate in order to go in. For those with their own claims to the site, for example that their ancestors are buried there, this is very strongly felt. Some local characters appear on a NMMZ list allowing them free entry, but this is restricted largely it seems to those with the loudest voices. Their conduct is restricted – to conduct a ceremony for example, requires permission from the regional director, who himself refers the request up to his own superiors in Harare. From before independence to the present day, National Museums and Monuments have had problems with squatters, fence cutting, cattle grazing, wood cutting and poaching, which, taken as a whole, reflect the distancing of Great Zimbabwe from local communities.

The history of Great Zimbabwe's management since its 'discovery' by European explorers in the late nineteenth century mirrors, to a certain extent, the development of academic discourses about its past. If the early years of the representation of Great Zimbabwe was dominated by European explorers and Rhodesian antiquarians, with their absurd ideas of biblical origins, ancient civilisations and rumours of gold, then this is mirrored by what actually happened at the site. Like many ruins across Zimbabwe, the site fell victim to treasure and relic hunting that has been described in one tourist brochure as the 'rape of Great Zimbabwe' (PhotoSafari (Pvt.) Ltd. 1999: 9).

Before 1900, other ruins in the country fared worse at the hands of the B.S.A.Co. authorised Ancient Ruins Company than Great Zimbabwe (Kuklick 1991: 142). But Great Zimbabwe by no means escaped, and William Posselt's dubious acquisition of one the Zimbabwe Birds in 1889 from Chief Mugabe, then residing on the hill

(Matenga 1998: 22), was certainly followed by other unreported cases of 'Indiana Jones' or 'Tomb-raider-style' theft. By 1902 however, Rhodesian public opinion had turned against the destruction caused by treasure hunters. That year the Ancient Monuments Protection Ordinance became law and laid 'the foundation of the present heritage management system in Zimbabwe' (Pwiti 1996; Pwiti and Ndoro 1999; Ndoro 2001: 15; see also Ndoro and Pwiti 2001;). Ironically, some of the same characters previously involved with the Ancient Ruins Company re-emerged as 'bona fide archaeologists' (Kuklick 1991: 142). In 1902 Richard N. Hall managed to get employed by the BSA Company for 'not scientific research but the preservation of the building' (Garlake 1973: 72). But as Kuklick noted, 'Hall exceeded his charge, recovering a goodly number of relics – and disturbing the site so as to make the stratigraphical reading of the archaeological record even more difficult' (1991: 143). Randal-MacIver, often heralded as the first professional archaeologist to study the ruins, was very critical of Richard Hall, not just for his 'ancient/exotic origins' ideas, but also his suitability for excavation work. In a letter to Sir Lewis Michell, Randal-MacIver wrote,

> Mr Hall is not only unqualified for excavation work he is positively disqualified by temperamental incapacity. Not only does he blunder in interpretation, not only is he incapable of giving a straightforward account of a plain thing, but he is not conscientious in executing his duties on the spot. (Letter from D. Randal-MacIver to Sir Lewis Michell, 23 November 1905. National Archives A11/2/18/66)

And further on he continued,

> The reckless blundering of his excavation of the Elliptical Temple is almost indescribable (see figure 5.3). Except in one tiny corner he has lifted up and carted bodily away everything above foundation level. That is to say he has destroyed all the original huts, and removed all the objects found in them.

> Even a trained man should hardly have been allowed to do so much on a site which is unique. And if he did he would have been considered obliged to publish the most scrupulously exact and minute account of every little detail. Mr Hall has published no such account; his *Great Zimbabwe* may be considered a standard example of what an excavator's report should not be. (Letter from D. Randal-MacIver to Sir Lewis Michell, 23 November 1905. National Archives A11/2/18/66)

The debate between Randal-MacIver and Richard Hall was perhaps the fiercest of the 'Zimbabwe Controversy', and this letter indicates that this extended debate did not only concern the representation of Great Zimbabwe's past, or who built it. It was minutely concerned with what happened at Great Zimbabwe itself. Hall's belief that Africans could not possibly have built Great Zimbabwe 'led to his destruction of most of the Dhaka structures and artefacts, which clearly indicated the indigenous origin of the site' (Ndoro 2001: 40). This supports the argument that place is often dealt with according to the view of its past held by those doing the managing. The professional archaeologists of the time strongly criticised the destruction of archaeological, stratigraphical evidence, and continue to do so today. The explicit opposition that was created and reinforced through the 'Zimbabwe Controversy', between 'scientific' and 'amateur' ways of studying the past, was therefore reflected in how the place itself was treated.

Other people are also now blamed for their 'unprofessional' conduct in the management of Great Zimbabwe. In 1911, having already been in charge of the Police

Camp for several years, Corporal Wallace was seconded from the B.S.A. Police, and began work as 'caretaker' of Great Zimbabwe. Later, when concern over the survival of the ruins reached a peak – partly in anticipation of a rush of visitors accompanying the opening of an expected new railway line – Wallace took up full time employment under the Department of Public Works.[6] After some initial tension with Hall who had managed to get himself re-appointed by the B.S.A.Co.,[7] Wallace began what turned out to be his life's work – rebuilding huge sections of walling at Great Zimbabwe. By 1915, under the instruction of Douslin from the Department of Public Works, Wallace had also managed to destroy most of the archaeological deposits in the western enclosure of the Hill Complex, which continues to be lamented by archaeologists today alongside Hall's destruction of deposits in the Great Enclosure. But it was the unguided and undocumented rebuilding of the vast walls that has left perhaps the most visible mark on Great Zimbabwe. From close reading of early descriptions of the ruins, it becomes very apparent how different the site looked a hundred years ago. As Edward Matenga, the former regional director, put it during our walk around the site in July 2001 'all the directors that have worked here have left their own mark', but Wallace's is unrivalled.

Wallace's liberal (though it should be admitted structurally sound) rebuilding did eventually raise some eyebrows in the Rhodesian administration and the Historical Monuments Commission (formed after the 1936 Monuments and Relics Act). After Wallace's retirement in 1948, his successors did not take long to condemn his 'inauthentic' re-constructions. Indeed by then an important offshoot of the 'Zimbabwe Controversy' was the notion that the preservation of archaeological remains should be the responsibility of someone with archaeological training, which Wallace lacked. Minimal rebuilding of fallen walls continued at Great Zimbabwe after Wallace, but not anywhere close to the scale he achieved. At least until independence in 1980, when for a brief period under Ken Mufuka, some extensive and undocumented rebuilding took place among the eastern enclosures. But that was again quickly condemned after Mufuka's departure, when NMMZ began to find its feet and the direction it was to take. It took the 'professional' route, and with assistance of UNESCO, Great Zimbabwe became a World Heritage Site in 1986, and site preservation and reconstruction became the primary functions of the archaeologists managing Great Zimbabwe.

To its credit, since the early 1980s, NMMZ has become a very capable and 'professional' heritage organisation, with a great deal of expertise in the continual surveillance, monitoring and preservation of dry stone walls. Furthermore, it has achieved some quite impressive reconstruction works, the 1995 reconstruction of the Western Entrance of the Great Enclosure (Matenga 1996) being perhaps its pinnacle achievement. Yet it is significant to note that this institutional development mirrors the professionalisation of discourses about the past which have sidelined different ways through which the past is constructed. It is no surprise that the top positions within NMMZ go to trained archaeologists, most of whom have come through the same channel – an archaeology degree from the History Department at the University of Zimbabwe. In a sense, archaeologists now dominate the management of the past at Great Zimbabwe, as much as they dominate the representation of it. The two are, perhaps, inseparable. Local clans and individuals are as ignored in the site's management as they are in terms of representation. In VaChainda's words

Thats the truth … they are using Kuchenjera. Kuchenjera is ruling … Its not from the masvikiro.
(Interview with VaChainda, 30/12/2000)

Thus, perhaps as an indirect consequence of the *Zimbabwe Controversy*, the professionalisation of the representation of Great Zimbabwe's past has come hand-in-hand with the gradual professionalisation of the site's management, and in the process the appropriation of the site has continued, distancing, separating and alienating it from the wider landscape and the communities that surround it. In this sense, the management of Great Zimbabwe did not witness a decisive break with its colonial past after independence in 1980.[8] Interested stakeholders, both local and 'national', particularly chiefs, *masvikiro* and, more recently, war veterans – as well as, most crucially of all, the ancestral spirits, and the *Voice* to whom Great Zimbabwe ultimately belongs – are no more meaningfully consulted about the management of the site today than they were before the name 'Zimbabwe' was bestowed upon the country in 1980.[9] As I will show in the second part of this book, herein lies the kernel of the *silence of anger* at Great Zimbabwe; the inability of the postcolonial state, and NMMZ in particular, to overcome the colonial past in order to engage with other, non-archaeological ways of imagining and managing the past, place, heritage and landscape. As Aiden Nemanwa once explained to Chenjerai Hove (Hove and Trojanov 1996:79–82) 'life has changed …

… The new education and wisdom have brought strange ways and learning. The ancestors are not fools. So they said: you have brought your ways and new wisdoms. The new education came, wanting to dig into holy places, searching for the voices of our ancestors which spoke from the caves. They wanted to know why it was that the implements and tools of the ancestors had to be returned to the shrine at certain times without fail. Where did those tools and implements come from? The new wisdom yearned to know. The newcomers took those holy implements to strange lands without asking anyone. Defilement. That was the way to defile the stone shrines. All they saw and envied they took away. That brought the anger of our ancestors.

That was the beginning of the silence of our ancestors. The silence of anger, not happiness. From that day, they hid all those tools and wares which had remained behind. They hid them in their ways, they took them away from the eyes of those who remained. If the people came back and confessed their ways and repented, promising to follow the ways of our people, the ways of respecting holy shrines like the Dzimbabwe, the ancestors will return the revelations which they took away in anger and frustration.

When the silence came, it was the Year of Silence. Not many things were said. There were many things which we could not figure out from the silence. It was the silence of anger, not happiness or rest. The soil is sick now. The earth cannot smile at us.

Why? When we fought to rule the land, there were elders who were sent by the politicians to request that the name of these shrines be the name of the country. The shrines, this home, was not for the people living near here. Messengers came to bring voices, to receive voices from these shrines. We, the spirit mediums, did it as messengers. We made the requests in the ways of our ancestors.

[…]

The war to free our land was a war of the ancestors, to restore the respect and holiness of our land, our ancestors. The children who went to fight were called *Children of the Soil*.

The laws and rituals of the land have not been followed up to now, since the return of the leaders. The only thing is that they gave to the country the name which they had requested from the ancestors. The other rituals which they were supposed to perform, they did not. The spiritual leaders of this country do not know what happened. The soil got annoyed. It was not the fault of the ancestors. It was not only them who got angry. Even the spirits of the children who died in the land of strangers are also angry. They cry to be returned to the soil of their birth. (Aiden Nemanwa, quoted in Hove and Trojanov 1996: 79–82)

The silence of the ancestors and the *Voice* at Great Zimbabwe – this silence of anger – represents the desecration and alienation of the site which began with the arrival of Europeans at the end of the nineteenth century, and continues today. While we have seen that members of surrounding clans put forward competing claims over Great Zimbabwe, their narratives of the processes which have led to its 'desecration' are remarkably similar. As I will show in the second part of the book, they illustrate how the processes through which a place becomes a national and international heritage site can alienate local communities and thereby undermine any 'spiritual values' associated with it.

Notes

1. Zawaira and Malorera in their 'Foreword' to Mufuka's (1983) *DZIMBAHWE: Life and Politics in the Golden Age 1100–1500 A.D.*
2. Like Walsh, Pierre Nora (1989) also constructs an artificial, and problematic, opposition between 'modern' and 'traditional' ways of perceiving the past. McGregor (2004: 22) argues in relation to stone ruins in north west Zimbabwe, that this opposition between 'memory' and 'history', 'tradition' and 'modernity', provokes an 'anti-historicist, apolitical understanding of pre-modern memory'. Her paper emphasises the politics and 'historicity of traditional means of relating to the past' which defies simple oppositions between 'memory' and 'history'. I make a similar point in Chapter 2; local discourses of the site's past are, in part, based on a sense of place and memory, but also borrow and utilise ideas from archaeological and historical discourses. My specific point here is to turn this problematic opposition around; while the past can be constructed through a sense of place, 'place' itself is also often perceived of and constructed through a sense (or multiplicity of senses) of its past.
3. This argument relates to a broader academic debate about the 'ambiguity' of the landscape concept (Gosden and Head 1994), which has been focused around a central, polemic distinction between the physical and the conceptual, or in Layton and Ucko's terms (1999) the 'environment' and 'landscape'. While one view (e.g. Daniels and Cosgrove 1988: 1) of landscape is as 'a cultural image' or representation, phenomenologists (e.g. Ingold 1992 and 1993; see also Tilley 1994) have attempted to shift the focus onto an understanding of landscape based on action and movement in it. Despite Ingold's very deliberate efforts, this debate has tended to reify the modernist, mind/body, culture/nature or 'understanding/ explanation' (Layton and Ucko 1999: 2–3) distinctions upon which it is based. Although a post modern approach would seek to overcome such distinctions,

according to Layton and Ucko (1999: 3) 'it is questionable whether either archaeology or anthropology has fully come to terms with the challenge'. For my purposes here it suffices to say that peoples' relationship to the past/place is based on the interplay of both their bodily existence ('dwelling') in and action ('management') on/upon the landscape, and their mindful representations and discourses of/about it.

4.   See National Archives files S917/a312/802/1-3, S914/12/7, S917/A312/800, S533T312/79. The golf course was maintained right up to the advent of the liberation struggle in the area in the late 1970s. The Van Reit plan of 1973 recommended that the golf course be closed and although this was met with disquiet from some white residents of Fort Victoria, alongwith plans to charge entrance fees, both happened in 1978 (NMMZ file 0/3 and National Archives file H15/10/1/3,10,20).

5.   Of course the close links between archaeology and nationalism that scholars have exposed (e.g. Gathercole and Lowenthal 1990; Diaz-Andreu 1995; Kohl and Fawcett 1995; Abu El-Haj 2001) further illustrate the close links between the discipline of archaeology and 'heritage' management.

6.   National Archives files L2/2/192 and W1/3/4/1-3.

7.   Apparently finding himself impoverished in 1913 Hall made various requests for employment to the B.S.A.Co. Eventually he was appointed in February 1914 as curator and guide only. At the same time Wallace had his temporary contract extended as he continued restoration work in 'Temple'. Soon the two men found themselves in disagreement, and in June Wallace resigned 'because this ridiculous old man will not leave him alone' (Letter from Douslin to Inskipp, 18 June 1914, National Archives file L2/2/192). With the support of his superiors in the Department of Public Works, Wallace was retained by the B.S.A.Co. board, and in a letter dated 27 July 1914, Richard Hall was told to move to Fort Victoria from where he would take visitors to the ruins until the end of his contract in March 1915. By August, Hall was lying sick in Bulawayo Memorial Hospital, where he died on 18 November 1914 (National Archives file L2/2/192).

8.   A similar argument about heritage management in Zimbabwe being a 'product of colonialism' has been made by Ndoro and Pwiti in various papers (Ndoro 1996; Pwiti 1996; Pwiti and Ndoro 1999, Ndoro and Pwiti 2001).

9.   Similar issues about the neglect of local communities' concerns in the management of heritage sites, and the importance of recognising the spiritual and intangible aspects associated with them have now begun to be raised in relation to other sites in Zimbabwe. One important example is Pwiti and Muvenge's paper (1996) on the conflicts and disputes around Domboshava Rock Art Site between NMMZ and local communities for whom it is an important place for conducting rain-making ceremonies.

PART II: THE SILENCE OF ANGER AT GREAT ZIMBABWE

SILENCE, DESTRUCTION AND CLOSURE AT GREAT ZIMBABWE: LOCAL NARRATIVES OF DESECRATION AND ALIENATION

When Mwari, Musikavanhu was creating this land, he gave an instruction 'let there be rivers, let there be trees let there be grass and these things began to appear, but there were some mountains he made holy/sacred, there are some rivers that were also made important, and some trees.
(Interview with Ambuya VaZarira, 19/11/2000)

So Great Zimbabwe' sacredness vanished because the whites fenced the area, and we lost or just despised our traditional rules. That's one big precaution/word of caution that I would give, that when you do your things do not despise or look down upon 'chikaranga' [tradition] that would assist you.
(Interview with VaMatanda, 8/1/2002)

SACRED LANDSCAPES

One of the most striking features of nearly all local accounts of Great Zimbabwe's sacredness is that they describe things which used to happen there, a long time ago, and no longer occur. Therefore, any discussion of Great Zimbabwe as a sacred site must also be a discussion of its desecration. But before we consider local narratives of Great Zimbabwe's desecration and alienation – this silence of anger at Great Zimbabwe – it may be useful to consider briefly how spirituality or sacredness can be embodied in the landscape, in much the same way that landscape can embody the past. Indeed, it is the link with the past that makes landscape sacred. As we have seen, for the Mugabe and Nemanwa clans who continue to dispute ownership of the site between themselves, Great Zimbabwe is of special significance because it is related to their respective pasts. Both claim their ancestors lived there, and for the Mugabe clan their ancestors were buried there, and for Nemanwa it is where they 'germinated'. But in this sense, Great Zimbabwe is not unique in the wider landscape; rather it is but one particular place in a landscape that is dotted with sacred places. Indeed, if it is the link with the past that makes a place sacred, then the entire landscape is sacred because of the close association of the land with the ancestors and *Mwari*.

This close association of the land with the spirit world has been noted and described in much of the academic literature on 'traditional religion' in Zimbabwe (Bourdillon 1976; Garbett 1966, 1977; Gelfand 1959; Lan 1985; Ranger 1967, 1985; Werbner 1989). Much of the more recent work has focused on roles played by the ancestors and *Mwari* as the *owners of the land*, during the liberation struggle (Lan 1985; Daneel 1995; Ranger 1967, 1985) and as *guardians of the soil* responsible for 'traditional' conservation and the fertility of the land (Daneel 1998; Makamuri 1991; Schoffeleers 1978; and Ranger 1999). It is the emphasis upon these two aspects which differentiates Shona concepts of ancestral relationships to the land from that reported in the ethnography of Australia and elsewhere, where landscape is often conceptualised and perceived of as a fixed embodiment of the past actions of the ancestral beings (Morphy 1995: 187). This way of conceiving the landscape is not appropriate to Shona perceptions of the relationship between the land and spirit world. While landscape can embody past events, the ancestors and *Mwari* do not, as Morphy (1995: 188) described it in Yolgnu terms, '*turn into* place'. Rather than in this sense *being* the very form of the landscape, the Shona spirit world shadows or parallels the human world, and as owners and guardians of the land, and the people on it, it exists separate from it.

However, at times the spirit world (the ancestors, *Mwari* and other spirits) does manifest itself in the landscape through rocks, caves, pools and trees. It can also appear as animals, especially lions and eagles, and perhaps most frequently of all, the spirit world emerges among people themselves, by possessing spirit mediums or appearing in dreams. In this way parts of the landscape, certain animals and certain people can act as vehicles for communication between the parallel existences, or worlds, of the spirits and people, particularly on those ritual and 'sacred' occasions when these separate 'worlds' do share temporal and spacial dimensions.

Stirrat (1984) has suggested that sacred places exist at the point of tension between Eliadean and Durkhiemian concepts of the sacred, which are complimentary and a feature of all religions. Using his formulation, these 'vehicles' (places, people and animals) through which the world of the spirits meets the social world of people occupy a liminal position between these worlds, and carry some of the attributes of the spirit world with them. In Stirrat's terms, they 'exist uneasily both within space and time and outside space and time' (Stirrat 1984: 209). As I discussed in Chapter 3, *masvikiro* (spirit mediums) are both distinct from the ancestors that possess them, and yet when they are possessed they almost *become* them. Even in their daily lives, and especially their own narrated pasts, the agency of the medium becomes intertwined with that of the ancestor. They are treated with respect, reverence and even fear by those around them, whether they are possessed or not. Similarly with sacred places, and none more so than Great Zimbabwe, they are both located in the present landscape, within the territories of a chiefdom (or contested between several), and yet they are governed by rules and customs that refer to the past and the spirit world, and which separate them from the rest of landscape.

The idea of 'liminality' grew out of the study of ritual in the work of Van Gennep (1909), and was developed by Turner (1969) among others. Here the sacred is viewed as, using Van Gennep's phrase, 'betwixt and between'; the point of confusion where the social and the divine meet. For Turner (1969), the liminal phase of a ritual is characterised by what he termed 'communitas'; a transcendental feeling of social unity, a state of 'direct unmediated contact' between individuals free of social structure. Yet ultimately this state of 'anti-structure', works to reaffirm basic social values and structures. And this is where a weakness lies that is repeated in much of the anthropological writing on ritual and religion. As anthropologists have recently noted (Asad 1983; Kertzer 1988; Kelly and Kaplan 1990; Mitchell 1996: 490–3), while functions, structures, and symbolic meanings of rituals and symbols have been emphasised, practise and power have been ignored. There has been an emphasis (which I want to avoid) on how values, structures and meanings are reproduced, reaffirmed or (in Marxist analysis) concealed through ritual, rather than on how practise shapes the meanings of rituals for participants, and, importantly, how social structures, values and meanings can be altered through the conscious manipulation of rituals and symbols by actors themselves.[1] I have described how the authority of the spirit world can be used by *masvikiro* to serve their interests and political agendas. But this authority is not a given, they are dependent upon society, and the effectiveness of their performance during ceremonies. Therefore their authority is not just a function of their status, but also of their practise. And this is equally applicable to the sacred places on the landscape – the caves and rocks, springs and rivers, mountains and trees – that also act as 'vehicles' through which the social and sacred worlds meet. Not only are these liminal places on the landscape *marked* by specific customs, taboos, and rules regarding access and use it is through these practises themselves that their sacredness is achieved. A place is not sacred unless treated as such.

This then provides the context within which we can now consider how Great Zimbabwe is viewed as a desecrated site by the local communities surrounding it. But desecration may be the wrong term, because it implies permanence. Many people did suggest that the sounds, voices, and other 'miraculous' features could return if *chikaranga* or *tsika dzechivanhu*, traditional rules and customs, were once again followed, and the spirits themselves consulted. Indeed the 'silence of anger' referred to by Aiden Nemanwa implies that Great Zimbabwe's sacredness has been suspended by the angered spirits until such a time that their authority is once again respected.

SOUNDS, VOICES AND LOCAL DISPUTES

J.F.: What are your earliest memories of Great Zimbabwe?

Aiden Nemanwa: It used to have a voice

J.F.: What kind of voice?

Aiden Nemanwa: There were the voices of people. They sounded like people talking to each other. Also these voices used to talk to visitors that came there.

J.F.: When did the voices stop?

Aiden Nemanwa: They finished with the coming of the white man. There used to be long back, some other white people that came to trade, Indians and others, but the voices only stopped when the white colonisers came.

J.F.: Was it the Voice of Mwari?

Aiden Nemanwa: There were voices in Shona. These people were not physically there, they were 'spiritual people'. You used to be able to see the big wealth of these people physically, but not the people who were only spiritually there. There were sounds of cows, goats, and bows and arrows which could be seen. Those spirits could communicate with people, but were not seen. The ruins were not constructed by people but by spiritual beings. Among the things that could be seen were utensils for building and hunting. The cows and goats could only be heard but not seen. (Interview notes, Interview with Aiden Nemanwa, 21/10/2000)

Most accounts of Great Zimbabwe's past sacredness referred to sounds and voices that used to be heard among the stone walls and on the hill. Some people described these sounds and voices as the general sounds of unseen people and animals going about their daily business: people ululating, grinding corn, playing the drums, whistling, and cows bellowing, sheep and goats bleating and so on. But others were very specific, emphasising a single *voice* in a way that resembles descriptions of the *Voice of Mwari* that speaks at various shrines in the Matopos hills (see Daneel 1970; Ranger 1999; Nyathi 2003). Indeed it was quite common to hear the suggestion that the two are the same voice; that *Mwari* (God) or *Musikavanhu* (lit. 'creator of people') used to speak at Great Zimbabwe before moving to the Matopos.

> *The place that is now called Matopos, the voice was not there long back, the voice was speaking at Great Zimbabwe, but when the whites came into this country the voice transferred to Matopos.* (Chief Nemanwa, Group discussion with Nemanwa elders, 18/7/2001)

Some people, however, seemed less keen to draw a direct comparison with the Matopos shrines or to state overtly that the *Voice* was that of *Mwari*. The Nemanwa clan's most articulate spokesman, Aiden Nemanwa, was very careful to avoid suggesting that the *Voice* that speaks in the Matopos hills is the same as that which used to speak at Great Zimbabwe. This may be related to his concern to secure Nemanwa's claim for the custodianship of the site, making him wary of the possibility of outside interference from people connected with the Matopos shrines.[2] This reminds us that local narratives about Great Zimbabwe's past sacredness and desecration are located within local disputes and claims over the site's custodianship. For some prominent members of the rival Mugabe clan, the link between Great Zimbabwe and the *Voice of Mwari* has been used to directly promote their claims to the site. VaHaruzvivishe, a elder of the Mugabe clan and descendant of the last pre-colonial resident of Great Zimbabwe, claimed that Ambuya VaZarira, a senior *svikiro* of all the VaDuma clans, had been granted authority by *Mabwe aDziva* ('the stones of the pool' – which refers to the Matonjeni shrines) to '*gadzirira*'[3] Great Zimbabwe.

> *VaHaruzvivishe: In 1993 I was asked by Ambuya VaZarira to go with her to Mabwe aDziva*
>
> *JF: There at Matopos?*
>
> *VaHaruzvivishe: Yes. So there Ambuya VaZarira was told that: 'It is you, who is supposed to be the custodian of that Mountain* [Great Zimbabwe]. *When you have prepared/cleansed that hill there, you should come to Mabwe aDziva and do the same thing here at Mabwe aDziva'*
> *Because Mabwe aDziva is a mountain that came from here* [Great Zimbabwe]. *It helps the mountain here, it is like a branch of Great Zimbabwe.* (Interview with VaHaruzvivishe, 4/3/2001)

Whether the *Voice* of the Matonjeni shrines has actually given its support to efforts by Ambuya VaZarira and others to revitalise Great Zimbabwe is hard to fathom, although she certainly does have connections with the Dzilo shrine and makes fairly frequent visits in relation to her various wider efforts to revive the role of the *masvikiro*, respect for the ancestors, and *chikaranga* (tradition) in general. Her claims do, however, help us to place local stories about the sounds and voices at Great Zimbabwe, into a wider discourse about the importance of Great Zimbabwe as a regional sacred site in Masvingo province, and even as a national sacred site.[4] Daneel has described how the 'traditionalist' environmental organisation AZTREC – which he helped set up – became part of a much wider 'religio-cultural movement and an outlet for political frustration' (1998: 153), which involved

> attempts of Shona traditionalists in post-independence Zimbabwe to move the old cult centre from the Matopos to Great Zimbabwe, or at least to a new site in Masvingo Province not far from the ruins. To many *masvikiro* [such] attempts … are evidence of a fairly common belief that, prior to the emergence of the Matonjeni oracle, Mwari's voice was heard at Great Zimbabwe, possibly during the Rozvi reign. (Daneel 1998: 152–53)

This idea that Great Zimbabwe was once a previous centre for the *Mwari* cult has been dismissed by some academic historians (Beach 1973b; Mtetwa 1976: 114). Despite this, it is clearly an idea that is very much alive and gaining ground across Zimbabwe today. In local narratives such wider discourses on Great Zimbabwe's sacredness and national significance are interacted and negotiated with in a politically engaged manner through the prism of local claims and disputes. Therefore for most members of the Mugabe and Nemanwa clans Great Zimbabwe is sacred both in terms of their own ancestors, and much 'higher' spirits such as *Mwari* himself/herself.

OTHER SACRED FEATURES AND THE SILENCE OF RESISTANCE

Sounds and voices are by no means the only 'sacred features' that have been ascribed to Great Zimbabwe. Other features include secret underground passages and caves, *nhare* or *ninga*, that link up with other sacred places across the country including the Matopos shrines and Chinoyi caves, and where it is said people used to hide from Ndebele raiders. There are also many stories about endless supplies of ripe tomatoes and *chechete* fruit at the ruins, as well riches of gold and diamonds, secret hoards of iron hoes and tools in the Conical Tower, and, of course, the mysterious pot with four legs *pfuko yeNevanji*, which bit off the hand of the man who tried to remove the gold it contained. Still other stories refer to animals. The *mhondoro* lions associated with the ancestors, leopards, pythons, and even the baboons and monkeys that now swarm all over the site – all feature in accounts of Great Zimbabwe's sacredness.[5] It is clear that besides the clan differences, Great Zimbabwe can mean different things for different people within these local communities. Or in other words, within the limits of widely known stories about sounds and voices at Great Zimbabwe, people are able to narrate and construct their own experiences and stories about Great Zimbabwe as a sacred place. Therefore a profound dream I had early on in my fieldwork was explained by Mai Rukasha (in whose household I stayed) in terms of Great Zimbabwe's sacredness (see below). Similarly one *n'anga* ('traditional

---

Extract from field notes: 'Kumusha, 26 August 2000'

It had been a good night though next door at Sekuru Nemanwa's household, there was a party that went on all night, which kept a lot of us awake. I had a very profound dream, which shook me up quite a lot. I told it Amai to this morning, and her response was interesting.

I was with several people from Kumusha, like Amai, Baba and others who were not directly clear. We were on the back of a pick-up, or next to one, at a growth point, or at some shops, but in a rural area, and one of us became sick with a mudzimu, possessed. A nearby n'anga or svikiro was got, and some reddish soup with medicine 'mushonga' put in it and offered to the 'sick' person. Then a small puppy, just like the one they have here started lapping at the bowl, and even standing in it, when it turned and looked at me, with deep piercing eyes.

Suddenly it was like something jumped into my head, through my ears or the back of my head, almost from the eyes of the little dog, and seemed to try and take over my head. I fell off a chair I was sitting on, and the voices of those around me became very distant. They seemed to be saying that I had a spirit that was trying to come out and indeed that was what I was thinking, as it felt like something was trying to take over my head. Then I woke up with a start.

When I told Amai this story of my dream, she told me it was because I was back here near Great Zimbabwe. She said that I am always telling her that I have 'full-on' dreams when I am here – referring to a series of unusually vivid and startling dreams I had during my last visit here in 1997, when she had placed the burning coals of African sandalwood in my room at night, to prevent bad dreams. As she had done in 1997, today she told me, it was because I am near 'to that place, the hill'.

healer') and dancer who regularly performs for tourists within the ruins, described it as a place where people used to come to be healed, and outlined what happened when she accompanied a film crew around the ruins

People used to go to that mountain to ask for help curing their illness or disease ... it would be healed. That mountain could help you. Say if you couldn't get children, you didn't know what to do, you could go there. You would become very sleepy, and in your dreams you would be told what to do and whether you were going to have a child or not. If you have bad things, you would not be able to sleep in that mountain, you would be chased away by a leopard. If you are a good person who does not kill or use witchcraft, you will sleep well, and have good dreams you wouldn't find any problems, but if you are bad you couldn't spend a night there.

Even now it happens, I have seen it. There were some men who came to do some filming in that mountain. Then we met a big baboon, and we had to take some snuff to put down, and clapping, we said 'Aiwa, please can we pass, kukumbira kufambe zvakanaka, [requesting to pass safely] *we are only looking for food'* [laughing] *that's when the baboon disappeared.*

It went mad!! Really!

There were seven men with us.

So it is a sacred place even now. (Interview with Mai Mafodya, 23/1/2001)

Apart from illustrating how personal experience can be woven into narratives of Great Zimbabwe's sacredness, this account also emphasises the importance of correct conduct, of being a 'good person' at Great Zimbabwe, which is common to many of the stories of its sacredness. I was often told that people disappeared or were lost in Great Zimbabwe if they failed to follow the rules.

Maiwa! yes, we used to go there and eat tomatoes and chechete. How then would I know that people were climbing there clapping hands first? If you did not clap hands when climbing, then there were some spirits that made people disappear. Then when you were in that situation of not knowing where you are going, the only thing you could do was to dig a little hole, and then you spit into it, then you would start to see the vision of where you have come from. (Interview with Ambuya VaZvitii, 7/11/2000)

The importance of following the right procedure reveals the liminality of a sacred place; set apart from the rest of the landscape, it has its own rules and customs that refer not to the world of people but to that of the spirits. These rules and customs both indicate that a place is sacred, setting it apart in the wider landscape, and are essential for that very sacredness; a place, after all, is not sacred unless treated as such.

We can say that even us, we can destabilise the status of the winds that are found there, because if we are told to enter some places, and **not** *enter some other places, if we disobey that we can actually destabilise the sacredness. At the same time, there used to be people, old people who are allowed to enter some of the sacred places, only the old women and the old men, who are no longer young.* (Interview with Ambuya Chibira, 4/12/2000)

What the rules and customs should be, and *chikaranga* itself, are of course debateable, as is the required status, age, gender and of course clan/totem of the 'proper' custodian, and I am sure that elders of the Nemanwa, Mugabe and Charumbira clans would have much to disagree on here. But all accounts do agree that at Great Zimbabwe the sounds and *Voice* have become silent because the correct rules and

Figure 5.1: The Hill Complex at Great Zimbabwe, showing fire damage in the fore ground. (Author 2001)

customs, that is *chikaranga*, have not and are not being followed. The spirits have not been respected; the place is not treated as sacred. As one Mugabe *svikiro* put it

> *Musikavanhu who was speaking in Great Zimbabwe has decided to keep silent because he is angered by what is happening at Great Zimbabwe, so Musikavanhu is quiet. But he is still there, not willing to talk because it hurts to see what is happening at Great Zimbabwe at this moment ... .*

> *It's like if my mudzimu is hurt by what I am doing, it will just get out of me, and stay somewhere because it will be hurting, that's the same as what has happened at Great Zimbabwe, Mwari was speaking in the stones, he just left the monument, and decided to keep quiet, but he is still there at Great Zimbabwe* (Interview with Ambuya Jowanny, 12/3/01)

In this sense the silence of the voices and sounds at Great Zimbabwe is also a silence of resistance; the resistance of the spirit world against the refusal to follow *chikaranga*. Focusing on the silence of anger as a form of resistance reminds us that there is always the potential for the voices and sounds to return, if the spirits and *chikaranga* are once again respected. And silence is not the only form of resistance the ancestors use. Bush fires (see figures 5.1 and 5.2), drought and winds are also signs of the anger of the spirits.[6] Ambuya VaZarira gave me a clear explanation for a fire she witnessed at Great Zimbabwe.

Figure 5.2: NMMZ staff and volunteers fighting a fire on the Great Zimbabwe estate in September 2001. (Author 2001)

Ambuya VaZarira: Great Zimbabwe got burnt. The spirits once told the Ministers, that the mountain at Great Zimbabwe would catch fire ... because they have failed to abide by the rules of the stones. And of course the fire broke out, and the mountain was burnt, but they could not stop the fire by any means. It could only be stopped by rain, it was rain that stopped the fire. [Rain is often associated with the benevolence of the ancestors and Mwari]

J.F: When was this?

Ambuya VaZarira: I do not quite remember, some of these things are written down, probably 1986. I went to the offices to tell them, that we should abide by the rules of chikaranga but they do refuse, because we never had fires when we used to follow the rules, but now because we do not follow the rules, the hill catches fire nearly every year. The masvikiro [referring her not to mediums but the ancestors themselves] *claim that it is 'us ourselves that burn the mountain, out of anger because people do not want to listen'*

The first year when the mountain caught fire, we were at a meeting at Nemanwa Growth Point. CIO [Central Intelligence Organisation] *was there, CID* [Central Investigation Department] *was there, the ZRP* [Zimbabwe Republic Police] *was there, the ministers were there, and the DA* [District Administrator] *was there. As we were at the meeting, the mountain roared, and then came a wind, and that wind did not go further than where we were, just stopped there.*

As we were there, there came another wind, a stronger one, it came exactly to where we were, and everyone had to disperse. And Minister Zvobgo's car was almost lifted up, and also Minister Mudenge's. No one said a word about it, except me. I just called out Hokoyi!

It was only then that the big wind went, then the mountain caught fire.

J.F: Do you know why the mountain caught fire?

Ambuya VaZarira: It is the midzimu that were punishing the people because they do not want to follow the rules that are set by the spirits. They don't want to make our place important/sacred. So as a punishment they burnt it. (Interview with Ambuya VaZarira, 27/11/2000)

This explanation for a fire at Great Zimbabwe reveals the extent of the anger of the spirits at the failure of those in charge at Great Zimbabwe to follow the 'correct' rules and customs. It also reveals that the silence of Great Zimbabwe is not only understood to be the result of what happened during the colonial period, though it is usually agreed that this is when the site's destruction and desecration began. The silence of Great Zimbabwe continues today; the rules of the spirits are still not being followed.

DESTRUCTION, THIEVERY AND LOOTING

The silence of Great Zimbabwe is frequently stated to have begun with the arrival of Europeans at the end of the nineteenth Century, and the destruction they caused as they dug for gold and relics, or for the source of the mysterious sounds and voices.

The main problem which caused the Voice to transfer, is that when the whites came into this country they came and dug on the spots where the Voice was speaking, getting gold and diamonds and other things, valuable things from that place. So that is why the Voice decided to transfer to Matopos. (VaTogaripi, Group discussion with Nemanwa elders, 18/7/2001)

Some of these descriptions mention the use of 'binoculars' or 'machines' or even explosives in the search for gold or money, to illustrate the kind of disruption the intrusion of these uninvited foreign visitors caused.

There used to be gold inside the monuments, and so they [white people] *used to speculate, using maybe binoculars, so they were seeing gold in it, using their own machines, and because of that they started destroying the stones.* (Interview with VaChinomwe, 9/11/00)

The sounds went away because they were using explosives which produced sounds like 'paooooooo!', so the Vakaranga were afraid of that. (Interview with Ambuya Sophia Marisa, 8/3/2001)

These accounts often add that the whites were unable to actually get the gold or money they sought, which always disappeared mysteriously, implying that despite their use of 'alien technologies', Europeans were no match for the powers of the spirit world.[7] Nevertheless the disruption and destruction caused by the whites gracelessly 'clumping with their shoes' was enough to upset the balance required by the spirits.

And the year when the white man came, they used binoculars and they saw that there was a clay pot that was full of gold. Then they took some picks and shovels, and some of our brothers were taken from here to go and dig and the white man was seen using binoculars, maybe their machine, because they were seeing that there was a clay pot that was full of gold. Then they destroyed all the stone walls which are facing to the east, thinking that they would see this clay pot full of gold, but when they had completed their destruction, they found nothing. Up to then the Great Zimbabwe ruins had never been climbed with shoes, it was always climbed with bare feet. Then the white man came and started clumping with their shoes and everything started to be lost. (Interview with VaMututuvari, 4/11/2000)

The emphasis upon the destruction caused by whites hunting for gold or for the source of the mysterious sounds and voices in local narratives, mirrors the condemnation that the early 'amateur' antiquarians and explorers, particularly Hall, have received (and continue to) from later 'professional' archaeologists lamenting the destruction of archaeological deposits and stratigraphical evidence at Great Zimbabwe and related sites.[8] But it is very significant to note that local narratives draw absolutely no distinction between the treasure-hunting antics of the early European explorers and antiquarians and the careful, 'scientific' excavations of the later archaeologists. This confirms what Linda Tuhiwai Smith has written

> And, of course, most indigenous peoples and their communities do not differentiate scientific or 'proper' research from the forms of amateur collecting, journalistic approaches, film-making or other ways of 'taking' indigenous knowledge that have occurred so casually over the centuries. The effect of travellers' tales, as pointed out by French philosopher Foucault, has contributed as much to the West's knowledge of itself as has systematic gathering of scientific data. From some indigenous perspectives the gathering of information by scientists was as random, ad hoc and damaging as that undertaken by amateurs. There was no difference, from these perspectives, between 'real' scientific research and any other visits by inquisitive and acquisitive strangers. (Smith 1999: 2–3)

Aiden Nemanwa described how the excavations worried the spirits, who closed the secret caves, roads and *ninga* to prevent their secret contents from being discovered.

> *Throughout all the excavations, the spirits were very worried, as well as being worried about their sons being killed. The 'ninga' and caves and roads were closed by the spirits because the white people were trying to find out what was inside, so they were closed. The spirits of the land are no longer seen because of that harassment being done to the land.* (Interview notes, Interview with Aiden Nemanwa, 21/10/2000)

Similarly, Samuel Haruzvivishe made it clear in his account that there is no meaningful differentiation between the reckless pillaging of the early antiquarians and the 'scientific' excavations of the 'professional' archaeologists, because neither made any attempt to consult the custodians of the place.

> *J.F.: There has been a lot of archaeological digging at the ruins. What did the spirits think about this?*
>
> *VaHaruzvivishe: The spirits were not happy about this, even the chiefs were not happy. They knew where the voice of Mwari used to be heard, and the sounds of sheep and goats and cattle, as well as the sounds of people pounding grain. We were much pained that these things stopped, that the white people stopped paying respect to those people, as custodians of that place. Perhaps they could have consulted the custodians about how the place should be looked after.* (Interview notes, Interview with VaHaruzvivishe in Ambuya VaZarira's presence, 28/10/2000)

The worry, anguish and pain that such accounts describe is brought into sharper relief when we consider that the manual work for many, if not all, of the early diggings and excavations was carried out largely by men from the surrounding communities (see figures 5.3 and 5.4). This is well documented in the accounts of early explorations written by Bent (1896: 31,66,69), Willoughby (1893: 5–6), Hall (1905: 31–50) and others. Indeed most of the manual work carried out at Great Zimbabwe, be it excavations, the clearing of vegetation, rebuilding of walls and so on, has always been, and continues to be carried out by members of local communities. No doubt many did so willingly[9] – lured by meagre wages in the form of blankets or cups of salt (Bent 1896: 68–69; Hall 1905: 43) – although there is also evidence that chiefs negotiated on the behalf of their people in attempts to gain European support in their disputes with other clans (Willoughby 1893: 6 and Mtetwa 1976: 195). On other occasions, the threat of force was used to ensure local co-operation.[10]

Figure 5.3: Hall's excavations and clearances in the Great Enclosure, 1902–04. (National Archives of Zimbabwe 1734)

Instances described by both Bent (1893: 79–80) and Hall 1905: 19, 43) clearly demonstrate that local workers and elders were often deeply troubled by the work being carried out. Bent described the annoyance he caused when he uncovered the remains of a sacrificed goat (1893: 79–80), and Hall how he persuaded and bribed 'my native boy' to accompany him into the 'Elliptical temple' [Great Enclosure] one night at full moon, despite the man's obvious unwillingness[11] (Hall 1905: 19). Moreover, some of these insensitive diggings involved the disturbance and removal of recent graves on the hill (Bent 1896: 79; Hall 1905: 43, 94). Bent described how they 'unwittingly' opened several graves, one of which was that of the man who had died the year before, whose brother came to complain (Bent 1896: 79). Hall wrote that he came across about fifty graves,[12] of which the 'remains in a score of instances were removed' (Hall 1905: 94). Interestingly, while people I spoke to often lamented the destruction caused by diggings and excavations at Great Zimbabwe, no one referred to this desecration of graves on the hill.[13]

Apart from destruction, local stories about the desecration of Great Zimbabwe by Europeans often referred to the looting of artefacts that occurred widely, and is very well documented by historical sources.

In that mountain there were so many things. Wild tomatoes were germinating, even hoes were found there and so many other things including Pfuko YaKuvanji, [and] the Zimbabwe Bird (see figures 5.5 and 5.6)

[...]

That's where the mistake came from, and after that mistake, they did not take measures to repair those things. They took the Zimbabwe Bird and Pfuko YaKuvanji, and people were going there thrashing those tomato plants with slashers, and then everything went away. (Interview with VaMakasva, 13/2/2001)

Of all the relics that were stolen, and disappeared from Great Zimbabwe during the early excavations and diggings at the site, the case of the soapstone carvings of birds

Figure 5.4: Locals employed as labourers for R. Hall's excavations, 1902–04. (Taken in 1903 by Mrs G.N. Flemming)

known as the Zimbabwe Birds (see figure 5.5), is certainly the most documented and widely known (see Matenga 1998). The story of their removal from Great Zimbabwe at the end of the nineteenth century and beginning of the twentieth century, and the final return to Zimbabwe of most of the birds in the early 1980s has come almost to symbolise the cultural colonisation of Zimbabwe, and the ultimate victory of its independence in 1980. The last stages in this process was the recent re-unification of two halves of one of the birds at a nationally televised ceremony held at Great Zimbabwe in May 2003 (Ranger 2004: 226), after the return of one half by the Museum für Völkerkunde in Berlin (Matenga 1998: 38). The bird was subsequently 'officially' handed over to the 'Masvingo chiefs' (who then passed it on to NMMZ to keep for the nation) at the National Assembly of Chiefs held at the site in May 2004.[14] The amount of state pomp invested in the 2003 re-unification ceremony, in particular, reflects the national status of the Zimbabwe Birds today. These birds take pride of place in the site museum, and images of them feature on the national flag, coat of arms, the currency, as well as numerous logos of both public and private companies in Zimbabwe. Apart from Great Zimbabwe itself, there is nothing which has been invested with as much national symbolic sentiment as these birds. They have become an intricate part of the spiritual and historical basis of the state of Zimbabwe. As Matenga put it,

> The determination by the Zimbabwean government to reclaim the birds stemmed from a desire to rehabilitate Great Zimbabwe as a cultural symbol of the African people. The desire was inspired by the belief that the potency of Great Zimbabwe as the guardian spirit of the nation lies in its possession of sacred artefacts such as the conical tower and the Zimbabwe Birds. It was imperative to bring back the bird emblems in order to re-equip and revive the shrine of Great Zimbabwe. (Matenga 1998: 57)

Given the national iconic importance of the birds, it is surprising that local narratives about Great Zimbabwe's sacredness and desecration had relatively little to say about them. When the Zimbabwe Bird was mentioned (it is usually mentioned in the singular – 'the Zimbabwe Bird' – reflecting the fact that one bird in particular has become the national icon) its removal from Great Zimbabwe, along with a variety of other objects, was the focus. Discussion with locals about the Zimbabwe Bird rarely went beyond the wider debates that Matenga has carefully discussed (Matenga 1998: 63–83) about what they were meant to represent. Common ideas that Matenga wrote of – about it representing either the bateleur eagle, *chapungu*, which is often considered to be a messenger from *Mwari* and ancestors, or that it represents the fish eagle, *hungwe*, a totemic bird which the controversial historian, and now Minister of Education, Aenus Chigwedere (1985: 70, 78) has suggested was the totem of the original Shona ancestors – were often repeated to me by people surrounding Great Zimbabwe. But there was no, or very little, elaboration upon these ideas that are so widely known across Zimbabwe.[15] Perhaps the only really interesting reference to the Zimbabwe Bird that I heard was from Aiden Nemanwa, who refused to divulge its secrets, claiming the appropriate time had not yet come.[16]

There is a similarity here to the Conical Tower in the Great Enclosure at Great Zimbabwe, which also receives wide national representation on, for example, the dollar coin, the ruling party's (ZANU PF) logo and as the backdrop of various ZBC programmes. Apart from a common claim that it holds or once held secret contents such as hoes, spears and other tools, the Conical Tower does not receive a great deal of special attention in local narratives.[17] Indeed many of the features of the site that receive the most attention in national, and also international, heritage representations of Great Zimbabwe, such as the Zimbabwe Bird and the Conical Tower, as well as the walls of the Great Enclosure, are not the features that local narratives tend to focus upon, such as the hill, the secret passages and caves beneath it, or the Chisikana spring (see also Ucko 1994: 274). One really good example that is mentioned far more frequently in local narratives than the Zimbabwe Bird is the mysterious striped pot with legs, *Pfuko yaNevanji* (see figure 5.6).

There are many different versions of the story about *Pfuko yaNevanji* (some people call it *Pfuko yaNebandge* or *Pfuko yaKuvanji*), which was not actually found at Great Zimbabwe, but on a hill in the Charumbira lands called Mupfurawasha, in 1900, by William Posselt's brother Harry (Hall 1905: 87). Both Carl Mauch and Harry Posselt described how the pot was regarded with reverence, fear and awe (Hall 1905: 86–87). The stories that Posselt and Mauch were told about *Pfuko yaNevanji* have not changed

Figure 5.5: One of the Zimbabwe Birds. This one in particular has become the National icon of Zimbabwe, and features on the flag, the currency and the National Coat of Arms. (NMMZ)

much. People I spoke to also described how it moved between places, carrying water, or gold or beads. Apart from its ability to walk on its four legs, *Pfuko yaNevanji* is also well known for its tendency to bite off any hand that was put inside trying to steal its contents.

One difference between the stories I was told and those recorded by Posselt and Mauch, is that people now often emphasise *Pfuko yaNevanji's* ability to elude (or bite off) the covetous hands of white explorers.

> *Pfuko yeNevanji was taken away by the whites. One of the whites placed his hand inside the pot, and he was bitten and his hand was cut off, then they came for a second time, and took the pot, with out putting their hand inside, and they took it away and later it came back again. It came back on its own, and then they came back and took it again.* (Interview with Ambuya Sophia Marisa, 8/3/2001)

A second difference is that *Pfuko yaNevanji* has become directly associated with Great Zimbabwe and the mysterious things which used to happen there, a link that is not apparent in Mauch's or Posselt's accounts. It is not that everyone told me *Pfuko yaNevanji* used to actually stay at Great Zimbabwe – some mentioned Bingura Hill next to

Figure 5.6: Pfuko yaNevanji. (NMMZ)

Nemanwa Growth Point, and others emphasised the pot's movement between places rather than a place itself – but rather that *Pfuko yaNevanji* was always brought into our discussions in relation to the desecration of Great Zimbabwe caused by Europeans through their appropriation of relics. VaChokoto made this very clear,

Now these white people, when they came, they started to enjoy the place. They started to turn the place into their own museum, and started to make people pay to go into Great Zimbabwe. Then they started to see which are the things that were sacred during that time , and then we, the Nemanwa people we had our own person who was called Mupfura Washa. Long ago there was once a ceremony, and this person was bitten by the clay pot, and this clay pot is now known as, Pfuko yakuvanje. Then the Europeans took that Pfuko as a symbol of the old things that used to be found here, and they took it and put it into a museum. So those are some of the first things that they managed to take from the people, because they took that Pfuko yakuvanje, they also took some spears, some hoes and some axes which were anciently made by the local people, and they took all these things and put them in the museums. (Interview with VaChokoto, 6/11/2000)

Apart from showing us how the removal of relics is seen to have contributed, along with physical destruction of the site itself, to Great Zimbabwe's desecration, what is also made clear is that the way in which 'they started to turn the place into their own museum' was deeply problematic. Just as local narratives do not differentiate between the destruction caused by antiquarian diggings, and that caused by 'scientific' excavations, similarly no distinction is made between the reckless looting of relics, and their careful display in a museum. Beyond the material destruction of the site, or removal of objects, it is the process by which the site was turned into a heritage site that is perceived to have caused, above all else, the anger of the spirits. This is even more clearly revealed in statements made about the reconstruction of walls at the site.

When I began my fieldwork in October 2000 I had expected to find that local 'traditionalists' would be against the very principle of reconstruction of walls at

Great Zimbabwe. I was wrong. There is very little objection against rebuilding *per se*, rather the objections fit into a scheme of wider grievances that revolve around power and authority. As Aiden Nemanwa, put it,

> *Personally, I think it is good to reconstruct something that has fallen down. The spirits are worried that the reconstruction should have been discussed; they should have consulted with the spirits. The spirits are unhappy because they were ignored. They own the place so whatever happens should be led by the spirits.*
> (Interview notes of interview with Aiden Nemanwa, 21/10/2000)

VaChainda went one step further to suggest that the reason the walls were falling down was because the ancestors were not being listened to.[18] The falling of walls, in this sense, is another form of the resistance of the ancestors.

> *J.F.: Did you see any change I how Great Zimbabwe was run before Independence, and after independence?*
>
> *VaChainda: There is no change, as I see it because I see that they are restoring and rebuilding the ruins but as I understand it, for a building to fall apart then there is something going wrong.*
>
> *J.F.: OK, so you are saying, if it was going right, then they wouldn't need to rebuild it, because it wouldn't be falling down.*
>
> *VaChainda: They should have consulted … they should have consulted over the falling of the rocks from the mhondoro dzenyika* [senior/tribal/clan ancestors of the country]. (Interview with VaChainda, 30/12/00)

VaHaruzvivishe put it in the following terms,

> *Everything changes but when things change those that have their things being changed should be informed and when they are informed, then I don't think there would be any problem. If the mhepo dzenyika* [literally 'Winds of the land'] *that are in the mountain* [that is the hill in Great Zimbabwe] *are informed of any change or new developments that may be done at the mountain, I don't think there would be a problem, if they agree.*
>
> *You see, us, the Mugabe people, we claim it is our place and of course it is our place but right now we have nothing to do with that mountain.*
>
> *You can see that our Government which is there, what it is doing, what it is following. It follows the customs that came with the white people, that's what it follows. It does not follow some of our tsika dzechivanhu* [customs] *… if it wants to follow white customs, then they should have informed the Vadzimu* [ancestors] *that: 'We have adopted these white customs because they are good for this, this and that' Then there would have been no problems but they have just totally ignored our way of doing things. …'.*
>
> *Do you understand?*
>
> *J.F.: yes …* (Interview with Samuel Haruzvivishe, 4/3/01)

The silence of anger and the resistance of the spirits at Great Zimbabwe, therefore relates directly to the fact that the spirits, be it *Mwari*, the ancestors or both, are considered the *owners of the land*. In this sense the silence of Great Zimbabwe is ultimately caused not so much by the failure to follow the 'correct' rules or customs, but rather by the appropriation and alienation of the site, and the refusal to consult the spiritual owners of it. In local narratives, this appropriation, alienation and loss of control of the site was often discussed in terms of closure – the closure of the site from the surrounding landscape, local communities and the spirit world which ultimately owns it.

CLOSURE, APPROPRIATION AND ALIENATION

We have already discussed closure in terms of the caves and tunnels which were closed by the spirits who were attempting to protect the objects that were kept inside them. The closure of those tunnels which connected Great Zimbabwe with the shrines at the Matopos and elsewhere, also obviously relates to the perceived disruption of communications between these sacred places. But people also discussed the physical 'closure' of the landscape, caused not by the spirits, but by those who manage Great Zimbabwe. They frequently spoke of the closure of the Chisikana spring[19] with cement, which occurred in the 1950s, as 'part of the landscape modification done at Great Zimbabwe to accommodate a golf course' (Matenga 2000: 15). The closure of this spring fits into a wider list of grievances concerning the physical management of the whole landscape, which include new roads and buildings put up among the ruins; the planting of exotic eucalyptus trees; the felling of sacred *muchakata* trees; the rebuilding of the walls with cement; and perhaps most significantly for my argument, the fencing of the site which effectively closed it off from local communities. This fence embodies the appropriation, distancing and alienation of Great Zimbabwe, and is seen in itself as a major cause of the anger of the spirits. As Ambuya VaZarira so forcefully put it

> *It's because they refuse to respect the mountain, to make it important. They have put a wire around the ruins, they have put electricity there. They put a wire fence around the place and a wire to us means you are claiming that as your place.*

> *That's the only thing that we blame the government for, they have put a wire fence around the ruins, and a wire to us claims ownership of a place; without our knowledge, we were not informed when they put the wire there. So to us its like they have taken it, it now belongs to them, and the midzimu, because they know the real owners they have to punish them.* (Interview with Ambuya VaZarira, 27/12/00)

The history of the fencing of the site goes back to the Masey Report of 1909 (Masey Report 1909: 13) which recommended, in the interest of economy, that only the ruins themselves should be fenced. It was only later, in the 1950s, when the Department of National Parks took over management, that the entire site was fenced. The intervening period saw the gradual appropriation of the site,[20] and when the last local residents were evicted access to it was increasingly denied. Older people frequently described how, as children, they used to go into the site to collect wild tomatoes, and other fruits, or to sell crafts and vegetables to tourists. Others spoke of hunting there, or the fields that used to be cultivated amongst the stone walls, until they were chased away by the 'museums people'.[21] But such issues about denied access to natural resources at Great Zimbabwe culminate, for most of the elders of the surrounding clans, into a single complaint about the loss of control/ownership of the site. As Samuel Haruzvivishe put it,

> *The white people stopped the masvikiro from coming here. Refusing to allow the spirit mediums to enter, and even the chiefs were removed. So we started to live around the area, but not in Great Zimbabwe. We lived all around, but not inside. We were no longer having any control over Great Zimbabwe.* (Interview notes, Interview with VaHaruzvivishe in Ambuya VaZarira's presence, 28/10/2000)

One particularly sensitive aspect of this loss of control, and ownership of the ruins are the entrance fees that are charged at the gate. Ambuya VaZarira narrated a story about

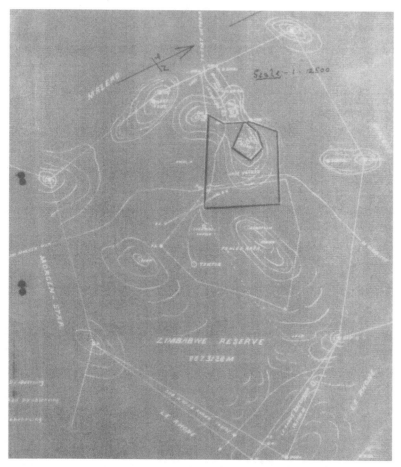

Figure 5.7: Photograph of a blueprint attached to memo, dated March 5, 1934, from Secretary of Mines to Chief Clerk Department of Mines (National Archives of Zimbabwe, file S917/a312/800). This shows the Great Zimbabwe Reserve in relation to the neighbouring farms of Morgenster, Mzero and Le Rhone; illustrating how much of the surrounding area was alienated from local communities. It also shows the area within the reserve that was originally fenced as a result of the Masey Report of 1909. In the 1950s, when the estate came under the control of the Department of National Parks, the whole estate was fenced. Also shown (dark lines) are the golf course, and the curator's plot within it, and above them the site of the Great Zimbabwe Hotel which remains there today.

an occasion when she refused to pay entry fees, and then forced her way into the site. The perceived anger of the ancestors, the owners of the land, is made chillingly clear in her claim that the white person she argued with that day at the gate, died a week later because she had been denied entrance to Great Zimbabwe.

> *The week after the brawl at the ruins, I went back to my home in Chief Shiku in Zvishavane. I received a letter that the white man I had argued with had died a week later. Because I had told him, if the mountain was his, yes I was going to pay, but it wasn't his, I wasn't going to pay, I refused.*

[...]

There is one thing I want to tell you Joe, our midzimu are very powerful, in this land they are very powerful because this soil you see here, is theirs, given to them by Musikavanhu (Interview with Ambuya VaZarira, 27/11/2000)

The entry fees relate directly, in local narratives, to the fencing of the site as both are markers of the loss of ownership and control of the site by the ancestors, the owners of the soil. Furthermore, the fences act as a 'technology of power' on the landscape, by forcing people who wish to go in, or pass through the ruins, through certain gates where entry fees are paid. The fences are therefore seen as integral to the alienation of Great Zimbabwe caused by its management as a business and NMMZ are often blamed for increasing entry fees, and treating the site like a business (like their Rhodesian predecessors) rather than, as many people had expected/hoped at independence, as a sacred site belonging to the ancestors.

We can say that the old people who used to stay in there, those old men, they used to communicate with Musikavanhu directly. But now you can find that the people who are managing it right now, they are making it a business, whereby anyone who wants to see the monuments will have to pay. Even to reach some of the places, there are now some policemen, some guards who are there, and things like that. So it makes the place not function very well, because the soil, and the midzimu yenyika, they get angry. (Interview with Ambuya Chibira, 4/12/2000)

Related to the idea that Great Zimbabwe is run like is a business is the issue of tourism, though tourists themselves are rarely blamed or disliked.[22] What people are concerned about is the lack of respect for the ancestors that accompanies tourism, and ignorant tourists: inappropriate dressing; people 'kissing their girlfriends'; and the signs, buildings, paths and roads that are part of the paraphernalia of tourism. Ambuya VaZarira (Interview 27/12/00) described the rubbish that tourists leave behind as 'things we have not seen in our lives', illustrating the incompatibility of the ancestral spirits with some of the unpleasant aspects of tourism.

But there is another way in which Great Zimbabwe has been 'closed', or alienated from the surrounding communities and the spirits themselves. Beyond the (mis-)conduct of tourists at the site, or the physical distancing of the site through fencing and entry charges (or even the destruction of excavations and the removal of relics), the most devastating and continuing impact on the sacredness of Great Zimbabwe is caused by the prevention of ceremonies at the site. This is because the prevention of ceremonies, in effect, closes a place of communication between the world of people and that of spirits. As I have already argued, a place is not sacred unless treated as such, and 'traditional' *bira* ceremonies are probably the most profound way of 'treating a place as sacred'.

CEREMONIES AT GREAT ZIMBABWE

Academic historical debates aside, it is clear from the writings of early explorers that before the colonisation of what became Rhodesia, ceremonies of some kind did occur at Great Zimbabwe.[23] Both Mauch (Burke 1969: 215) and Hall (1905: 93–4) referred to seasonal sacrifices and offerings which occurred during their stays at Great Zimbabwe. Since the early twentieth century ceremonies have been disallowed, as access to the site became increasingly restricted. The last recorded ceremony, until after independence at least, was a 'sacrifice' in the Great Enclosure recorded by Hall in 1904 (Hall 1905: 93). It seems likely, however, that ceremonies continued to be held in secret. In 1997, Aiden Nemanwa told me that ceremonies were still being held at Great Zimbabwe, but in secret, at night.[24] I was also told by the Nemanwa elders that in the 1960s some African nationalists including Simon Muzenda (the late vice-president) conducted a brief ceremony a Great Zimbabwe, when a pot of beer was offered to the ancestral spirits. This relates to a widely held belief that the nationalist liberation movements, and the freedom fighters themselves, the *vana vevhu* (children of the soil) were provided with ancestral legitimacy from the ancestors at Great Zimbabwe (see Chapters 6 and 7). Local concerns about the need to hold ceremonies at Great Zimbabwe reverberate within these wider discourses about the need to hold a national ceremony there to thank the ancestors for their help in the liberation struggle. Among local clans there is a strong belief that were beer to be offered to the ancestors and such a ceremony carried out, Great Zimbabwe's sacred features could be restored. But access to the site for 'traditional' ceremonies continues to be strictly controlled by NMMZ, despite widening calls from across the country. As an elder of the Murinye clan (and also a close associate of Ambuya VaZarira) put it

> *The chiefs and other traditional leaders have made several attempts, to the extent of preparing about 4 biras, in the past, but the rituals are supposed to be done at the top of the mountain, and the Museum Authorities, the responsible authority of the Monument, refuses them that chance, to get up to the top of the mountain to carryout their rituals. Only Government officials the president and probably his ministers are allowed to get there, so how can then the voices return, when the traditional leaders are refused the chance to do their rituals on the top of the mountain.*

> *Surely if we play our drums on top of the mountain, we are trying to call our ancestors so that they can come, and probably the things that used to occur can return. But then because we are only given permission to do our rituals away from the mountain, how would the sacred things that used to happen there come back?*

> *A black bull should be killed on top of the mountain up there, to soak the land with blood.* (Tungamiri Murinye, Group interview with Chief Murinye and his *Dare*, 10/1/2001)

This is, then, the starkest point of continuity between the management of Great Zimbabwe before independence, under colonial rule, and after independence, in the post-colonial state of Zimbabwe today.[25] The fences are still there, (and are frequently replaced or mended), the entry prices have gone up, and most important of all, the ceremonies that could restore Great Zimbabwe are still prevented, or very tightly controlled. Whilst much of the damage and disruption that is seen to have caused the disappearance of the sounds and voices, was a result of excavations, looting, and general (mis-) management of Great Zimbabwe by white colonials, local elders and traditionalists equally apportion blame to the government of postcolonial Zimbabwe, for failing to take the necessary steps so that the *Voice* and the sounds might return. In particular, the government refuses to consult with the chiefs and the *masvikiro* over the management of the site.

> *Great Zimbabwe is no longer seen as that important, or seen as a sacred place, to the extent that nobody actually cares, the government does not consult the proper people, that should be consulted to arrange for*

the return of the voices. They do not consult the chiefs, neither do they consult the svikiros and other responsible elders. So it's like once these people are consulted, and they actually sit down and consulted one another then probably the voices may come back. (Matopos Murinye, Group interview with Chief Murinye and his court, 10/1/2001)

It is in this context that Great Zimbabwe is the apex of (at least for elders of surrounding communities) a much wider array of grievances that revolve around the marginalisation of the ancestors and 'tradition' in Zimbabwe today. These grievances are often framed in terms of an opposition between the knowledge of books versus the knowledge of the ancestors, or education versus 'tradition'; or even the knowledge of 'whites' versus *chikaranga*. And in these terms, NMMZ stand accused of following the 'understanding of the book', not that of the ancestors.

I would say that there are two antagonistic groups, that of the traditional leaders, including the chiefs and the svikiros, and that of the national museums that follows the understanding of the Book or the school. They only read of information that was written by people, that's all they know, which disregards what traditional leaders such as chiefs and masvikiro claim to know about the ruins. So there is a clash between this school that belongs to the traditional leaders, made up of chiefs and masvikiro, and the other school, that includes the government and the responsible authority of the monument, this National Museums.

What happens here is that the government and its National Museums claim ownership or heritage of Great Zimbabwe by way of ruling over or looking after the ruins, and us traditional people also have our own truth or understanding of the ruins, so there is this controversial situation that prevails between the traditional leaders and the government. So there is a problem now since we have told you that the chiefs or the traditional leaders had arranged biras and were not permitted to take a pot of beer on top of the mountain, playing their drums there, and actually maybe things would change. But now National Museums, because they have their own beliefs and understanding and truths about Great Zimbabwe, which are different from what traditional leaders have, so we have clashes. (Matopos Murinye, Group interview with Chief Murinye and his court, 10/1/2001)

For their part, NMMZ often cite the continuing local disputes over Great Zimbabwe, particularly between the Mugabe and Nemanwa clans, as the main reason for their tight control on *bira* ceremonies at the site. The issue of ceremonies is taken very seriously by NMMZ, and decisions are always referred to the head office in Harare. Part of the sensitivity of the issue, from NMMZ's point of view, relates to the desire to avoid negative publicity, and to remain seen as 'impartial' in local disputes; as Ndoro (2001: 60) put it 'the fear that the granting of such permission would involve National Museums in what they consider "petty local politics" (Matenga *per.comm*)'. In 1984, after a request was received by NMMZ from chiefs Nemanwa and Mugabe to 'return artefacts to Great Zimbabwe in a ritual ceremony' (Minutes from meeting of Local Board, 8/8/84, NMMZ file C1a), a ceremony was held that turned into a 'fiasco' (Matenga 2000: 15) when representatives of these local clans became involved in a heated dispute. After this event it became an unwritten policy, or in Matenga's words 'modus vivendi', that ceremonies would not be allowed at Great Zimbabwe.

In 1984 a *bira* was held at the site which ended in a fiasco with the elders trading blows and hot words. The question of course was 'who was who?' between the two communities. After the fracas it was deemed practical not to hold such ceremonies; certainly not the best thing to do, but some kind of 'modus vivendi'. (Matenga 2000: 15)

This 'modus vivendi' has operated since that time, despite increasing numbers of requests for ceremonies from the early 1990s, when Zimbabwe was hit by a very severe drought. Letters in the 'chief's file' (NMMZ file G1(I)) at the NMMZ conservation centre contains a great deal of correspondence, from both local chiefs and traditionalists across the country, calling for ceremonies to be held at Great Zimbabwe. Between 1993 and 1998 one particularly active *nyusa* (*Mwari* cult messenger) called Mhukayesango, travelled across Zimbabwe raising support for a national ceremony at Great Zimbabwe to request for *mukombe wemvura* – rainfall. It appears that this event was scuppered by objections from Headman Nemanwa (letter dated 16/1/98, NMMZ file G1(I)), as Mhukayesango was working closely with the VaDuma clans including Chief Masungunye, Ambuya Vazarira and Chief Mugabe.[26]

However, it is not just *local* politics which prevents NMMZ from permitting ceremonies. With Great Zimbabwe's elevation to the status of 'world heritage site' in 1986, the 'fear' of being involved in 'petty local politics' was multiplied, and at same time NMMZ was provided with a very powerful means of justifying its refusal to permit ceremonies (see Chapter 9; also Fontein 2000). This was demonstrated in October 2000, when Chief Mugabe applied for permission to hold a ceremony at Great Zimbabwe with war veterans in order to honour and thank the spirits there for independence. Referring to the political, social and economic turmoil that Zimbabwe was facing at the time, Chief Mugabe argued that,

> since independence we had nothing to honour them. And now we think all this chaos, seeds of polarisation, conflict, instability and disintegration in our country. We think this [is] caused by the negligence we have done to our spirit mediums. Mugabe people unanimously agreed to have their traditional ceremony together with war veterans right at Great Zimbabwe, so now we call for your urgent intervention in this wrangle with a view to solve urgently. (Ref G/1:EM/wcm, NMMZ file G1(I))

The official NMMZ reply to, and indeed rejection of Chief Mugabe's request is very informative about NMMZ's position. Dawson Munjeri, the executive director, was clearly worried about NMMZ's international image, especially in relation to the very poor press reports that Zimbabwe was receiving in the international media at the time. Concern also seems to have been focused on the UNESCO World Heritage Committee, which was meeting the following month in Australia.

> Great Zimbabwe is not only a national monument but is also a UNESCO World Heritage Site which must abide by provisions of the UNESCO World Heritage Convention. One of those rules is that a World Heritage Site shall not be used for purposes that may bring disrepute to the World Heritage Convention. Zimbabwe is a signatory to that Convention.

> The possibilities of adverse publicity arising from the nature, the format and levels of representation of the planned ceremony is such that the element of disrepute may creep in and Zimbabwe will be sanctioned by the World Heritage Committee, which incidentally is meeting from 23rd November to 2 December 2000 in Australia. Zimbabwe is again a member of the governing council (the Bureau of that Committee) and thus is expected to lead by example. (Letter from D. Munjeri letter to Chief Mugabe, Ref G/1:EM/wcm, NMMZ file G1(I))

Clearly the requirements of UNESCO, the World Heritage System (Fontein 2000) and international heritage standards inform NMMZ's deliberations as much as, if not more than local concerns about the neglect of sacred places and the role of the ancestors, the owners of the land. This reveals the tension between Great Zimbabwe's various local, national and international roles, and the understandings and interests that accompany these different statuses; in particular the tension between Great Zimbabwe as a sacred site (locally and nationally) and as a national and international heritage site. This tension is reflected in the claims of 'traditional connoisseurs' about a 'conflict of wisdoms'[27] or forms of knowledge: the knowledge of books or education or of the 'whites'; and the knowledge of the ancestors, the soil and *chikaranga* or *tsika dzechivanhu*. Of course, the distinction between 'wisdoms' is not, and does not have to be binary or absolute; there does not have to be a contradiction between respecting the ancestors and the wishes of local communities, and Great Zimbabwe's role as a national and international heritage site. But for NMMZ to overcome the tensions that exist between different ways of understanding Great Zimbabwe's past and managing its remains, it must first acknowledge its own complicit role in the processes of appropriation, distancing and alienation by which, according to local communities, Great Zimbabwe's voices and sounds became and remain silent. Only when this is understood can efforts be made to respect and act upon the wishes of local elders, and the spirits themselves.

ACKNOWLEDGING GREAT ZIMBABWE AS 'LIVING', 'SPIRITUAL' HERITAGE

Things are changing at NMMZ. As we shall see in later chapters, since the mid-1990s the issue of local community participation has been taken increasingly seriously, and there have been renewed efforts to accommodate the wishes of local communities.[28] This came about within a wider context of increasing international recognition of the importance of local community involvement in any 'top-down' intervention, be it development or 'heritage management'. While it is unlikely that NMMZ will 'hand over' control to local communities (even if they were to resolve their disputes), there is a sense in which NMMZ is trying to broaden the recognition of the values associated with the site to include 'local sacred values' alongside the more established national and international 'heritage' values for which the site gave its name to the new country, and for which it was originally inscribed on the World Heritage List. This change in NMMZ's approach mirrors recent changes in the World Heritage System itself, and is perhaps best epitomised by the formal recognition of the 'living' and 'spiritual' values associated with the Matopos Hills (Munjeri et al. 1995; Ranger 1999) which culminated in their inclusion on the World Heritage List as a cultural landscape in July 2003 (Ranger 2004: 228). At Great Zimbabwe itself, this new emphasis on local participation and the 'intangible values' associated with heritage has resulted in increasing efforts to accommodate the wishes of local communities, and has had some quite dramatic results. The most important of these was a ceremony held within Great Zimbabwe in July 2000, just before I began my fieldwork.

On July 2 2000 NMMZ sponsored a small 'traditional ceremony' at Great Zimbabwe to mark the reopening of the Chisikana spring, which had been closed with cement in the landscaping for a golf course on the estate in the 1950s. The elders, chiefs and *masvikiro* of the three surrounding clans were invited; beer was brewed; and according to Matenga (2000: 15), nine mediums went into trance. In many ways this ceremony marked a turning point. It was a public marking of attempts to rid the site of some of the more conspicuous landscaping features and management decisions that were part of the colonial appropriation, and for locals 'desecration', of Great Zimbabwe. The ceremony also passed peacefully, with no squabbles about order of precedence by competing clans, despite 'one small incident that nearly scuttled proceedings' (Matenga 2000: 15). Perhaps most of all it marked, for locals at least, a conscious change in policy by NMMZ, and thereby re-energised the possibility that Great Zimbabwe could be restored as a major sacred site in Zimbabwe today. This optimism was especially apparent among members of the Nemanwa clan.

> *When the Europeans came they started saying that this mountain should be paid for, for someone to see, that is when the collapse of everything began. We couldn't even find our ancestral spirits, they are no longer heard, even all that used to make sounds at Great Zimbabwe is no longer heard. And even some of our rivers were closed, like we can talk of that well/spring, that Chisikana spring, where that small girl was found, it was also closed. Now since it is now ruled by the black people, they actually tried to find out what really is the cause of the problems, and they discovered that it was necessary for us to do this ceremony, and we did the ceremony at the well/spring, and it was revived. Now they know that there were a lot of traditions that were not followed properly.* (Interview with VaChokoto, 6/11/2000)

Aiden Nemanwa, himself, suggested that this ceremony marked a turning point in the history of Great Zimbabwe. I asked him whether, as he had told me in 1997, they still carried out ceremonies in secret. He informed me that since the Chisikana ceremony they no longer have to do so, because 'they are given permission, so they just do it' (Notes of interview with Aiden Nemanwa, 21/10/2000).

This claim that they have now been given permission to do these ceremonies is not entirely true. Chief Mugabe's request for a ceremony in October of the same year was turned down, because of the potential for international repercussions. More recently, a *svikiro* from Zaka turned up at Great Zimbabwe with a high profile entourage of chiefs, government officials and security agents, in order to carry out a ceremony to 'unlock the mystery of Great Zimbabwe' (*The Herald* 10 May/2003). They were turned away by NMMZ who 'felt the self proclaimed spirit medium would tamper with physical structures at the national shrine' (*Daily News* 14 May/2003).[29] The optimism displayed by some of the Nemanwa clan was certainly not repeated by members of the Mugabe clan, or by Ambuya VaZarira herself. Various descriptions of events at the Chisikana ceremony itself reveal that the tensions between the different clans were only barely suppressed, and that each clan formed its own group, separate from the others. In particular, Chief Charumbira's presence seems to have been resented, given that it is widely felt that he has no claim upon the site at all. Furthermore, Matenga told me that while NMMZ would in principle be happy to hold and sponsor such ceremonies more regularly, this is very dependent upon the situation of the disputes between the local clans; it is by no means certain.

The optimism that I came across early on in my field work, only a few months after this ceremony, may disappear once it becomes obvious that not much will change at Great Zimbabwe. Entry fees, the fence, and restoration and conservation work will, of course, continue. As Chief Mugabe made clear, NMMZ may need the local communities in order for them to hold ceremonies, but it will never relinquish control of the site to them.

> *All those things are being done by the Government. Like we can see it right now, they are now building their houses, they are now making the place for their business, maybe they are building houses for rent, with people paying money, but we the Duma people, they are just respecting us in terms of when they want some ceremony to be done, but all in all it is now becoming their own particular place.* (Interview with Chief Mugabe, 20/11/2000)

And it is clear that NMMZ's motivations are not entirely selfless either. There is a sense in which it is in NMMZ's interest to encourage ceremonies 'to enrich the heritage of Great Zimbabwe' (Matenga 2000: 15). Recognising Great Zimbabwe as a 'shrine' or 'sacred place' increases its 'heritage value', especially in the context of the World Heritage system's adoption of concepts such as 'living' and 'spiritual' heritage, and 'cultural landscapes' (see Chapter 9; also Fontein 2000). For NMMZ, local community support for their conservation and management programme at Great Zimbabwe is much desired, and they may be willing to encourage ceremonies at Great Zimbabwe to this end, without significantly changing the management of the site itself, as local elders and other 'traditionalists' desire. This was revealed to me in a comment the regional director made as we walked around the site on a Sunday morning in July 2000. Passing a recently reconstructed wall along the 'modern ascent' up the hill, I mentioned that many local elders I had spoken to felt that before such a reconstruction could take place it is necessary that a *bira* ceremony be held and the ancestors asked directly for their permission. In my notes of our walk around the monuments (which he later checked and made minor additions/corrections) I wrote down Matenga's reply as follows,

> From the NMMZ point of view, he says, it is important that we can persuade people so that our proposals or plans are accepted, and indeed happen. I can envisage despite the obvious problem of money, a short bira being held before such a project, and then again a short bira when it is finished, as an 'opening'. (Walk around Great Zimbabwe with the Regional Director Edward Matenga. Fieldnotes 15/7/2001)

This is quite informative about NMMZ's position. They are willing to hold ceremonies, as long as their proposals are accepted, and they retain control. Would this be acceptable to people like Aiden Nemanwa, VaHaruzvivishe or Ambuya VaZarira, for whom the meaningful consultation with, and the ultimate authority of the spirits is essential if Great Zimbabwe is to be restored? In Aiden Nemanwa's words, there will not be 'spiritual independence' until Great Zimbabwe is again ruled by the traditional custodians, and he will not be able to reveal his secret knowledge until such a time.

> *Before the colonials came, the surrounding people were responsible for Great Zimbabwe's management. White people came and started managing it. The only people who were allowed to be seen there were those working there.*
> *During the Colonial regime, white people controlled Great Zimbabwe. After independence, people are not yet independent, they are still following the ways of the white people. As there is not yet independence, spiritual independence, I have knowledge that I cannot yet divulge until independence. When Great Zimbabwe is again ruled by the traditional custodians, then there will be independence. Great Zimbabwe and the other shrines are still ruled in the white man's strategic ways.* (Interview notes, Interview with Aiden Nemanwa, 21/10/2000)

I mentioned at the beginning of this chapter that any discussion of Great Zimbabwe's sacredness must also be a discussion of its desecration. I suggested that 'desecration' may not be the appropriate term because of the permanence that it implies. Many people stated that the sounds and voices, and other 'miraculous features' could return if the place was again treated as sacred – if the 'traditional' rules and customs, *chikaranga*, were followed, ceremonies permitted, and the spirits, the ancestors and *Mwari*, were again respected. Despite the recent efforts by NMMZ to involve local communities in a sponsored ceremony at the site, it seems clear that in terms of a commonly held distinction between different 'wisdoms' or truths – the rules of the soil, the ancestors, the owners of the land; and those of the book or education – it is the latter that maintains its grip as strongly as ever, on the management of Great Zimbabwe. And therefore the 'restoration' of Great Zimbabwe's sacredness, as a marked, liminal spot on the landscape between the world of the spirits and that of humans, remains a challenge yet to be achieved. For many local elders it is unlikely that the silence of Great Zimbabwe can be successfully filled. 'Impossible, it cannot change' states one local elder firmly:

> *Azviti, impossible, it cannot change, because it was spoiled completely. It needs people who can know better who can know, not anyone we can see right now, for this thing to be revived. Muzukuru, let me tell you, Zimbabwe will not change and become like it used to be, because whoever talks the truth, they don't want to hear him. Whoever talks the truth, they will say you are lying because you are not educated, and this place will become worse and worse, because it is like someone who has put his faeces in a well.*

> *Milk is something that is white, but if you see flies inside it, you won't eat it. That is what was actually done to Zimbabwe. Zimbabwe was milk but flies fell into it.* (Interview with VaMututuvari, 4/11/2000)

Notes

1. As Abner Cohen (1993) showed in his writings on the Notting Hill Carnival, the chaos of the carnival atmosphere does not necessarily represent socially conservative and structure re-affirming 'communitas', but quite possibly genuine potential dissent with real political consequences.

2. There is a parallel here with Aiden Nemanwa's strong denial that the Rozvi had anything to do with the building of Great Zimbabwe which, as I explained in Chapter 2, relates to his experiences of Rozvi *masvikiro* turning up at Great Zimbabwe to claim it as their own.

3. *Kugadzirira*: to prepare/repair – referring here to brewing beer and conducting a cleansing ceremony; and related to the *kugadzira* ceremonies that are essential for the successful transformation of a dead man's spirit into a *mudzimu* (family ancestor).

4. Great Zimbabwe's importance as a national sacred site is discussed in greater detail in Chapters 6 and 7 which focus on the different ways in which the site featured in Zimbabwean nationalism.

5. This plethora of stories relates not only to the number of different people interviewed, but also to their sources (which often seem to include written 'historical sources', for instance in the case of the existence of gold and relics like the Zimbabwe Birds) and, importantly, people's individual experiences and ability to tell stories.

6.   Bush fires are quite common at Great Zimbabwe, especially during the dry winter months as the weather gets to its hottest before the first rains in October/November. I witnessed (and helped people from NMMZ to fight) a bush fire at Great Zimbabwe in September 2001 which burnt for three days and left great, though temporary, black scars over the landscape (see figures 5.1 and 5.2). Often NMMZ blame local communities for starting these fires to clear illegally occupied areas of bush for farming or hunting in the neighbouring game park.

7.   It is interesting to speculate how these local stories about Europeans coming to Great Zimbabwe in search of gold and money, and ultimately failing to do so, could be related to the discourses of European explorers and colonialists of that period who were extremely 'fired up' with the possibilities for gold prospecting in the unexplored lands north of the Limpopo, and rumours of King Solomon's mines (Burke 1969: 4; Kuklick 1991: 138–9). These dreams of vast gold reserves were not realised and while quite large amounts of gold were removed from Great Zimbabwe and other similar sites, it was a great deal less than most had anticipated (Kuklick 1991: 139).

8.   For example, Randal-MacIver, letter to Sir Lewis Michell, National Archives of Zimbabwe A11/2/18/66. See also Summers 1963; Garlake 1973; Ndoro 2001: 39–40; Pikirayi 2001: 13–14.

9.   It is, perhaps, not surprising that very few people I spoke to mentioned that it was their forefathers who had been employed in the excavations that destroyed the sacred features they described; those that did emphasised that their relatives were forced to work there (Interview with VaMututuvari, 4/11/2000).

10. On one occasion, after diggings caused the collapse of granaries on the hill at Great Zimbabwe, Bent found himself surrounded 'by a screaming crowd of angry men and women, with Ikomo at their head, brandishing assegais and other terrible weapons of war' (1896: 74). Chief Mugabe's brother, 'the rascal Ikomo' (Bent 1896: 73), who was living on the hill at the time, was later warned by Sir John Willoughby 'that if such a thing happened again, his kraal would be burnt to the ground and his tribe driven from the hill' (1896: 74).

11. Hall added that he 'only discovered two natives and these elderly men, who would willingly go into any of the ruins, especially the temple, after darkness had settled down' (1905: 19), which suggests that there used to be age restrictions over who could safely enter the ruins.

12. He claimed that unlike Bent, he got permission from Haruzvivishe Mugabe, who was then living on the hill, 'on the understanding that he is given half a cup of salt, that the remains are to be properly re-interred and that the boys who did the work should be allowed to go to their kraals to purify themselves' (Hall 1905: 43).

13. It may be that this is because stories about the desecration of graves undermine local claims over Great Zimbabwe, especially for the Mugabe clan, whose central claim is based on the existence of the *mapa* of their ancestors on the hill. It may also be that they are not aware of the excavations, or that there is confidence that the diggings did not effect the graves of their most senior ancestors, like Chipfuno himself. For the Nemanwa clan, who don't (or rarely) claim to have buried their own dead at Great Zimbabwe, focus on the desecration of Mugabe graves is hardly welcome if it attracts attention to Mugabe claims over the site. Edward Matenga,

the former Regional Director, (Field Notes, Sunday 15 July 2001) described this as a very sensitive issue, but added that it was very unlikely that Chipfuno's grave had been excavated, and that Mugabe people still carried out rituals at these *mapa*, though he himself had deliberately never enquired as to their exact position.

14. I discuss these events in more detail in chapter ten. For newspaper reports of the May 2003 event see *The Standard* 6/6/03; *The Sunday Mail* 25/5/03; *The Daily News* 31/5/03. For the May 2004 event see *The Herald* 6/5/04 & 8/5/04.

15. Yet William Posselt (1924: 74) reported that in 1889, Haruzvivishe Mugabe protested so fiercely when Posselt first attempted to remove the birds that he only narrowly avoided being physically attacked. This makes it clear that, as Matenga has put it 'at least according to the custodian Chief Mugabe, … the birds were so dear to the site that they should not have been removed, [and] … it is a reasonable deduction that the stone birds were sacred objects' (Matenga 1998: 64). This makes it all the more striking that more than a century later, this national icon seems not to receive any particular special mention among locals, beyond the fact that its removal from Great Zimbabwe, along with other objects, was part of a list of disturbances at the site which caused the anger of the spirits, the disappearance of the sounds and voices, and the closure of the sacred *Nhare*, caves and tunnels.

16. It would be interesting to see if Aiden Nemanwa has maintained this self-imposed silence about the Zimbabwe Bird, following the national event held in May 2003, and the official return of the bird to 'Masvingo chiefs' the following year. While it is likely that these events will 'feedback' into local stories about Great Zimbabwe and the Zimbabwe Bird, it also important to note that the 'multi-million dollar' re-unification ceremony provoked much criticism within Zimbabwe, as well as in the South African press (Ranger 2004: 226–7), for being a waste of money at a time when the whole country was facing enormous economic problems. Such criticism is equally likely to find its way into local discourses and stories about Great Zimbabwe and the Zimbabwe Birds. As I discuss in chapter ten, Chief Charumbira's prominence at the second event in the ambiguous role of both deputy minister of local government and 'head' of the Masvingo chiefs was obviously controversial for other clans in the district who have their own claims over Great Zimbabwe.

17. The only important exception here is a very specific story that Nemanwa people told about a person called Marange, who successfully climbed the conical tower and thereby demonstrated that the site belongs to the Nemanwa clan.

18. Ucko (1994: 271, 272), citing Mabvadya (1990), makes a similar point; that 'in a Zimbabwean context … the collapse of ancient walls at least at some sites is seen to be the intentional actions of relevant ancestors'.

19. As discussed in Chapter 2, this spring plays an important role in the historical narratives of the surrounding clans, each of whom put forward different versions of its importance. In July 2000 NMMZ held a ceremony at this spring to 'revitalise' it, to which members of all three clans were invited. It is likely that as a result of this ceremony, the local importance of this particular feature on the landscape of Great Zimbabwe has indeed been 'revitalised', which is reflected in the variety of stories local people now tell about it.

20. It was during this period that the last local occupants, the Haruzvivishes, were first removed from the Hill Complex onto another part of the site (Hall 1905: 6, 10), then to Mutuzu hill, until according to VaHaruzvivishe (Interview 4/3/01), his grandfather died, at which point the sons were removed as the site became a national monument in 1936.

21. The loss of the land for cultivation, for collecting wood, or hunting animals, is still a contentious issue. In a similar way that the lands next to Great Zimbabwe, formerly Mzero farm, have continued to be used for illegal grazing, firewood collection and hunting/snaring, these things have also continued at Great Zimbabwe itself despite the efforts of NMMZ to prevent them. But unlike the 'game park' that has recently been re-occupied by members of the Mugabe clan (see fontein 2005), illegal 'squatting' at Great Zimbabwe has not been tolerated at all, bar a tense period shortly after independence. Attempts to re-occupy the estate have been relatively few since the mid 1980s. But NMMZ does continue to have problems with poaching/snaring, fence-cutting, cattle grazing and wood collection on the site.

22. Much of the local rural economy is based on the meagre proceeds of tourism that are allowed to filter to the ground level. Furthermore, according to Aiden Nemanwa, there have been visitors at Great Zimbabwe for a very long time, and the spirits 'are happy to have visitors' (Notes of interview with Aiden Nemanwa, 21/10/00).

23. There has been some academic debate about pre-colonial ceremonies at Great Zimbabwe. Some writers have suggested that Mauch's description is evidence of an 'annual ritual occasion held at Great Zimbabwe after the harvest in honour of *Mwari*' (Matenga 1998: 17). Mauch's account has frequently been cited as evidence of Great Zimbabwe's past role as an *Mwari* cult centre (e.g. Abraham 1966: 33 and Aquina 1969: 389–405). However, Beach (1973) has argued that Mauch's description bears more resemblance to Shona rituals in honour of the ancestors. Local accounts today are slightly ambivalent about which part of the spirit world Great Zimbabwe refers to; most people suggested or implied both that *Mwari* used to speak at Great Zimbabwe, and that there are ancestral spirits associated with the site.

24. This was later confirmed by Aston Sinamai, then curator at Great Zimbabwe, who told me that sometimes traditional beer pots and offerings of *rapoko* [finger millet] were found on the hill or in the Great Enclosure (Fontein 1997: 25–6).

25. This continuity between the alienation of the site and the prevention of ceremonies there under Rhodesian management, and what has happened after independence, is perhaps best illustrated by the case of Sophia Muchini (Garlake 1983: 16–17), which I discuss in Chapter 7. Claiming to be a *svikiro* for the great war-hero ancestor Ambuya Nehanda, she came to live within the ruins during the liberation struggle and shortly afterwards, calling for a national ceremony of reconciliation there. Her treatment, and ultimately, her violent eviction from Great Zimbabwe, by NMMZ in the early 1980s in what was one of the ugliest moments of NMMZ's history, not only illuminates the strong vein of continuity in the management of Great Zimbabwe, but also some of the contradictions and discontinuities of African nationalism's utilisation of Great Zimbabwe for the purposes of fighting for liberation, and post-independence nation-building.

26.   In May 1998, the Executive Director of NMMZ, Dawson Munjeri requested more information from the Regional Director, Edward Matenga about a ceremony proposed by Ambuya VaZarira and VaHaruzvivishe, because 'we would not like NMMZ to be involved in political struggles between local traditional leadership' (Memo from Exe. Director to Reg. Director, dated 27/5/1998, NMMZ file G1(I)). After organising a consultation meeting with representatives of all three clans, Matenga replied 'It was regrettably clear at the meeting of 19 June that the three were far from achieving unity of goals and purpose on this issue. I therefore made it clear to them that in these circumstances NMMZ would not support the project for fear of a repeat of the events of the early 80's' (Memo from Reg. Director to Exe. Director, undated, NMMZ file G1 (I)).

27.   As Aiden Nemanwa (Hove and Trojanov 1996: 84) put it, again using a metaphor of 'closure', 'There is a conflict of wisdoms. The new wisdom is the wisdom of defeat, of conquering other wisdoms. The new wisdom fought to gain its space. The old wisdom does not fight for its space. It withdrew and looked forward to the day when it will be sought once more. Our ancestors closed the doors of their wisdom. The key to unlock the doors of wisdom is to follow the rites and rituals'.

28.   This also the case at other sites, such as at Domboshava where NMMZ was advised to ' "cool" the anger of the ancestral spirits' by providing three head of cattle to be slaughtered and offered to the ancestors (Pwiti and Mvenge 1996: 821).

29.   I am grateful to Professor Terence Ranger for making me aware of these events, and for providing me with copies of the newspaper reports.

CHAPTER 6

'LET US FIGHT AND REBUILD': NATIONALISM AND GREAT ZIMBABWE

In this chapter and the next I wish to consider how Great Zimbabwe was employed for the purposes of African nationalism – for the imagination (Anderson 1983) of a nation with which, and for which, to fight for liberation from colonial rule. I will trace some of the multifaceted meanings that Great Zimbabwe came to embody through this process. In particular I will focus on the blurring that emerged between a relatively crisp view of Great Zimbabwe as a symbol of national *historical* identity – in the words of one veteran nationalist, 'an eternal heritage for people of all tribes in this country'; and another view of Great Zimbabwe's role, which became persuasive among many guerrillas and 'traditionalists', as a national *sacred* site, and the basis of the spiritual and ancestral authority of the liberation movement, and later the newly independent state.

GREAT ZIMBABWE: 'AN ETERNAL HERITAGE FOR PEOPLE OF ALL TRIBES IN THIS COUNTRY'

At midnight on 17/18 April 1980, in Rufaro Stadium, Mbare township, Salisbury (soon to become Harare), the Zimbabwean flag was raised, and independent Zimbabwe was born, becoming Africa's 50th independent state. Along with the name 'Zimbabwe' itself, the new state's national icons were derived from Great Zimbabwe in the form of both the image of the conical tower from the Great Enclosure, and, more definitively, the image of one of the Zimbabwe Birds (see figures 2.2 and 5.5). The former features on the dollar coin, the ruling party's logo, and as the backdrop of ZBC news programmes, whilst the latter is proudly perched on the national flag (see figure 6.1), the national coat of arms (see figure 6.2) and the currency, representing national sovereignty, and in the words of the Minister of Foreign Affairs Dr S. Mudenge, 'embodying the body spirit of the modern nation state of Zimbabwe' (Mudenge 1998: viii).

It was not the first time that the Zimbabwe Birds had featured in state iconography. In line with the argument that the colonisation and settlement of what became Rhodesia was inspired and justified in terms of the precedence provided by settlers' 'foreign-origins' mythology of Great Zimbabwe (Kuklick 1991: 139), images of the Zimbabwe Birds had first appeared on the coat of arms of Southern Rhodesia in 1924, and from 1932 had featured almost continuously, except for the brief period of the Federation of Rhodesia and Nyasaland in the 1950s, on the currency, the coat of arms and other state insignia right up to independence in 1980 (Kuklick 1991: 136; Matenga 1998: 51, 58). As the 'most precious symbols of the new state' (Garlake 1983: 16), little time was lost after independence in negotiating for, and finally securing the return of five of the birds from South Africa in 1981 (Matenga 1998: 57–60). In 1988, the

Figure 6.1: The National Flag of the Republic of Zimbabwe, showing the Zimbabwe Bird. (Author 1997)

Figure 6.2: The National Coat of Arms of the Republic of Zimbabwe, showing the Zimbabwe Bird and the Conical Tower. (Author 1997)

Great Zimbabwe Site Museum extension, built to provide room for the public display of the birds, was opened by President Robert Mugabe. He ended his speech with the following words, indicating clearly the importance of Great Zimbabwe as a symbol of past African achievement and Zimbabwean national unity.

Great Zimbabwe is an important symbol for it shows this generation what we as a people were capable of achieving. It encourages us to reach for greater heights in our fight to rebuild Zimbabwe. Great Zimbabwe will remain an important unifying symbol that should inspire us to defend our national sovereignty and hard-won independence so that we can continue to affirm and promote our cultural identity. (Draft of President's speech, Opening of Shona Village and Museum Extension, 25 November 1988, NMMZ File H4)

Apart from the obvious historical precedence that Great Zimbabwe provided African nationalism, the choice of the name 'Zimbabwe' for the new nation has often been explained by commentators as a reaction against the 'cultural aggression' of Rhodesian settlers' 'foreign-origins' myths. As the archaeologist Peter Garlake put it,

There has been universal recognition that the Zimbabwean cultural heritage was despised, denigrated and stolen by settler propagandists. It was in part a deliberate reaction to this cultural aggression that led to the choice of Zimbabwe as the name to be given to the new nation by all the political parties committed to its birth. It was also chosen to give a historical validity to the nation. (Garlake 1983: 15)

Similarly, in a paper on white colonial attitudes to landscape in Southern Rhodesia, Ranger has suggested that Great Zimbabwe 'assumed a disproportionate significance to black nationalists' because of the attempts by white settlers to claim it as part of the heritage of their own race. Therefore 'it became inevitable that the independent state which has now succeeded Southern Rhodesia should be called Zimbabwe' (Ranger 1987: 159).

This view was substantiated by Lawrence Vambe, an early African nationalist, in interviews carried out by I.J. Johnstone of the National Archives of Zimbabwe in 1983. Vambe identified and credited the late Michael Mawema with first utilising 'Zimbabwe' for the purposes of African nationalism in Rhodesia, when he was president of the NDP in the early 1960s.

I have no clear idea as to the exact moment it was decided to name our country after the Zimbabwe Ruins. You may be right in your supposition that a definite decision was made in this respect very possibly in about 1960 and Michael Mawema, then a very militant personality and president for a brief period at about this time, would probably have been making an official announcement of this name shortly after the party had decided to adopt it. Otherwise I have no true recollection or record of exactly when this name was chosen.

If I may make a loose interpretation of this decision, I would say that this name was arrived at by us, among other reasons, as an act of defiance to our white master who argued that we were far too primitive to have been capable of constructing such a sophisticated structure as the Great Zimbabwe edifice. We had to emphasise that our certain past and certain future as a free people was enshrined permanently in that 'big house of stone' as the name literally means, which was built by our ancestors. The Rhodesians did not like that, particularly later when Smith and his fellow rebels declared themselves independent. [interviews carried out on 1, 8 and 13 June 1983 by I. Johnstone, National Archives of Zimbabwe, Oral History index, Oral/2333 p.60]

This was further corroborated by the veteran nationalist and Masvingo MP, the late Dr. Eddison Zvobgo, when I interviewed him in August 2001. He implied that the choice of the name Zimbabwe first emerged, almost accidentally, in a comment by Michael Mawema at a political rally, after which 'it caught hold'.

J.F.: As you know, I'm doing research about Great Zimbabwe. How was it that Zimbabwe became such an icon for the nationalist movement?

Dr Eddison Zvobgo: During the time of the NDP [National Democratic Party – a precursor to ZAPU] *the guy we were governed by, Dr Mawema, at a rally, he simply said that this country was going to be called Zimbabwe, and it caught hold, and that was that. So in a sense, he gave this country its name.*

J.F.: Just as simple as that? And from there it got its own momentum?

Dr Eddison Zvobgo: Ya, so when we formed, after the NDP, the next party was called the Zimbabwe African People's Union. And when we formed ZANU, it was the Zimbabwe African National Union. So it was now clear that Zimbabwe would be the name of the new country. (Interview with Dr Eddison Zvobgo, 18/8/01)

During the same interview I asked Dr. Zvobgo about the ZANU logo of the war years, 'Let us fight and rebuild' below a picture of the conical tower and the Zimbabwe Bird. His answer further illustrates how Great Zimbabwe became a 'refreshing' source of inspiration and African nationalist pride in their own past, during a period when the denial of the very existence of an African past was only beginning to be challenged by the work of oral historians.

J.F.: One of the things that struck me, is that if you look at the logo for ZANU PF now, it says 'people, unity and development'. During the war it was 'let us fight and rebuild'; there is a sense in which the idea of Zimbabwe, as a great African achievement, which was destroyed, should be rebuilt. Is there any truth in that?

Dr Eddison Zvobgo: Not rebuilding the ruins. The pride stemmed from the fact that if our ancestors could do this then obviously it was an eternal heritage for people of all tribes in this country. It gave a us a reference point, something to be proud of, and it has remained a national shrine.

The historians have obviously done a lot of work, and in your profession, the anthropologists and political scientists have put forward various [histories] *going back to the Mutapa period. Like Stan Mudenge, who wrote his PhD on the politics of the kingdom of Munhumutapa.*

It was a very refreshing thing, that we had empires before the British came, and so we were not just people of yesterday.

J.F.: Like something from the past inspiring something for the future.

Dr Eddison Zvobgo: yes.

J.F.: So obviously, you must have been aware of the Rhodesian stubborn refusal to accept Zimbabwe as having been created by Africans, to what extent did that play a part?

Dr Eddison Zvobgo: A very great deal, I mean I majored in African history, all the books that had been written until that time simply said 'these people couldn't have done it. It must be some ancient, some old, foreign empire, which came and built this' but when you go from area to area, Mapungubwe, Khami and so on, you will find that these were widespread here, so the idea of foreigners having done it is rubbish. (Interview with Dr Eddison Zvobgo, 18/8/01)

It was during the same period, the 1960s, that historians and archaeologists began to take oral traditions seriously as a means through which to glimpse the African past. Many of the books that were written by historians during this period fed directly into nationalist discourse. Some of these (for example, Gelfand 1959; Abraham 1966; Ranger 1967) became partly/largely responsible for raising the profiles of several key ancestors, such as Chaminuka, Ambuya Nehanda, and Sekuru Kaguvi, who later became nationalist icons of the *chimurenga*, the liberation war. As far as Great Zimbabwe and the other 'Zimbabwe style' ruins were concerned, these new histories emphasised their African origins, and specifically associated them with the Changamire Rozvi, which 'emerged as a favourite' out of all 'the great political units in southern Zambezia known to the 1960s historians – the Mutapa, Changamire, Gaza and Ndebele' (Beach 1980: 223–4). The work of the archaeologists Whitty (1961), Summers (1963b) and Robinson (1966), who were already convinced of Great Zimbabwe's African origins, built on that of the oral historians in linking Great Zimbabwe with the Rozvi, as did the later writings of Aquina (1969–70) and Daneel (1970) at the turn of the decade.

Such links between Great Zimbabwe and the Rozvi that academic historians and archaeologists were putting forward during the 1960s related to, and fed into the histories that had been collected and constructed by various African historians/activists of Rozvi revival movements since the 1920s (Beach 1994a: 193–207). There were several attempts to recreate the old title of *Mambo* of all the Rozvi, and one activist, Noah Washaya even managed to gain the permission of the Historical Monuments Commission to 'hold a ceremony and sacrifice at Great Zimbabwe and to remove six stones from the ruins for foundation stones for his temple and college' (Beach 1994a: 205). A ceremony did occur in Chidoma in September 1961, but it is not clear whether there was one at Great Zimbabwe. According to Beach, from 1959 Washaya had been able to get some support for his efforts from white Rhodesians, and in particular from one Andrew Dunlop, a Federal MP and later Rhodesian Front minister, who 'hoped that the rise of the Mambo movement would divert popular African interest from the rise of the African National Congress' (Beach 1994a: 204). In July 1962, at a large, and heavily publicised gathering in Chidoma, Washaya delivered a strong 'anti-Zimbabwe African People's Union' speech, and was crowned 'Mambo Rodia Musasa' (Beach 1994a: 206). But soon Rhodesian government attitudes to the idea of restoring the title Mambo hardened, and in 1963 Dunlop was 'pointedly told to stop pushing the Washaya case' and thereafter 'Washaya finally vanished into obscurity' (Beach 1994a: 206).

While many of these various Rozvi activists were partly or wholly motivated by their own immediate interests (which led in Washaya's case to his courting of white Rhodesian officials), it also clear that their efforts, in conjunction with the increasing academic interest in the oral history of the Rozvi, and their links to Great Zimbabwe, must be seen in the context of the rising nationalist fervour of the 1950s and 60s. 'The role of the Rozvi in history' as Pikirayi (2001: 29) has put it, 'was exaggerated to meet the demands of modern nationalism'. It is therefore not surprising that during the 1970s it was Beach (1973a, 1973b), then working for Rhodesian Ministry of Internal Affairs (Pikirayi 2001: 30), who spearheaded the subsequent backlash against this exaggeration of the Rozvi past. But this academic backlash against the 'myth' of the Rozvi did not affect to any great extent the usefulness of Great Zimbabwe for nationalism. After independence, when the role of the Rozvi in relation to Great Zimbabwe had been thoroughly questioned by historians such as Beach (1973a, 1973b,1980, 1984, 1994a) and Mtetwa (1976), writers then began to explore, in various ways, and with varying degrees of academic 'approval' (see below), links between the Mutapa dynasty and Great Zimbabwe (Chigwedere 1980, 1985; Mufuka 1983; Mudenge 1988). Away from academia, the link between the Rozvi and Great Zimbabwe has remained in popular consciousness, where, in any case, distinctions between the Rozvi, and the Mutapa state are rarely rigorously maintained.

The political advantage that the African nationalist movement derived from the presentation of Great Zimbabwe as an African site is lucidly revealed by the evidence of Rhodesian concerns about the use of the name 'Zimbabwe' by nationalist parties. In fact, in an ironic twist, the nationalist use of the name 'Zimbabwe' – partly adopted in defiance of the exotic-origins fantasies of Rhodesian settlers – 'reactivated' the Zimbabwe Controversy (Kuklick 1991: 158). This was clearly illustrated in a letter, dated 6 June 1969, from Edmund Layland, who later co-authored Gayre's *The Origin of Zimbabwean Civilisation* (1972), to the Minister of Information Mr N.J. Brendon.

> The name Zimbabwe has become synonymous to the outside world with the political ambitions of African Nationalists to establish a 'black' government. The militant Rhodesian Africans in exile at U.N.O. Headquarters in New York have even gone as far as to describe themselves as 'the dispossessed people of Zimbabwe' a name which is intended to take the place of Rhodesia in the same way that Northern Rhodesia became 'Zambia'. In line with British long term planning for a similar 'hand-over' in Rhodesia, the Zimbabwe 'Bantu theory' has been supported by British Authorities and used by them to enhance the claims of Africans to govern this country. This propaganda continues through the medium of books, articles., broadcasts and telecasts both in and outside Rhodesia; it is therefore imperative that a Government Ministry should not give support to this subversive movement by publicising their claims in a Government brochure designed to be distributed throughout the world. The origin of Zimbabwe as described in the new brochure is <u>misleading</u> as it gives the impression that the mystery has been solved, and that the other theories and opinions of some of the world's leading authorities on the subject are no longer of any importance. The part played by Sabeans, Phonecians, Arabs and Indians in establishing the earlier civilisation is dismissed in <u>one sentence</u>, and the views of three persons little known outside Rhodesia are described as representing <u>modern scientific opinion</u> when in fact scientists of <u>world status</u> such as Professor Dart, Professor Gayre, and the late Abbe Breul and many other eminent authorities do not subscribe to those opinions.
>
> The Bantu 'Roswi' people are described as builders of Zimbabwe from 1450–1833 when they were finally driven out by invading Nguni from the south: an unreal and historically inaccurate theory based on unreliable native legend. (National Archives of Zimbabwe, Historical Manuscripts HA17/1/2 p.1)

Edmund Layland was in close and frequent correspondence with Colonel Hartley, the MP from Fort Victoria District, who stood up in parliament on 5 September that year to address the Minister of Home Affairs on this issue of the presentation of Great Zimbabwe's origins.

> I rise briefly to draw the attention of the minister to a seeming trend which is developing among Government Officials and quasi-Government Officials in relation to the history of the Great Zimbabwe Ruins. This is causing some concern in my constituency knowing as you well do, Sir, that the ruins lie there. There is one trend running through the whole presentation of the image of the ruins which apparently is being directed to promoting the notion that these ruins were originally erected by the indigenous people of Rhodesia. This may be a very popular notion for adherents to the Zimbabwe African People's Union and Zimbabwe African National Union and the Organisation of African Unity but I wish to make the suggestion that this notion is nothing but sheer conjecture. I feel that it is quite wrong that this trend should be allowed to continue to develop, particularly in Government circles themselves. I suggest to the Minister that he should take note of the trend that I have described because I believe that it is time that some publicity was given to some other theories as to what the origin of the ruins may, in fact, have been. (Frederikse 1982: 11)

The *Rhodesian Herald* (September 6 1969) reported the following day that the Minister had agreed with Colonel Hartley and that he 'had told personnel of the country's archaeological institutions that it would be "more correct" to convey to the public that there was no incontrovertible evidence about the origin of Great Zimbabwe' (Kuklick 1991: 159). Censorship was introduced at Great Zimbabwe; employees were told they would lose their jobs if they told visitors the ruins had been built by African people; they were not supposed to discuss radio-carbon dates; and guidebooks were 'physically censored' by order of the Minister (Frederikse 1982: 10–11). Finding that 'they could no longer work under the Rhodesian regime and sustain their intellectual integrity' (Kuklick 1991: 158), the archaeologists Peter Garlake and Roger Summers left Rhodesia in 1970. But the letters column of the *Rhodesian Herald* indicate, if taken as a fairly reliable indicator of Rhodesian public opinion, that there was widespread support for these moves among Rhodesians (Kuklick 1991: 159). Furthermore, in what Pikirayi (2001: 23) has labelled 'the new revisionism', new books by apologists of the foreign-origins myths that had begun emerging in the late 1950s (for example, Dart 1955; Wainright 1949; Jeffreys 1954), were followed by yet more additions from the 1960s onwards (Bruwer 1965; Gayre 1972) and continued to appear well after independence (Hromnik 1981; Mallows 1986; Parfit 1992).[1]

With the benefit of hindsight, the value of Great Zimbabwe for African nationalism, as a symbol of past African achievement, and as the focus of nationalist aspirations for the future, may seem obvious, especially in the context of the 'cultural aggression' of the attempts by Rhodesian colonists to claim it as their own. But there is some circumstantial evidence to suggest that this was not actually the case. At the risk of reading too much into an individual's use of phrase, Eddison Zvobgo's assertion that Michael Mawema '*at a rally, simply said that this country was going to be called Zimbabwe, and it caught hold, and that was that*' carries the implication that the name was initially just stumbled upon, after which it was met with almost universal agreement among African nationalists. This ignores two important 'footnotes'. First, as Kuklick has diligently pointed out,

> The meaning of the ruins, however, had not always been the same for African nationalists. In the 1930s, there were Shona partisans who accepted the ancient-exotic interpretation of Great Zimbabwe's origins, and described the early use of their ancestors as forced labourers by non-Africans as emblematic of the indignities since suffered by their people under colonialism. (Kuklick 1991: 160)

The more familiar and conventional use of Great Zimbabwe, as a symbol of past achievement and future aspirations, only came in the later, and wider context of Pan-Africanism that emerged after Second World War. As Kuklick continues,

> But in the Pan-African political argument that captured the nationalist imagination all over sub-Saharan Africa after World War II, Zimbabwe was one of the most powerful African states that had been suppressed by colonialism, the vanished glories of which would return when the descendants of their citizens created new states with the old names. (Kuklick 1991: 160)

Thus the African nationalist use of Great Zimbabwe 'to give historical validity to the nation' (Garlake 1983: 15) had not always been immediately obvious, and, far from being 'original' and innovative, needs to be seen in relation to the choice of names for other African states that had already achieved their independence, like Ghana and Malawi (Garlake 1983: 15).

The second issue relates to the general acceptance with which Great Zimbabwe was received as a symbol of African *national unity*, binding together, along specifically racial lines, under the label 'Zimbabwe', the plethora of different ethnic groups and identities but most significantly the Shona and Ndebele, into one nation, with which, and for which to fight for liberation. The quotes from both President Mugabe and Eddison Zvobgo mentioned above, carry this common presumption that the name 'Zimbabwe' did effectively unify Africans under the vision of one nation, or at least, was not responsible for any of the factionalism that did occur at various stages during the nationalist movement and struggle for liberation of the 1960s and 70s.[2] Given that my fieldwork was carried out mainly in Masvingo province, among Shona people of the Karanga dialect, it is not surprising that people I spoke to confirmed the role of Great Zimbabwe as a unifying force, as exemplified by the words of VaKanda, the deputy chairman of the Masvingo War Veterans Association.

> *So the fact that this country was called Zimbabwe unified everyone. Even the Ndebele people recognised the importance of Zimbabwe when they were fighting for this country. For example, Joshua Nkomo was called the father of Zimbabwe. So it was a unifying force.* (Interview with VaKanda, VaMuchina, MaDiri, 16/3/01)

Of course VaKanda'a argument is well taken; nationalists across ethnic divides did utilise the name 'Zimbabwe', despite the obvious point that, while it was recognised generally as an African creation, it has also been quite specifically related to the Shona past. It should not be surprising to note, as Ranger has done (1999: 212), that there were some murmurs of dissent from cultural groups in Matabeleland allied to the nationalist cause, for whom the name 'Zimbabwe' was not the obvious choice.

> There was no debate in the MHS [Matabele Home Society] leadership about their alliance with nationalism – though there was some debate about the exact ways in which the new African nation should be defined by its past. In August 1960, for instance, the MHS publicly rebuked Salisbury leaders of the NDP for unilaterally choosing the name 'Zimbabwe' for the projected nation. This was a name, they objected, which 'promoted tribal feelings'. The Matopos made a better symbol of national unity: 'the Matopos are both historically and traditionally of greater significance and attempts to belittle it would be resisted in Matabeleland'. (Ranger 1999: 212, with references to the *Bantu Mirror* 20 August 1960).

I have found very little else to indicate any further dissent, from among cultural nationalists in Matabeleland or elsewhere, against the choice of the name 'Zimbabwe' for the new state, though I dare say it did exist in some quarters. Lack of evidence here may in fact be an indication of the extent to which the idea of a unified nation, under the banner of 'Zimbabwe', took hold among the African nationalist elite, concerned as they were to avoid the pitfalls of regionalism/tribalism, so that whatever was seen as 'tribalist' dissent was, in effect, silenced. The emphasis on Great Zimbabwe as the heritage of all Zimbabweans is clear in Dr Ushewokunze's, then Minister of Home Affairs, introduction to the first edition of Garlake's (1982) *Great Zimbabwe: Described and explained*, published two years after independence. Describing Great Zimbabwe as the 'precious cornerstone of our culture', he finished his 'Introduction' with:

> In spite of what hired cynics would like the world to otherwise believe, Great Zimbabwe
> was built by the **Great People of Zimbabwe**. So be it known. (my emphasis, Garlake
> 1982: 4–5)

Given that from the early 1960s, after Mawema's alleged utterance at a political rally,
the name 'Zimbabwe' did become widely, even internationally, accepted as the name
for the new country yet to be born, or maybe reborn, it is startling to note that there
seems to have been an apparent lack of direct, politically motivated, nationalist
mythologising about Great Zimbabwe during the period leading up to independence.
It seems that beyond asserting that Great Zimbabwe represented the glorious past
achievement of Africans, that was then denied and stolen by colonial settlers,
the nationalist propaganda machine was strangely silent. Perhaps it was left to the
archaeologists and historians, or perhaps Great Zimbabwe spoke for itself, as the
symbol for the nation to be liberated. This stands in marked contrast to the extended
mythology about the first *chimurenga* of 1896, and the activities of ancestors like
Ambuya Nehanda, Sekuru Kaguvi and Chaminuka, which was fed, as 'political
education', to aspirant guerrilla fighters during their training in camps in Zambia and
Mozambique, and through them, to the 'masses' inside Zimbabwe at all-night *pungwe*
sessions. I will develop this in the next chapter, where I argue that this apparent vacancy
was filled by the discourses of guerrillas and 'traditionalists' in rural Zimbabwe, far
away from the exiled politicians spending the war in other countries. They combined
this other mythology, or even theology, of liberation (inspired by the works of oral
historians such as Gelfand 1959, Abraham 1966 and Ranger 1967), with their newly
created identities as 'Zimbabweans' fighting for the freedom of their nation, so that
Great Zimbabwe became viewed as a national sacred site, and closely associated with
the ancestral forces that guided the struggle.

The 'vacancy' of specifically nationalist historical/mythological writing on Great
Zimbabwe during the period between the first adoption of the name by nationalists in
1961 and independence in 1980, also stands in marked contrast to the nationwide
calls for new histories that followed independence. In his speech to the Zimbabwean
nation on the eve of independence, the Prime Minister Robert Mugabe made the call
to arms explicit: 'Independence will bestow on us a new ... perspective, and indeed, a
new history and a new past' (Robert Mugabe cited in Garlake 1983: 15). The
emphasis of this call for new histories was to bury the foreign-origins myths of
Rhodesian settlers. It was, as its first respondent later put it, 'an indication that the
Zimbabwean African did not accept what the settler regime and its historians
brandished to us as our history. Alternatively, it was a call for the reinterpretation of
that history' (Chigwedere 1985: 5).

The aim was not only to establish Great Zimbabwe solidly as African heritage, but
also to challenge the very means by which history was written; to provide an African
alternative to the study of the past. In the words of the second respondent to the call
to arms, Ken Mufuka, the first African director at Great Zimbabwe (1983: 7),

> The study of Great Zimbabwe is therefore an attempt to find our self-identity. The
> search for our roots is not an easy one. The bulk of European scholarship, beginning with

Theodore Bent in 1891, R.N. Hall, W.G. Neal, Professor Raymond Dart, Mr Gayre of Gayre and a host of others have conspired to deny us that which by right belongs to us. To write an African viewpoint is to attract the ire of the 'learned world' and to risk being laughed out of court. But in Zimbabwe, this is the second year (1983) of national transformation and we deem it permissible to throw away old skins and to don new ones.

In this brief treatize, we have completely departed from the well-worn paths of proven scholarship. Proven scholarship after all sought to deny us what is ours. Even those who were sympathetic to our cause, because of their inadequate grasp of African cultures, did not quite appreciate the significance of the facts at their disposal. (Mufuka 1983: 7)

While the response from these two nationalistic writers to the call for a new national history was quick,[3] with the benefit of hindsight, it was also brief. The works of both authors have been badly received by academic historians who questioned both the content and the methodological use of oral traditions by their authors. Chigwedere's work was accused of encouraging tribalism, because of its 'new emphasis on ethnic origins as the explanation of history [which] clearly encourages tribalism and divides the new nation' (Garlake 1983: 15), whilst Mufuka's work, with its focus on Great Zimbabwe's 'egalitarian society' (Mufuka 1983: 28, 30), was charged with 'romanticism', as well as the unreliable use of oral traditions (Beach 1984b). Mufuka's response to these criticisms in his *Foreword* to the 1984 reprint is informative because it highlights his view of what differentiates the alternative African history that he espoused, from that of the white 'learned world', the 'professional connoisseurs' of historians and archaeologists. Betraying his determination to be a nationalist writer, he also emphasised that it had been his intention to be 'romantic'.

Two criticisms have been levelled at DZIMBAHWE. The first one is that 'any historian getting as many as 12 generations before 1900 is comparatively lucky' when using oral evidence. Oral evidence is by its very nature unreliable. Secondly I was told that the book is 'romantic'. The first criticism is based on cultural judgement. White people, who have a long written tradition, are naturally suspicious of oral evidence. If I accepted that premise, I would have to conclude that Zimbabweans have no reliable history before the coming of the white man in 1890. That I have chosen to believe the evidence given to me by my people is natural and congruent with the nature of DZIMBAHWE.

Dzimbahwe was written as a romantic account of a Great Shona-Karanga civilisation as seen through the eyes of the natives whose civilisation it was. It is in the nature of political testament. It is an attempt at recreating a self identity and bringing to light the DZIMBAHWE civilisation of which the monument is silent witness.

I make no apologies for this romanticism. (Mukfuka 1984 (1983): 3)

It seems clear that Mufuka's work, which was the only 'nationalistic' work to focus exclusively on Great Zimbabwe, embodies some of the euphoria of the post-independence period during which it was written. His assistant, VaNemerai, Education Officer at Great Zimbabwe, suggested that for Mufuka himself, this was closely linked to his personal desire to make a contribution after a long period abroad. VaNemerai also suggested that since then, his own views have changed.

*There was also a feeling of him as someone who had been away for a long time, and now he was back home, to an independent Zimbabwe. He wanted to make a contribution. And the euphoria was there, such as 'now we are independent, and we can write about Great Zimbabwe from **our own** point of view, as Zimbabweans'. That aspect comes in very clearly in his book.*

But you find that, well for me it was a new experience. I was young, he was more experienced, so that my view now is slightly different to what it was at that time. But we were influenced by being free, it being independent Zimbabwe, being free. So it was some kind of euphoria, the joy of being independent. (Interview with VaNemerai, 20/11/01)

Mufuka's contribution was quite well received at the time by the 'lay' audience, and his work gained some popularity, but not among 'academic' audiences for whom he was not, in any case, writing. In a sense he was responding to what Garlake (1983: 14) has described as the 'progressive alienation from the general population' of archaeology. But as a struggle seemed to develop over the right to represent Great Zimbabwe's past, between the professional archaeologists who had for so long kept alive the flame of Great Zimbabwe as African heritage, and these new nationalist/populist historians, Garlake and Mufuka were situated on opposite sides of the fence. In a letter, dated 28 September 1984, to Dawson Munjeri, then Regional Director at Great Zimbabwe, Garlake made his displeasure known about the fact that it was Mufuka's *DZIMBAHWE* (1983), and not his *Great Zimbabwe: Described and Explained* (1982) that was being promoted by staff at Great Zimbabwe.

For a long time my ZPH pamphlet was not on sale at the site museum despite invitations to the museum to order it. Only last month, when I asked, without giving my name ... which of the various booklets I should buy, I was informed by the museum official that Mufuka's work was recommended as 'all Garlake's work are copied from Mufuka' – a clear untruth given the dates of publication & one that caused some amusement to the group of foreign academics I was taking round.

I have a local press cutting of this year ... stating that your tourist officers now tell all visitors that Great Zimbabwe is the work of Marengu or Rusvingo or one of Mufuka's others characters. Certainly from the little I have overheard of your guides at the site, they all do preach Mufuka's gospel. (Letter from P.S. Garlake to Dawson Munjeri, 28 September 1984, NMMZ Files H2)

This 'struggle' over the right to represent Great Zimbabwe's past also emerged during a visit to Great Zimbabwe by Prince Charles in March that year, which later led to a very public debate in the pages of the *Sunday Mail* between Mufuka, and the then Permanent Secretary to the Ministry of Foreign Affairs, Stan Mudenge. In a letter to the *Sunday Mail* on the 6 January 1985, Mufuka accused Mudenge of making 'himself absolutely obnoxious' during the visit by Prince Charles. In his long reply, Mudenge described how,

instead of giving way to Dr Mudenge as official escort to the Prince – as previously arranged and agreed upon by all parties – Dr Mufuka stepped to the fore and proceeded to impose himself upon the party. This unhappy and really quite embarrassing situation continued all the way to the top of the ruins [...] Dr Mufuka began to explain that stone building at Great Zimbabwe had in fact begun on the hill and not in the valley. He pointed to the building style over head the present main entrance and identified it as typical of the Hill Ruin style from which the rest of the Great Zimbabwe building tradition evolved ....

It was only at this point that I intervened ... to point out that the particular stone work ... was not in fact part of the original style. I stated that it was part of the reconstruction work effected earlier in the century by Mr Wallace ... and therefore did not represent a typical example of the original Zimbabwe Ruins Hill style.

Immediately following this intervention, Dr Mufuka proceeded to explain to the royal visitor the significance of the entrance to the Hill ruin as now used. Once again I felt compelled to intervene and to explain that the original entrance, which is now closed, was to be found on the edge of the cliff ....

[...]

However, in complete defiance of the instruction of his superior, Dr Mufuka hastened to rejoin the royal party when it visited the Great Enclosure and was later heard to be to be indicating what he claimed to be the exact position where, in the 13th century, the first wife of one the great rulers at Zimbabwe used to sleep.

[...]

Dr Mufuka claims he ... overheard Dr Mangwenda accusing him of misleading or lying to the visitors. Unfortunately neither Dr Zengeni nor any other member of the entourage has been able to corroborate this statement. Had such a remark been made, however, it would have been justified, for Dr Mufuka was in fact misleading our VIP visitors on more than one occasion. (Dr S. Mudenge, Secretary for Foreign Affairs, in *Sunday Mail* January 13 1985)

Stan Mudenge's book *A Political history of Munhumutapa c1400–1902* (1988) was, in contrast to the efforts of Mufuka and Chigwedere, very well received by the academic establishment of historians and archaeologists, and can be taken, for the purposes of my argument here at least, to represent the moment when efforts to establish 'an authentic national history' (Mudenge 1998: vii) by Zimbabweans working from within the sphere of the established academia took over from those working on its periphery, such as Mufuka, and Chigwedere. Much like the 'professional connoisseurs' managed to hold sway over the very tenacious 'amateur' Rhodesian foreign-origins theorists for much of the twentieth century, so by the mid 1990s there was a new breed of Zimbabwean 'professional connoisseur' of the past, based at the History Department of the University of Zimbabwe, who were beginning to hold intellectual sway in the debates on Great Zimbabwe's past. It is a far cry from when Garlake (1983: 14) lamented 'with the only black Zimbabwean to have had a formal academic training as an archaeologist killed in the war, ... the image of Zimbabwean prehistory has been entirely determined by foreign and settler archaeologists'. Recent publications such as Matenga's *The Soapstone Birds of Great Zimbabwe* (1998), and Pikirayi's *The Zimbabwe Culture: Origins and Decline of Southern Zambezian States* (2001) are testimony to the maturity of Zimbabwean archaeology.

That is not to say the 'nationalistic', and indeed populist theories of Mufuka and Chigwedere disappeared. For most of the 1990s they remained in a largely muted form in public discourse. Mufuka's work is often referred to as an example of the 'misuse of oral history' whilst Chigwedere's work gained some popular notoriety through the televised history programme *Mitupo neMadzinza* [literally 'Totems and Clans'], on which he appears with fellow 'TV historian' VaDzova, discussing the oral traditions of Zimbabwe. More recently, Chigwedere's fortunes have seen a revival through his efforts, as Minister for Education and Culture, to make history a compulsory part of the school syllabus. While this has been seen by some as a cynical attempt to promote his own work,[4] Ranger (2004: 224–5) has identified Chigwedere as a key player in a new, ZANU PF-driven discourse of state history that is 'explicitly antagonistic to academic historiography' (Ranger 2004: 218), which he labels 'patriotic history'.[5] In Chapter 10, I explore whether this 'patriotic history' represents a revitalisation of the attempts of the early 1980s to not only rewrite African history, but also the means by which it is written, drawing attention to the involvement and complicity of more 'sophisticated historians' such as Stan Mudenge (Ranger 2004: 226), and even NMMZ itself, to highlight its complexity.

Despite Chigwedere's influence at government level (and possibly now, in part, because of it) the chasm between 'nationalistic'/populist history and that of the 'professional connoisseurs' at the University remains wide. For most of the post-independence period, it is the latter that has had influence in the representation, and, crucially, in the management of Great Zimbabwe National Monument. Indeed NMMZ, the organisation that took over responsibility for Zimbabwe's monuments and archaeological sites from its Rhodesian predecessor NMMR, has become intertwined with the History Department at the University of Zimbabwe. Since Mufuka's departure from Great Zimbabwe in 1984, the majority of the senior posts within NMMZ have been occupied by former archaeology students from that department. So not only has the representation of Great Zimbabwe's past been firmly re-appropriated by 'professional connoisseurs', so too has its management. As I will develop in a later chapter, the management of Great Zimbabwe as Zimbabwe's prime national heritage site, has, since independence, seen increasing professionalisation, as NMMZ has become an internationally recognised authority on the scientific monitoring, preservation and restoration of dry stone walls. Furthermore, with Great Zimbabwe's elevation to the status of a World Heritage Site in 1986, its management has been increasingly determined by the requirements of the World Heritage Convention (UNESCO 1972), and the international discourses on heritage management that surround it, as well as by the demands of international tourism.

In sum, therefore, Great Zimbabwe's centrality to nationalism's imagination of the nation, as the '*eternal heritage for people of all tribes in this country*', did not lead to a very dramatic reformulation of how its history should be constructed. Despite efforts by Chigwedere, Mufuka and others, the re-appropriation of Great Zimbabwe's past for Africans has come most successfully from within the academic sphere of the 'professional connoisseurs', which should really be seen as a continuation of the efforts of archaeologists under Rhodesian rule, such as Summers, Robinson, Huffman and Garlake, who worked for the Historical Monuments Commission and its successor NMMR, before the 'new revisionism' and indeed censorship began in response to nationalism's use of the name 'Zimbabwe'. Rather than representing a radical break with the past, as President Mugabe's call for new histories seemed to imply, this dominant post-independence view is better seen as a continuation of a discourse established long before independence. In this way, the period of censorship and 'new revisionism' of the latter years of the Rhodesian Front can be seen in a similar light to the nationalistic works written in immediate confrontation to them after independence by Mufuka and Chigwedere – as the last ditch efforts to respond to and counter the historical inevitability of the predominance of academic studies of the past, which then reasserted itself through the History Department of the University of Zimbabwe, and its cousin NMMZ during the 1980s and 1990s.

It remains to be seen where 'patriotic history' will fit into this scheme, though as I suggest in Chapter 10, in one respect it is unlikely to provide much of an alternative to any of the approaches to Great Zimbabwe's past so far discussed – it does not seem to promote local 'history-scapes' any more than any of its predecessors. Similarly, while both Chigwedere and Mufuka deliberately set up their approach to Zimbabwe's past in opposition to both the fanciful and disparate theories of Rhodesian settlers, and the 'professional', 'white' discourse of academic historians and archaeologists, neither actually strayed that far from the latter's position. Local perspectives on Great Zimbabwe do not feature in either author's work to any greater depth than they do in the 'conventional' pro-African origins accounts of Great Zimbabwe's past. Despite the very public, and quite angry exchange of views between Mudenge and Mufuka that occurred in the pages of the *Sunday Mail* in 1985, both have focused on the oral traditions of the Munhumutapa period from northern Zimbabwe, rather than local oral traditions. Mufuka mentions the oral traditions of local clans in no more detail than the only pre-independence oral historian who worked in the area (Aquina 1965, 1969–70) did in the late 1960s. From a local perspective, Mufuka's radical departure from the 'normal' views on oral history about Great Zimbabwe's past does not really appear that 'radical' at all. This was revealed when I discussed this issue with VaNemerai, who assisted Mufuka in researching and writing *DZIMBAHWE*:

> *VaNemerai: The reason is that it is believed that the people who were living here deserted the place, the main group went up northwards to the Dande area, Zambesi valley, that's Mutota. They say that the people who were living here, didn't build the place, the main group went northwards to the Zambezi, a smaller group went westwards. ...*
>
> *J.F.: to the Khami ruins ...*
>
> *VaNemerai: So that's why he talks about Mutota because he thinks that the descendants of these people, Mutota and so on, are found there. You are not going to find a great deal of information about Great Zimbabwe here. Also when you talk to the people here, the Manwa people, the Duma people, they seem to have come after the original builders of Great Zimbabwe, as keepers of the monuments, and even today they feel they are the keepers of Great Zimbabwe.* (Interview with VaNemerai 20/11/01)

Thus, even those authors who made the most determined and conscious efforts to overturn the biases of past antiquarians, and 'professional' historians and archaeologists, have not, from a local viewpoint, delivered much of an alternative in the form of a chance to be heard.

SOME THOUGHTS ON HERITAGE, NATIONALIST MIMICRY, AND DERIVED DISCOURSES

The past, in the words of Lowenthal (1990: 302), 'is everywhere a battleground of rival attachments'. In recent years , the 'politics of the past' (Gathercole and Lowenthal 1990) has come under increasing scrutiny by historians, archaeologists and anthropologists, who, having become aware of the historical complicity of their disciplines with the colonial project (for example, Asad 1973, Said 1978), and the institutionalised racism it embodied, have also come to recognise the subjectivity involved in their interpretations of the past (Ucko 1990: xii), and the heavy load of ideological implications they often incarnate. Gero and Root (1990: 19) put it as follows, in their analysis of the *National Geographic*.

> The past we construct, then, is more than passively conditioned by our political and economic system; it is a direct product of, and an effective vehicle for, that system's ideological messages. As a product of Western practice, archaeology reduces the cultural distance between past and present by reifying a commoditised view of the world and the values that support that view. Archaeology as an enterprise legitimises the hegemony of Western culture and western imperialism and imposes a congruent view on the past, one that is ably promoted by successful media such as the National Geographic. (Gero and Root 1990: 35)

These academic disciplines have also begun to investigate the role that their constructions of the past, and the physical remains of it, that is, 'heritage', have played in the formulation of differing, and often conflicting, national, ethnic and religious identities. As Kaiser has noted in the opening passage of his paper, the salience of the past in competing ethnic nationalisms is particularly obvious in the political transformations and fragmentations that occurred in the Balkans in the 1990s.

> Nowhere has it been made more horrifyingly clear that the past is a prize, a resource to covet and for which to contend, than in the west Balkans today. When towers and walls of ancient towns are shelled for no purpose, when medieval churches and mosques become targets, and when the call to arms unfurl histories like banners, then it is starkly apparent to what extent the past can intertwine with the present – and to what effect. … The wars of the Balkans unequivocally show that possession of the past is no trifling matter, and that the construction of the past is fraught with consequence. (Kaiser 1995: 99)

The recent wars in the Balkans illustrate the violence that can be inflicted in the name of the past, against not only people, but also against the very remains of the past themselves. This was further illustrated more recently in Afghanistan with the deliberate destruction of the Bamiyan Buddhist statutes by the Taliban regime.[6]

It should be evident that Great Zimbabwe is a very good example of how differing constructions of the past can be used to justify and legitimise both a military invasion/conquest, that is, colonisation, and an anti-colonial nationalist liberation movement. Furthermore its historiography illustrates how archaeology 'has been shaped as an academic discipline in Africa by its role in either supporting or countering the dominant assumptions concerning African history' (Holl 1990: 296). I have argued elsewhere (Fontein 2000: 8) with reference to the work of Kevin Walsh (1992), that the 'modern' academic disciplines of history and archaeology emerged as a result of the European enlightenment, through a combination of the idea of progress through linear time, an enchantment with a distant and separate past induced by a period of massive social change, and the quest for objective 'truths' and 'facts' about the world. As 'disembedding mechanisms' (Giddens 1990) these disciplines have appropriated knowledge of the past away from a past based on memory, a sense of place – and specifically to my case study here, oral traditions and the words of the ancestors – and have gained an almost hegemonic authority in terms of both the representation and the management of 'heritage'. This argument has particular salience in a post-colonial context, as it is increasingly realised that historical and archaeological discourses are eurocentric, in the sense that they are based upon perceptions of the past and time that originated as a result of the European enlightenment, which arrived in Africa on the back of colonialism. As Gathercole has put it, 'it is undeniable that archaeology's biases derive from its western origins and perceptions' (Gathercole and Lowenthal 1990: 3).

The debate about the eurocentricity of, or perhaps better put as the European/colonial origins of archaeology and history, as 'professional' disciplines of the past that appropriate to themselves both the authority to construct the past, and manage its remains, reaches a different level if we consider the role that these disciplines often play in the construction of national identity. Nowhere is this more apparent than Israel/Palestine where archaeologists report to the president and graduating army recruits swear oaths of allegiance on the summit of Masada (Paine 1994: 390; see also Zerubavel 1995; Abu El-Haj 1998, 2001; Said 2002). Kohl and Fawcett have referred to both Israel and Zimbabwe to illustrate the close connection which often exists between archaeology and nationalism.

> Archaeological sites are such potent symbols of national identity (e.g., Masada in Israel, or Zimbabwe in, significantly, Zimbabwe) that peoples today are frequently willing to fight over them. Archaeology and ancient history help define people as distinct and occupying (or claiming) territories that were historically theirs. (Kohl and Fawcett 1995: 11)

Of course, Zimbabwe is not the only country in sub-Saharan Africa where use has been made of the past, as 'heritage', for the purposes of African nationalist revival. Both Kaplan (1994) and Willet (1990) show how museums in Nigeria have been used to foster national identity in that country.

> As an artefact of colonial rule with artificial frontiers, evolving Nigeria needs to promote a sense of national identity, pride, and unity. It has begun to use museums to help do so by redistributing material from all parts of the country to museums throughout the land. In this way, Nigerians of different languages, cultures, religions, and allegiances can begin to appreciate how the past has made them one as well as many peoples. (Willet 1990: 181)

It is clear, therefore, that Zimbabwean nationalism was by no means exceptional in its use of Great Zimbabwe as national heritage; as a tool or 'useful rallying point' with which to 'imagine' a nation with deep historical precedence into existence. Indeed it may be much more appropriate to consider the extent to which Zimbabwean nationalism's use of Great Zimbabwe has followed an almost standardised model for nationalism, that is orientated partly around a need to reach into the past for primordial legitimacy in the present. In her work on nationalism in nineteenth century Spain, Diaz-Andreu (1995) has analysed the close relationship that existed in the development of archaeology as a discipline, and the emergence in the nineteenth century of that other curious product of European enlightenment, the nation-state. While Diaz Andreu's argument is particularly focused on Spain, her argument is clearly applicable in a wider context.

> the development of archaeology as a scientific discipline in the nineteenth century can only be understood in the context of the creation of a national history: that is to say a history directed at legitimising the existence of a nation and, therefore, its right to constitute an independent state. (Diaz -Andreu 1995: 54)

Therefore the disciplines of archaeology and history may have been implicated in the project of the nation-state right from the outset. In this light, the use of archaeology and history for the purposes of anti-colonial nationalisms, such as in Zimbabwe, adds significant fuel to the argument that has frequently appeared in the discussions on nationalism, that 'third-world nationalisms' are of a 'profoundly "modular" character' (Chatterjee 1986: 21) and reliant upon models of nationalism created in the 'west'.

Two of the most influential thinkers on nationalism in recent years (Gellner 1983 and Anderson 1983) have both linked its development to the European enlightenment, though their approaches and their positions are slightly different. Gellner, coming from the liberal-rationalist angle, argued that nationalism came about as a result of the needs of industrialisation, whilst Anderson, coming from a Marxist perspective, suggested that it came about as a result of the development of 'print-capitalism'. The latter's approach has been especially influential because Benedict Anderson 'demonstrated with much subtlety and originality that nations were not the determinate products of given sociological conditions such as language or race or religion; they had been, in Europe and everywhere else in the world, imagined into existence' (Chatterjee 1996: 216).

From this starting point Partha Chatterjee, a member of the influential Subaltern Studies Collective, has developed his own approach to the subject of nationalism. In particular he has focused on the links between the emergence of European nationalism and the processes of European colonial domination of the rest of the world. When anti-colonial nationalism later appeared in Asia and Africa, it was inevitably a 'derivative discourse' (Chatterjee 1986). In the words of one of Chatterjee's reviewers, anti-colonial nationalism has been successful in 'liberating the nation from colonialism, but not from the knowledge systems of the post-Enlightenment West, which continue to dominate, perhaps even more powerfully' (Ramaswamy 1994: 960). In Chatterjee's own words,

> Nationalism sought to demonstrate the falsity of the colonial claim that the backward peoples were culturally incapable of ruling themselves in the conditions of the modern world. Nationalism denied the alleged inferiority of the colonised people: it also asserted that a backward nation could 'modernise' itself while retaining cultural identity. It thus produced a discourse in which, even as it challenged the colonial claim to political domination, it also accepted the very intellectual premises of 'modernity' on which colonial domination was based. (Chatterjee 1986: 30)

In terms of the use of Great Zimbabwe by Zimbabwean nationalists that I have described above, Chatterjee's argument about 'derivative discourses' finds a remarkably strong echo. In claiming Great Zimbabwe as African heritage, in reaction to the 'cultural aggression' of Rhodesian settler myths of foreign-origins, Zimbabwean nationalism was indeed denying 'the alleged inferiority of the colonised people'. And similarly, the increasing professionalism of Zimbabwean archaeology and heritage management 'vanguarded' by NMMZ/History Department, University of Zimbabwe, could be seen as an assertion 'that a backward nation could "modernise" itself while retaining cultural identity'. Certainly its seems that the 'very intellectual premises of "modernity" ' were embraced by Zimbabwean nationalism, particularly in its post-independence approach to the representation and management of national heritage. In fact, given the explicit complicity of the Rhodesian state, in its final death throes, with the 'new revisionism' and censorship that saw the Rhodesian foreign origins myths re-emerge in opposition to African nationalist appropriation of Great Zimbabwe's past for themselves, we could even assert that the colonial state, on its very last legs, was far less aligned with 'modernism' than the nationalist movement itself.

Homi Bhabha (1994) reaches a similar, if slightly parallel position to that of Partha Chatterjee.[7] Describing 'mimicry' as 'one of the most elusive and effective strategies of colonial power and knowledge', Bhabha has developed his analysis of the 'mimicry' of colonial discourse to suggest that,

> colonial mimicry is the desire for a reformed, recognisable Other, *as a subject of a difference that is almost the same, but not quite.* Which is to say, that the discourse of mimicry is constructed around an ambivalence: in order to be effective, mimicry must continually produce its slippage, its excess, its difference. The authority of that mode of colonial discourse that I have called mimicry is therefore stricken by an indeterminacy: mimicry emerges as the representation of a difference that is itself a process of disavowal. Mimicry is, thus the sign of a double articulation; a complex strategy of reform, regulation and discipline, which 'appropriates' the Other as it visualises power. Mimicry is also the sign of the inappropriate, however, a difference or recalcitrance which coheres the dominant strategic function of colonial power, intensifies its surveillance, and poses an immanent threat to both 'normalised' knowledges and disciplinary powers. (Bhabha 1994: 86)

The 'mimicry' of the colonial master by the colonial subject, is therefore a technique of power over that subject – a means of both revealing, and creating the dominance of colonial discourse and knowledge. At the same time, however, it undermines that discourse of knowledge based on difference and becomes a means by which the domination of the coloniser is threatened, and ultimately overturned. Focusing on the 'mimic man' Bhabha continues,

> The line of descent of the mimic man can be traced through the works of Kipling, Forster, Orwell, Naipaul and to his emergence, most recently in Benedict Anderson's excellent work on nationalism, as the anomalous Bipin Chandra Pal. He is the effect of flawed colonial mimesis, in which to be Anglicised is emphatically not to be English.

> The figure of mimicry is locatable within what Anderson describes as the 'inner compatibility of empire and nation'. It problematizes the signs of racial and cultural priority so that the 'national' is no longer naturalizable. What emerges between mimesis and mimicry is a writing, a mode of representation, that marginalizes the monumentality of history, quite simply mocks its power to be a model, that power which supposedly makes it imitable. Mimicry repeats rather than re-presents .... (Bhabha 1994: 88)

Thus the roots of the anti-colonial effort and the form that is taken lie in the 'colonial mimicry' that is itself a technique of colonial domination, but provides the means of its own disposal. Hence colonialism, and its antithesis, anti-colonial nationalism, are '*almost the same but not quite*'. If we return to Zimbabwean nationalism, there should be little surprise therefore that the colonial appropriation of Great Zimbabwe, as witnessed, for example, by the use of the Zimbabwe Birds on colonial Rhodesian flags, currency and coat of arms, was mimicked by the determined effort of nationalists to re-appropriate these symbols for themselves. Shortly after independence a further measure was taken at Great Zimbabwe itself to rub away the very last traces of its colonial appropriation. Various parts of the ruins that had been burdened with the names of early Rhodesian explorers were renamed and a very conscious decision was taken that the term 'ruins' should be avoided. The most significant parts of the site received Shona names; like the Imba Huru (literally 'big house') for the Great Enclosure, and Nharira ya Mambo (literally 'sleeping quarters of the King') for the

Hill complex, to replace 'Acropolis'. In a revealing letter, dated 11 June 1981, to Thomas Huffman who was writing a new guide book for Great Zimbabwe, Cran Cooke, then Director at Great Zimbabwe, provided a list of the new names, and made his cynicism apparent.

> The word ruins is not considered diplomatic and the names of the early hunters and explorers are most definitely out. Even Ridge Ruins was suspect as it was thought to be a Colonial's name. Royal is not a good Marxist word. No doubt at some future time Karanga names will be invented or perhaps some areas named after important visitors since independence. We can only wait and see. (Letter from Regional Director Cran Cooke to T. Huffman, 11 June 1981, NMMZ File H2)

But contrary to Cran Cooke's comments, no further name changes have taken place. Even those Shona names that were introduced for the Great Enclosure and the Hill complex are only rarely used.

Norma Kriger has discussed how a similar effort of 'nationalist mimicry' – the destruction of 'colonial heroes' and symbols shortly after Independence, and their replacement by newly created 'national heroes', 'generated acute political controversy' (Kriger 1995: 140). She described how colonial symbols in the form of statues of Cecil Rhodes, and other Rhodesian heroes were removed from public places in urban landscapes, amid public protests not only from white former-Rhodesians, but also from differing black Zimbabweans. One such debate revolved around the appropriateness of Second World War memorials for the new state. The Harare City Council and the central government found themselves on opposite sides of the fence, the former arguing that 'it kept reminding them of colonial soldiers who had died for the British Empire', and the latter raising the counter-argument that the Second World War should be remembered 'as first and foremost a war against fascism' (Kriger 1995: 143). In what later became the norm, as is clear from the hindsight that emerges from the excellent work of Kriger (1995) and Werbner (1998), the government showed its determination to control the process of national memory building, and was in this case successful. The Cenotaph commemorating white Rhodesian war dead of the Second World War has remained standing in Harare Gardens (Kriger 1995: 143).

But if such controversy about what to do with colonial symbols had been unexpected, it paled into insignificance in relation to the debates that have raged about the creation of new national symbols in the form of National Heroes of Zimbabwe. Both Kriger and Werbner have illustrated in their complimentary analyses of the 'politics of creating national heroes' that the process of postcolonial, state memory building in Zimbabwe about the recent past, in particular the liberation struggle that brought independence, has been fraught with conflict and disputation. This has been particularly centred on the very centralised, and ruling ZANU (PF) party dominated process by which it is decided whose remains should be buried at the National Heroes Acre monument in Harare.

One thing that the vigorous debates about commemoration and memory building in Zimbabwe have revealed, is the 'plurality of political origins myths' (Werbner 1998: 75). Working with Balibar's argument (1991) that young and old nations alike 'resort to a myth of political origins for imagining the ongoing formation of the nation'

(Werbner 1998: 75), Werbner has argued that for 'nations whose political origins myths go back to a birth at the barrel of an anti-colonial gun ... such a birth readily opens out to a wide horizon of imagined nationhood'. Nationalism, and the 'myths of political origins' that are invoked in its name, are not homogeneous, and undifferentiated, but rather sites of 'contradictory appropriations and thus controversial memory-making' (Werbner 1998: 75). Unfortunately I do not have space here to consider in any greater detail the political issues and controversies of memory-making in post-colonial Zimbabwe that Werbner and Kriger have brought up in their work, except to add that attempts to control processes of commemoration and memory-making by the ruling party elite at the centre of the state have provoked a lively and varied popular discourse of counter-commemoration and memory-making by disgruntled and disaffected parts of Zimbabwe society on the periphery. Werbner's argument about the 'plurality of political origins myths' will tie in with what I expand upon in the next chapter, about that other mythology that fired nationalist imagination – the first *Chimurenga* of the 1890s. Its symbols, ancestral figures like Ambuya Nehanda and Sekuru Kaguvi, came to be linked to nationalist use of Great Zimbabwe by guerrilla fighters and traditionalists far away, and separate from, the really rather limited imaginations of the exiled nationalist elite.

Returning to the theme of nationalist mimicry and derivative discourses, it seems obvious that it is possible to view within both the perceived need to construct a National Heroes acre, and in the form that it has taken (for a good description see Werbner 1998: 82–6), evidence of postcolonial state mimicry of its 'western'/colonial fore-bearer. On the surface it seems that the post colonial Zimbabwean state has merely adopted what Werbner (1998: 72) described as 'the constitution of a whole, distinctively modern, complex for commemoration of the sacrifice of life in the cause of the nation-state'. This is not so. Werbner's own analysis argues that the memorial complex 'has been reworked in significant ways upon its reception in postcolonial Africa', and in Zimbabwe specifically,

> the postcolonial state has made no provision for the mass military cemetery. Instead it has created its national shrine near the capital around a cemetery for the elite, an inner circle even in the grandly monumental layout of their graves. Here a conflation of the cenotaph and the tomb of the unknown soldier has further reduced the emphasis on the common warrior in the modern memorial complex. Although Zimbabwe's national shrine has a main statue known as the Tomb of the Unknown Soldier, this is actually a cenotaph, empty of any body.

> In Zimbabwe the modern memorial complex has been given a distinctively postcolonial form, glorifying above all the individuality of great heroes of the nation. In itself a force for differentiation, it represents a centred nation triumphant in displacement of racist white settler domination, but it also registers, uneasily and contrary to official intent, the increasing disaffection between what Zimbabweans call the chefs and the povo, the people or the masses. (Werbner 1998: 73)

For Werbner it is this elitism, evident in the very construction of the shrine, let alone the official rituals of state pomp, or the partisan processes of selection that surround it, which lies behind the controversies that memory-making in Zimbabwe have engendered. For the purposes of my argument here, on the 'derivative' nature or 'mimicry' of nationalism, it is the postcolonial reworking of modern state commemoration that I want to focus on. If post-colonial states are able to utilise and rework, and in a sense 'make their own', aspects of state-craft that originated elsewhere, perhaps in the very backyards of their previous colonial masters, then surely this suggests more autonomy than talk of derivative discourses, and mimicry allows? On further reflection both 'mimicry' and 'derivative' imply, as Bhabha noted, *'almost the same but not quite'*, which must provide some space for authenticity and autonomy in non-European, anti-colonial nationalisms. It is very significant for my argument here, that Zimbabwe's monument to its own national heroes embodies many stylistic features and symbols that derive directly from Great Zimbabwe itself (see figures 6.3 and 6.4). As Werbner described it,

> The high walls are surmounted on each side by an emblematic figure from Great Zimbabwe known as the Zimbabwe Bird. Chevrons on the walls, in the style of those at the Great Enclosure of Great Zimbabwe heighten the resonance with the country's most famous and ancient African civilisation. (Werbner 1998: 85)

Surely therefore, in content, if not in form, all nationalisms must be somehow unique, or do non-European states have no historical alternative but to follow in the shadows of 'Western' thought? In a later reformulation of his own ideas, Partha Chatterjee has considered this problem. Taking as his cue a critique of Benedict Anderson's work, he asks

> If nationalisms in the rest of the world have to choose their imagined community from certain 'modular' forms already made available to them by Europe and the Americas, what do they have left to imagine? History, it would seem, has decreed that we in the postcolonial world shall only be perpetual consumers of modernity. Europe and the Americas, the only true subjects of history, have thought out on our behalf not only the script of colonial enlightenment and exploitation, but also that of our anti-colonial resistance and postcolonial misery. Even our imaginations must remain for ever colonised. (Chatterjee 1996: 216)

Figure 6.3: National Hero's Acre, Harare. The 'Zimbabwe-esque' stone walls with chevron patterns are visible in the background, to the right and left. (Author 1997)

Figure 6.4: One of two murals at the National Hero's acre, Harare, which carry depictions of different stages of the second *Chimurenga* underneath huge Zimbabwe Bird statues, visible in the top right hand corner. (Author 1997)

Partha Chatterjee has presented his reformulated ideas in his second book, 'The Nation and its Fragments' (1993), in which he argued that anti-colonial nationalism 'creates its own domain of sovereignty within colonial society well before it begins its political battle with the imperial power' (Chatterjee 1993: 6). He continues,

> It does this by dividing the world of social institutions and practices into two domains – the material and the spiritual. The material is the domain of the 'outside', of the economy and of state-craft, of science and technology, a domain where the West had proved its superiority and the East had succumbed. In this domain, then Western superiority had to be acknowledged and its accomplishments carefully studied and replicated. The spiritual, on the other hand, is an 'inner' domain bearing the 'essential' marks of cultural identity. The greater one's success in imitating Western skills in the material domain, therefore, the greater need to preserve the distinctness of one's spiritual culture. This formula is, I think, a fundamental feature of anti-colonial nationalisms in Asia and Africa. (Chatterjee 1993: 6, also 1996: 217)

By viewing anti-colonial nationalisms in this dualistic manner, Chatterjee is able to overcome the criticisms of his previous work by reviewers who questioned the 'fruitfulness of theorising within the autonomous/derivative discourse binary, where autonomy lies within the West, and the derivative in the rest' (Ramaswamy 1994: 961). Chatterjee is therefore able to return a greater level of agency to actors of anti-colonial nationalisms, by making this separation between the 'material' and 'spiritual' domains.

For it is in the latter that,

> Nationalism launches its most powerful, creative, and historically significant project: to fashion a 'modern' national culture that is nevertheless not western. If the nation is an imagined community, then this is where it is brought into being. In this, its true and essential domain, the nation is already sovereign, even when the state is in the hands of the colonial power. The dynamics of this historical project is completely missed in conventional histories in which the story of nationalism begins with the contest for political power. (Chatterjee 1993: 6, also 1996: 217–18)

As I now turn to consider a very important aspect of Zimbabwean nationalist mythology, based not so much on Great Zimbabwe, but on a more recent aspect of the past, the rebellions of 1896–97, the first *Chimurenga*, it will become apparent that Chatterjee's approach has considerable merit. Through another claim to legitimacy and authority, which was based not so much on past African achievement, as evidenced by the numerous Zimbabwe-style ruins that litter the landscape, but rather reached out to authority of 'national' ancestors such as Ambuya Nehanda, Sekuru Kaguvi, and Chaminuka, Zimbabwean nationalism was able to claim for itself an 'authenticity' that evades conclusively any accusations of 'mimicry' or 'derivative discourses'. But while nationalism may be a unifying project, its view and imagination must always be a multiplicity of views and imaginations. As I will develop below, the much celebrated and theorised collaboration between the ancestors and the guerrillas was only another 'useful' mythology for the pragmatic nationalist elite, exiled in foreign lands, but a lived experience for those actually fighting in, and suffering the effects of the war of liberation. And in this differentiation of perspective, as Chatterjee has put it, 'lies the root of our postcolonial misery' (Chatterjee 1993: 11), for in the movement from anti-colonial nationalism to postcolonial state, there has been an inability to create authentic forms; the 'authentic' spiritual domain surrenders to the 'derived' material domain. Thus at Great Zimbabwe it is not the rule of *chikaranga* or the words of the ancestors that dictate, but the needs of international and national heritage, informed by the 'professional connoisseurs' of the past, and the economic imperatives of international tourism.

> The result is that autonomous forms of imagination of community were, and continue to be, overwhelmed and swamped by the history of the postcolonial state. Here lies the root of our postcolonial misery: not in our inability to think out new forms of the modern community but in our surrender to the old forms of the modern state. If the nation is an imagined community and if nations must also take the form of states, then our theoretical language must allow us to talk about community and state at the same time. I do not think our present theoretical language allows us to do this. (Chatterjee 1993: 11; also 1996: 222)

Notes

1. Two very recent examples are Richard Ganter's (2003) *Zimbabwe's Heavenly Ruins: A mystery explained* and Robin Brown-Lowe's (2003) *The Lost City of Solomon and Sheba: An African Mystery*. One wonders about audiences, as Luise White (2003: 72–3) has done in relation to multiple Rhodesian confessions to the assassination of Herbert Chitepo, and whether these accounts are being written for an 'imagined Rhodesian

audience' so that 'a nation could be imagined as it had never been lived' (White 2003: 73, 74).

2. During the struggle, and shortly afterwards, there was much discussion and analysis of factionalism during the liberation movement (for example, Kapungu 1974; Saul 1979; Ranger 1979; M.Sithole 1980, 1984, 1985) which has focused on the role of class, ideology, and leadership as well as ethnicity. Masipula Sithole (1980) has suggested that the initial split between ZAPU and ZANU in 1963 was not ethnically motivated, but that leadership issues were really responsible. Those involved were aware that they could have mobilised along the Shona/Ndebele ethnic divide, but chose not to. In a later work (1985) he suggested that the wider ethnic factionalism that occurred in the second half of the 1970s – not just Shona/ZANU versus ZAPU/Ndebele, but also among other nationalist groups, and between ethnic 'sub-groups' within these larger ethnic groupings – were examples that supported Malcom Cross's argument that the 'salience of ethnic identity' (Sithole 1985: 185) increases if political change becomes more likely. Recently literature on ethnicity in Zimbabwe has re-emerged in the context of studies relating to the *Gukurahundi*, the Matabele massacres committed between 1983 and 1987 by the ZANU PF government in rural Matabeleland during their fight against ZAPU/Ndebele dissidents (Werbner 1991; Ranger et al. 2000).

3. Chigwedere published his first book *From Mutapa to Rhodes* (1980) only two months after independence, his second *The Birth of Bantu Africa* in 1982, and the third *the Karanga Empire* in 1985, Mufuka's *DZIMBAHWE Life and Politics in the golden age 1100–1500AD*, came out in 1983.

4. See, for example, *The Standard* 9–15 September 2001.

5. Chigwedere has published several new books recently including *The roots of the Bantu* (1998) and *British Betrayal of the Africans, land, Cattle and Human Rights* (2001). While the first is very much in the mode of his earlier writings, the latter is much more conspicuous for its extreme vilification of Britain, in which 'all the white characters … are painted with the treachery brush' (Mufuka 2002). Even Mufuka criticizes the book for being 'political', 'impartial' and unacademic, though he also suggests that reading it 'will help the reader understand the resentment towards whites now prevalent in ZANU PF' (Mufuka 2002).

6. After the destruction of these statutes, which was reported worldwide, and met with varies degrees of horror and condemnation (for example, CBC news website, 2/3/01 and 3/3/2001; CNN.com 12/3/01; BBC News website 12/2/01 and 26/3/01), calls for their reconstruction (for example, Newsweek Web Exclusive 31/12/01; BBC News website 30/1/02; CNN.com undated 'Karzai pledges to rebuild Afghan Buddhas') were met in some quarters with similar, if far less outraged, expressions of concern (ICOMOS Report 2002–03, Afghanistan). In 2003 the site was inscribed on both the World Heritage List and the World Heritage in Danger List.

7. I must credit this point to Dr. Paul Nugent as he initially directed me towards Homi Bhabha's work, suggesting the similarity of his position to that of Chatterjee. How it has been worked into my argument is, of course, my own responsibility.

'MAPFUPA EDU ACHAMUKA'
(OUR BONES WILL RISE):
NATIONALIST MYTHOLOGY OF THE FIRST
CHIMURENGA

One remarkable aspect of the Zimbabwean liberation struggle that has been debated a great deal since independence is the close collaboration between guerrillas and spirit mediums. David Lan's work (1985) in particular focused on the way in which guerrillas were able to harness popular support and legitimacy in the Dande, in Northern Zimbabwe, by working closely with spirit mediums of royal ancestors, the *mhondoro*. Ranger has made a similar argument, suggesting that there was a revival of 'traditional religion' and belief in the ancestors because 'above any other possible religious form the mediums symbolised peasant right to the land and their right to work it as they chose' (Ranger 1985: 189). The work of Daneel (1995, 1998), focused on Masvingo province, has also described in great detail the relationship that existed between *masvikiro* and the ancestors; *Mwari* and the Matopos shrines; and the *Vana Vevhu* ['children of the soil'] – the guerrillas themselves.

Several writers have emphasised that the mobilisation of the rural 'peasantry' in Zimbabwe was not always inspired by appeals to the ancestors. Norma Kriger (1988, 1992) has argued convincingly that in some places popular support for the liberation struggle was not willingly given, and more often the result of coercion by guerrillas. Her work also emphasised the importance of 'disaggregating the peasantry', and highlighted how struggles internal to rural society were played out during the liberation struggle. Maxwell (1999: 14) in turn has emphasised the importance of also looking more closely at differences among guerrillas themselves, in terms of elites versus rank and file, ethnic origin and year of recruitment. Both Maxwell (1999) and Daneel (1995, 1998) have added to Linden's work (1980), so that together they have showed how the Catholic church and several African churches were also involved in the mobilisation of popular rural support for nationalism. To add further to this 'ethnographic thickness' of Zimbabwe's liberation struggle, Ncube and Ranger (1995) have clearly demonstrated that contrary to popular stereotypes, it was not only ZANLA which co-operated with spirit mediums during the war, ZIPRA guerrillas operating in areas dominated by *mhondoro* mediums also made use of them; and similarly ZANLA as well as ZIPRA guerrillas visited the *Mwari* shrines in the Matopos, Matabeleland. Therefore differences between ZANU and ZAPU, and their armed wings, ZANLA, and ZIPRA respectively, that had previously been posited in terms of 'ZANLA as "Shona" and "spiritual" and of ZIPRA as "Ndebele" and "secular" ' carry no weight, and as Bhebe and Ranger (1995: 8) have put it 'we ought to see both guerrilla armies as responsive to the beliefs and institutions of the people among whom they were operating'.

What is clear from this lively debate,[1] is that there was a great deal of regional and local variation, and a complex multiplicity of factors involved in the relationships between 'traditional religion', Christian churches, rural people, guerrillas, and the nationalist elite during the years of war preceding independence. One of the key issues for the argument that I am now about to develop is the differences that existed in the ideology, and indeed education, of elite members of the nationalist movement, and guerrillas actually fighting in the struggle.

In her addition to the study of Zimbabwean nationalism and the struggle for liberation, Fay Chung (1995), who worked in the ZANU Education department in Mozambique during the war, has focused on the educational work done by ZANU during the struggle. In her conclusion she has clearly articulated the differences in ideology between the leading elite of ZANU, the political party, and members of its armed, military wing ZANLA.

> Because of the large number of intellectuals and professionals who joined it, ZANU regarded education as an essential part of the liberation struggle. One recalls that when ZANU was first formed in 1963 it was labelled as a party of intellectuals cut off from the masses. There was great emphasis within ZANU on intellectual and professional development which were seen as necessary to overthrow the settler regime. Since 1963, however, ZANU had come to realise the need for mass support for its guerrilla armies. ZANU and ZANLA were successful because they brought together intellectuals and peasantry.
>
> Tensions constantly existed between the two groups, of course, the one being more scientific in its analysis and orientation and the other more traditional. This tension was reflected in contrasting ideological manifestations, with one group asserting its adherence to democracy, nationalism, modernism and later Marxism–Leninism, while the larger group of peasants clung to their traditional ideology, dominated by the traditional resistance figures such as Nehanda, Kaguvi, Chaminuka and the ancestral spirits. But these different ideologies co-existed throughout the liberation struggle. ZANU did not try to destroy traditionalism in the way FRELIMO tried to do, but instead tried to win traditionalists' support for the liberation struggle. The traditional leaders, the spirit mediums, on the other hand, also tried to understand and accommodate modern trends [...] Whilst the spirit mediums remained upholders of traditional values, such as respect for life and preservation of the environment, they were able to accept modern trends such as the use of sophisticated modern weapons and education. (Chung 1995: 146)

Coming from someone who worked in the education department for ZANU during the war, and the Ministry of Education after independence, this indicates how Zimbabwean nationalism was not by any means internally undifferentiated ideologically. Rather, it emphasises how even within one theatre of nationalism (by which I mean, one nationalist party within a wider nationalist movement that consisted of at least two major parties, ZANU and ZAPU), there were very significant ideological differences. These were not divisive and disruptive to the execution of the war, even if after independence they came to be seen as the root of problems, and the unfulfilled expectations of war veterans and 'traditional connoisseurs'. My only point of disagreement with Chung is her implication that the ZANU political elite merely 'co-existed' with, or tolerated, 'traditionalism' and tried to win over the support of 'traditionalists' for their cause. This ignores the extent to which the nationalist, 'western-educated' political elite readily harnessed, exaggerated and inadvertently created some aspects of 'traditionalism'. In doing so, the spiritual authority of certain ancestors, for example those associated with the rebellions of the 1890s, such as

Ambuya Nehanda and Sekuru Kaguvi, became a means of harbouring the support of the masses. As political mythology/theology with which to imagine a nation, and, importantly, to provide historical/ancestral precedence for the use of violence as a means with which to fight for and ultimately establish an independent Zimbabwe, the rebellions of the 1890s, the first *Chimurenga*, became as important for Zimbabwean nationalism as Great Zimbabwe was in terms of providing a deep historical example of past African achievement and future aspirations. Thus prominent nationalists referred to their own struggle as the *second chimurenga* representing not a break from the past, but a continuation of it. In July 1962, Joshua Nkomo, leader of ZAPU, was met at Salisbury airport by a ninety year old veteran of the rebellions of 1896–97, who presented him with a ceremonial axe symbolising the ancestral authority for resistance, 'so that he might "fight to the bitter end" ' (Ranger 1967: 385).

Just as nationalism's use of Great Zimbabwe as a symbol of past achievement and future hopes was related to a growing body of work by archaeologists and oral historians, so too was nationalism's use of the mythology of the first *Chimurenga*, and the idea of 'national' ancestral figures, most importantly Chaminuka, Nehanda and Kaguvi, based on the enthused writings of oral historians. Ranger's early work *Revolt in Southern Rhodesia* (1967), with its heavy emphasis on the roles played by the mediums of Nehanda, and Kaguvi in Mashonaland, and the *Mwari* cult in Matabeleland, was particularly influential, as Eddison Zvobgo himself commented during our interview. Ranger even made it clear that he was aware that his book would feed directly into a nationalist discourse that had already begun to embrace 'traditionalism' in its search for popular support (1967: 384).[2] He described how in 1954 Bulawayo trade unionists visited the *Mwari* shrine at Matonjeni 'to seek the blessing of the god on a proposed strike'(Ranger 1967: 383) and he quoted Shamuyarira (1965: 28–31) to illustrate how 'the character of this new nationalism was profoundly modified by its discovery of the potentialities of the traditions of resistance of the rural masses' (Ranger 1967: 382).

> The point really is that what was happening was the involvement of the ordinary man and the encouragement of his morale to an extent which had not been paralleled since 1896. Mr Shamuyarira tells us something of these developments. The legends of Chaminuka 'took on an extra significance ... after the formation of the City Youth League', the first of the radical nationalist movements in Mashonaland. 'George Nyandoro particularly dwelt upon this memory in speeches as a binding factor in resisting the settlers'. Mr Shamuyarira goes on to describe how the National Democratic Party, founded in 1960, 'added one important factor that had been singularly missing in Rhodesian nationalism: emotion'. He described their mass meetings, the prayers to Chaminuka, 'thudding drums, ululation by women dressed in national costumes and ancestral prayers'. 'In rural areas meetings became political gatherings and more – social occasions where old friendships were renewed and new ones made, past heritage was revived through prayers and traditional singing with African instruments, ancestral spirits invoked to guide and lead the new nation. Christianity and civilisation took a back seat, and new forms of worship, new attitudes were thrust forward dramatically. Although all attendants wore western clothes ... the cars and loudspeakers were seen everywhere as signs of the scientific age, the spirit pervading the meetings was African, and the desire was to put the twentieth century in an African context'. These meetings he tells us, had an emotional impact 'that went far beyond claiming to rule the country – it was an

ordinary man's participation in creating something new, a new nation'. (Ranger 1967: 384–5, quoting N. Shamuyarira 1965: 28–31)

In a later work, Ranger (1982) has described how the profile of Chaminuka was raised to the status of 'national' ancestor through the works of Gelfand (1959) and Abraham (1966) which fed directly into nationalist discourse. The status of Chaminuka, the ancestor, has remained influential, and he is widely associated with having predicted the defeat of the Ndebele at the hand of white men described as 'men without knees' (National Archives 1984: 130). But of all the ancestral figures whose 'national' importance emerged through their association with the liberation struggle, it is Ambuya Nehanda in particular who became associated with war. The celebrated prophecy that '*Mapfupa edu achmuka*' – 'our bones will rise' – allegedly uttered by the spirit medium of Ambuya Nehanda before she was hanged for her part in the 1896–97 uprisings, was related as 'political education' to young guerrilla recruits in training camps in Mozambique and elsewhere, who were told that they were the very bones that Ambuya Nehanda had spoken of. VaKanda described to me the 'political education' he received in training camps in Mozambique.

> *When we were in the camps in Mozambique, we were given political education. And the starting point was how this country was colonised, and how the people suffered. How they were driven to violent areas by the first settlers, and how our forefathers resisted and fought the white colonisers during the first chimurenga. And we were told about how Sekuru Kaguvi, and Ambuya Nehanda led the struggle to fight against these new colonisers, and the heroics they performed. How they were overcome, because they were not adequately armed, the colonial settlers had superior weapons. And how they were captured and hanged. So to us it instilled a very big sense of admiration. If these people who were poorly armed could resist people who were armed with guns, whilst they were armed with spears. You know it actually inspired us, because they were very brave, and for the simple reason that they were fighting for their country. So there was a phrase that she [Ambuya Nehanda] said, when she was being hanged. 'Our bones will rise, you can kill me now, but our bones will rise against you' As I speak that phrase it sort of gives you an inexplicable feeling of wanting to take it from there and go forward, you see? So the inspiration was that, 'My bones will rise', and we were told that we were the bones, the very bones that Ambuya Nehanda was saying. So that inspired us to say, what ever happens, we will fight till the end. (Interview with VaKanda, VaMuchina, MaDiri, 16/3/2001)*

In this sense, Zimbabwean nationalism was able 'to preserve the distinctness of [its] spiritual culture' and managed 'to fashion a "modern" national culture that is nevertheless not western' (Chatterjee 1993: 6). Indeed nationalist use of this political mythology or even theology, based on the now legendary role played by spirit mediums and the *Mwari* shrines in the first *chimurenga*, was much more 'original' and 'creative' than the use of Great Zimbabwe as 'past African achievement'. The 'creativity' of this aspect of nationalist mythology is even more apparent if we consider how it was put into actual practise by guerrillas on the battle ground (e.g. Lan 1985; Daneel 1995). Indeed the great emphasis that scholars of the liberation struggle have put on the collaboration between the spirit world of the ancestors and *Mwari*, and guerrillas, is an indication that the uniqueness of this project has been recognised, even if it is now clear that the picture of 'peasant'/guerrilla collaboration in name of the ancestors that was often presented masked some of the brutalities that also featured in the relationship between guerrillas and rural folk. VaMhike, chairman of the Masvingo War Veterans Association, made explicit the link between the nationalist mythology of the first *chimurenga*, and the actual practical guidance and collaboration of the ancestors/spirit mediums with the guerrillas.

> *In as far as the purpose, or how we viewed the spirit mediums in our War of liberation, I understand now, and I firmly believe that all those who left during the armed struggle were inspired somehow by the fighting spirit of war. In Mozambique we did undergo political education. It was that orientation which brought to light to the recruits, made us understand why people had to fight the Regime and even to understand that it was not a war of liberation without guidance. The first heroes, Sekuru Kaguvi, Ambuya Nehanda and Chaminuka actually left the war as an incomplete battle. And it was thought and believed strongly that the sons of Zimbabwe should complete the war. And so we were in a situation whereby we had spirit mediums who we had to contact in order to get a way forward. Even in battle, in the field, in different areas where we were operating we had to consult the spirit mediums. Each chief in Zimbabwe has got a svikiro whom you consult when you operate in the area. And these used to tell us, or instruct us, or to order us to say when you are in this area, you don't do 1, 2, 3 things, you do this, that & that. Like you have to listen to the instructions from the spirit mediums, to say you occupy such type of hills or areas, and then you can go and operate in this way. We had things like birds of the spirit mediums which we believed were associated with the spirits, like the Chapungu. It would come, whilst we were camped, waiting for the enemy, it would come and even give us directions for retreat after the battle. Or it could signal that there is an enemy within the area you are operating and we would be made alert, and within minutes, there would be a battle. And you would now understand that even if helicopters or bomber planes would come, the chapungu* [Bateluer eagle] *would come and intervene. Yes, to give the warning and even interfere with the aeroplanes, and they would disperse, and we move out free. So we strongly believed that the spirit mediums played a role; even now we still believe. Consultations tell us that we still have a role to play as war veterans.* (Interview with VaMhike, 26/6/2001)

While this connection between the nationalist mythology of the first *chimurenga* and the ancestoral guidance of 'national' ancestors, with the well-reported collaboration of guerrillas and spirit mediums seems obvious, in fact it was a conceptual leap. The evidence suggests that it was actually quite rare that guerrillas worked with these 'national' ancestors. Rather they worked with local spirit mediums, who passed on the messages to the bigger ancestral spirits, what Daneel (1995: 204) has called the 'dare rechimurenga' – the 'ancestral war council'. Linking the ancestral efforts of the first *chimurenga* to the Zimbabwean liberation struggle of the 1960s and 70s, Comrade Mahiya described how the freedom fighters would daily consult the 'small *masvikiro*', who passed the messages and requests for help unto the 'big *masvikiro*', 'through their own channels':

> *Comrade Mahiya: So these Madzimambo were defeated, because Europe by then was quite advanced, and they already had the gun, so it is the same guns that helped the white people to defeat the people and take the country, and start to rule, under the rule of the Rhodes. He conquered Zimbabwe.*
>
> *So when they conquered there, those people who were defeated, the elders were saying 'No you have defeated us, but it is not yet over, Mapfupa edu achmuka* [Our bones will rise] *That was Ambuya Nehanda'.*
>
> *So truly their bones rose, because it was the question of land, it was the question of the people of Zimbabwe refusing to remain under bondage, under slavery, without social and economic independence.*
>
> *So as the war was being fought by the fighters, every morning we used to go under the tree to talk to them, the ancestors.*
>
> *JF: where?*
>
> *Comrade Mahiya: Everywhere, where we went, when we arrived would see the small masvikiro. Those spirits/winds would meet with the big masvikiro to discuss the issue that there is a war going on there, that war that we did not finish. So the freedom fighters, the excombatants, the war veterans that is what they were doing.*

J.F.: So they went to the masvikiro, the smallest ones first, after which they passed it on to the big masvikiro?

Comrade Mahiya: Yes, in their own channels, this is what was happening. (Interview with Comrade Mahiya, 1/8/2001)

Indeed, as a means of mobilising the rural 'masses', it was quite important that local ancestors were recognised, and therefore, as Comrade Nylon put it:

We would work with the chiefs, and the chiefs would guide us to the masvikiro. The masvikiro would tell us exactly what they wanted in their places/areas/land, at times they would say, we don't want any blood, and we would work with that in mind, and they would say give us the laws, for instance you are a man yourself, and at times you would want to get a woman, yet some of the demands were that you worked without having any kind of relationship with a woman, and that we would abide by. (Interview with Comrade Nylon, 8/8/2001)

Lan has discussed this prohibition against sexual relations with women as part of series of ritual prohibitions, similar to those of *masvikiro*, by which the ancestral legitimacy was bestowed upon guerrillas.

By observing the ancestral prohibitions the guerrillas were transformed from 'strangers', into 'royals', from members of lineages resident in other parts of Zimbabwe, into descendants of local *mhondoro* with rights to land. They had become 'at home' in Dande. (Lan 1985: 164)

It is not entirely clear how appropriate this was for guerrillas fighting in Masvingo province, but it is apparent from Daneel's work (1995: 1998) that through their close co-operation with local elders, *masvikiro* and the ancestors, guerrillas often gained authority locally as 'traditional connoisseurs' in their own right, despite their (often) young age, and the fact that their own *kumusha* ('rural home') was usually far from where they operated. They were often referred to during my conversations with *masvikiro* and other elders in Masvingo as *vana vevhu* (children of the soil), and they are often called to attend *bira* ceremonies, where they are treated with great respect. The ritual importance of *bute*, the snuff of the ancestors, which was often adopted by guerrillas whilst in the theatre of war, has been highlighted in Daneel's (1995) 'thinly fictionalised' (Ranger 1999: 218) account of the experiences of several guerrilla fighters in Masvingo province, appropriately titled *Guerrilla Snuff*. As 'traditional connoisseurs' in their own right, guerrillas often visited and hid in the sacred places like mountains and *marambatemwa* (holy groves of sacred trees) access to which is, and was, strictly restricted to elders and *masvikiro*. Comrade Nylon described how they used to build their bases on such places on the landscape because it was very difficult for others to go there.

There are quite a lot of places, especially mountains. There is one on the way to Ngundu, just before Ngundu, on Beitbridge road, which could actually burn sometimes, and you could see as far as Nyajena. I myself, and the other comrades, we could go into those mountains, and put our bases up there. But it was difficult for anyone else to do so, which symbolised just how much of a relationship we had with the ancestors. So if anybody else would go into those mountains, they would get lost or if anybody would try and climb there, even the enemy, if they tried to come, they would fall off, which just shows the connection between us and the ancestors. (Interview with Comrade Nylon, 8/8/2001)

It was through working with the local ancestral owners of the land, and through their worldly representatives, that the use of these local sacred places was made possible. It seems unlikely that appeals to the authority of Ambuya Nehanda would have persuaded members of the local 'traditional' leaderships to provide access, and show guerrillas the sacred sites of their ancestors. Indeed, it seems most likely that it was through their alliances with the local leadership that the political mythology of Nehanda and the other ancestral spirits was spread to the rural masses. The 'political education' that guerrillas had received was passed on during all night *pungwe* sessions, through which the rural people were politicised, and the guerrillas' own support network among the people was established (see Lan 1985: 127). In VaMhike's words,

> *We went through a number of battles but our main task was actually to politicise the masses, about the purpose of the war. We did not have any problem as far as operating from within the masses. We can say that our military camps were the masses. Our logistics were the masses. And our information network was actually the masses themselves. They could communicate with us, of the enemy whereabouts, advances and even retreats … .* (Interview with VaMhike, 26/6/2001)

Guerrillas also worked with their own personal ancestors, and in some cases they seemed of much more relevance than either 'local', 'regional' or 'national' ancestors. This should be added to the 'ethnographic thickness' of the Zimbabwean liberation war. Again Daneel's account reveals how the spiritual experiences of guerrillas were often related to their own particular ancestor's guidance first and foremost, over and above spectacular interventions by such great spirits as Nehanda herself. Comrade Nylon expressed this view himself,

> *My opinion is that we worked with our own personal ancestral spirits, that is the spirits that help you in particular are your family spirits, this is what I think we worked with. So that even if there were other stronger spirits, they could be counteracted with our own. The thing is that we were guided by our own spirits to go through and I think this is what helped us attack the Great Zimbabwe hotel.* (Interview with Comrade Nylon, 8/8/2001)

Important here is recognition that almost the whole range of the 'Shona spirit world' has some how been incorporated into experiences of the Zimbabwean liberation struggle. What began as a political origins mythology, which centred itself upon the activities of certain ancestral figures during the first *chimurenga*, became a means through which to capture the common rural imagination of the nation. Through this harking back to the memory of the rebellions of 1896–97 the guerrilla armies that formed the military wings of political nationalism were provided with historical precedence for their military struggle, and a means of politicising and co-opting the masses to that project. In this process, these 'original' and 'authentic' imaginings were taken a great deal further by guerrillas and the 'traditionalists' they co-opted than most of the 'western-educated' and thoroughly 'modern' nationalist elite themselves would have gone, as Fay Chung has herself made clear. For my purposes here, this is most transparent in the way Great Zimbabwe, already providing nationalism with historical legitimacy and 'ancient'/ archaeological precedence, as well as a future to aspire to, became linked to the 'theology/mythology' of the ancestors and the first *chimurenga*. Great Zimbabwe became very widely conceived of as a, if not the, 'national' sacred site, on par with the *Mwari* shrines of the Matopos, and thoroughly associated with the ancestral legitimacy of the armed struggle. For many guerrillas and 'traditionalists' I spoke to, 'Zimbabwe', as the name of the new country, was not an act of defiance against Rhodesian attempts to appropriate it for themselves, but because Great Zimbabwe was the place of Zimbabwe's greatest ancestral rulers whose authority was bestowed upon African Nationalism, and the liberation struggle. As the following quotes from war veterans illustrate, it is but a short step from saying Great Zimbabwe is an example of prime African achievement, to the suggestion Great Zimbabwe is the 'big place for our ancestors', though the implications of this short step are the basis of this entire thesis.

J.F.: Why do you think ZANU, ZAPU, ZANLA and ZIPRA all used the name Zimbabwe? What was the significance of Zimbabwe?

VaChuma: We used Zimbabwe, because Zimbabwe is our 'capital', or we can say 'inzvimbo huru yamatateguru' [the big place of the ancestors], *for all of Southern Africa, all of the ancestors. Even if we look in history, the old rulers used to come for their meetings at Great Zimbabwe.* (Interview with VaChuma, 17/02/2001)

Madiri: As I said I was born in 1936. When I grew up, I used to hear my parents giving the name Chishava, and I said 'what is chishava?'. They said 'That is where our great parents/ancestors came from.' 'Where is it?' They used to tell me 'Kudzimba dzemabwe dziri kuVictori [at the stone houses in Victoria district]'. *So that when I came to know the name Zimbabwe, Great Zimbabwe, I came to know it now when I was at school, being taught by the teachers. They said there is Great Zimbabwe in Fort Victoria and they used to tell us that is where the real Zimbabweans, and then Rhodesians at that time, African Rhodesians were born. That's what they used to tell us. But as I went on with school, that's when I came across the politicians. That is Zimbabwe, where our forefathers used to stay, and Great Zimbabwe is an English name , but it is Dzimbabwe, and it is where our spirits are, our Madzitateguru. That's what I know about it.* (Interview with VaKanda, VaMuchina, MaDiri, 16/3/2001)

Given the extent to which the work of the oral historians in the 1960s was heavily implicated in both nationalist use of Great Zimbabwe as a symbol, and in the raising of the profiles of ancestors such as Chaminuka, and those associated with the first *chimurenga*, it should be no surprise that oral historians also had their hand in the reformulation of Great Zimbabwe as a 'national' sacred site. Abraham (1966: 34) in particular made the link between the Rozvi, the *Mwari* cult, Chaminuka and Great Zimbabwe, suggesting that, they were all 'centred at Zimbabwe from the inception of the Rozvi monarchy until the Nguni invasions of the 19th century'. According to his informants,

It is within the 'Eastern Enclosure' ... that the medium of Chaminuka presided, and amongst other things, interpreted the squawkings of *Hungwe, Shirichena, Shiri ya Mwari* – The (Celestial) Fish-Eagle, The Bird of Bright Plumage, The Bird of Mwari, on its annual visit to the shrine, as pronouncements of the deity. (Abraham 1966: 35)

But while there would seem to be some correlation here with local perspectives on Great Zimbabwe as a place were *Mwari* used to speak before moving to the Matopos and, of course, the very common assertion that it was the Rozvi who built the stone walls, it does not seem that Abraham's ideas were directly related to those that war veterans had about the site's sacred role as a place of the ancestors providing legitimacy to the struggle. War veterans I spoke to did not clearly articulate one version of Great Zimbabwe's sacredness, rather, several different versions emerged. Most frequently, vague associations were made between Great Zimbabwe and the names Nehanda, Chaminuka and so on. VaMhike was able to articulate his views quite clearly, and made a very conscious distinction between Great Zimbabwe and Matonjeni, and then, rather ambiguously, linked Nehanda to Great Zimbabwe but also stressed that as a 'moving spirit' she could be consulted anywhere.

JF: I'm glad because you have just mentioned the issue that I wanted to talk about, which is Great Zimbabwe and this ceremony. Briefly, what role did Great Zimbabwe play? I mean people talk about it, it's not for no reason that this country was called Zimbabwe at independence.

Mhike: Yes, Great Zimbabwe is a place where the great rulers resided, and it is believed that the rulers were affected in their governance by the spirit mediums. Ambuya Nehanda and others, but we have got our great centre of all spirit mediums which is in Bulawayo at Matopos, Matonjeni. That's where all the spirit mediums go for advice and instruction. But Great Zimbabwe, even spirit mediums from Great Zimbabwe, needed to go to Matonjeni for consultation. But now Great Zimbabwe has been unique in the sense that it is actually the centre of the Great rulers, the fighters. They were planning their battles, if they were going out to fight but it is unique in the sense that it is a centre, a position of influence in terms of spirit mediums' consultation. All the spirits would gather at Great Zimbabwe before going to Matonjeni. So it is quite unique, and we still believe the spirit mediums residing at Great Zimbabwe are still very important in terms of influence, in terms of advice.

JF: Mhepo dzenyika?

Mhike: Ya, Mhepo dzenyika are centred at Great Zimbabwe, and all that happened, we believe, emanates from what is discussed or what is important for the spirit medium at Great Zimbabwe. Ambuya Nehanda at Great Zimbabwe, as a national spirit, is often consulted and that spirit can be consulted from any point in Zimbabwe, but we have very few masvikiro who actually say 'I am Ambuya Nehanda'.

Ambuya Nehanda is a moving spirit which does not have a permanent place. So she can come out through a medium in Harare, in any province, but in Great Zimbabwe, there is Ambuya Nehanda and we believe that is a very central place to instructions given in terms of governance. (Interview with VaMhike, 26/6/2001)

Great Zimbabwe being a 'central place … in terms of governance' is related to the idea of Great Zimbabwe as a point of origin for the ancestral authority of the nationalist movement, and indeed for the post-independent state. Another war veteran leader, this time based in Harare, expressed a similar view in his belief that nationalist leaders went to Great Zimbabwe to seek the advice of the ancestors there, for the liberation struggle. Speaking as a Provincial ZANU PF cadre, and in relation to the wider contemporary context of political turmoil in Zimbabwe, he also made chillingly clear his views about the divinely appointed authority of President Robert Mugabe himself.

Yah the nationalist leaders, as you know, erm, When the nationalist movement started, it started with these people, they went there. They went to Great Zimbabwe. They went to see the Masvikiro Makuru, and talked to them and told them, in fact they were now being advised that you have to fight. In matter of fact they went to receive military instruction. Hence when they went out to seek for military assistance, the masvikiro, because they are spirits, would intervene so that an agreement is struck, at least the support they had. Like I told you earlier on., when the children of Israel were in captivity in Egypt, Moses was told to go to Africa by then Jethro was the only man of God, who could tell Moses what to do. So it is Africa that is in the forefront before God. And for 400 years during the time of Jethro, there was only Jethro and there was no other person who could communicate with God. So when it happened, it is the same understanding that filtered into the people of Africa as they spread through out Africa, that it was their culture to talk to God. So it was the culture of the liberation struggle to talk to God, through the spirits asking for guidance. So the president of this country [President Mugabe] *became a president not by the British or the Dutch, or the American's approval but by the approval of the spirits, of the spirit mediums of this country. So he cannot be removed today by ordinary people. Anyone who does not approve of him, will be disapproving the existence of Zimbabwe 500 years ago, which is not true, which I think is the opposite of the truth.* (Interview with Comrade Mahiya, Harare Provincial ZANU PF Offices, 1/8/2001)

The idea is mirrored in the local discourses about Great Zimbabwe's 'national' sacred importance. This was revealed during an interview with Chief Mapanzure and his council, who unlike the neighbouring Mugabe and Nemanwa clans, do not make specific claims on Great Zimbabwe as their own.

Chief Mapanzure: First they wanted to call this country Nyanda [a mountain in Masvingo province] *but later they realised that Nyanda is too small, and it does not have a big value for the country like Great Zimbabwe does. So they decided to name the country Zimbabwe, from the Great Zimbabwe ruins. That's why they decided to call it Zimbabwe, because God used to stay at Great Zimbabwe.*

Chief Mapanzure's brother: The leaders, like President Mugabe, Joshua Nkomo, Chitepo and others, came to Great Zimbabwe and saw those wonderful features, and they said this country should not be called Nyanda, it should be called Zimbabwe, and they even went on to slaughter bulls, to thank the ancestors, as a sign to show them that they were happy with the name. (Interview with Chief Mapanzure and his *dare*, 18/2/2001)

As Chief Mapanzure's brother's comment indicates, in line with Comrade Mahiya's comments above, there is a belief that nationalist leaders visited Great Zimbabwe and made offerings to the ancestors there to request help with the war of liberation, and permission to use the name Zimbabwe. During our group discussion Chief Nemanwa himself claimed that he had been present when Simon Muzenda, the late vice-president, approached the Nemanwa elders in order to make such offerings to the ancestors at Great Zimbabwe. Aiden Nemanwa's contribution to the ensuing discussion characteristically emphasised that this demonstrated that Great Zimbabwe should be under the custodianship of the Nemanwa clan.

Chief Nemanwa: I was working in the temple, selling guide books, they came and asked me, Mr Muzenda and two others, they asked me saying 'Can we see your chiefs, and your elders?' The chief was then VaSiyavizwa. I said 'I can write the names of who you should go to, how to go to chief'. They said 'We want to go to war to take the country'. At that time they were with ZAPU. So we brought them a hari [traditional clay pot] *and we gave it to them ...*

They went with my father Muvenge, to the Chief and from there to buy a clay pot of beer, and we went to the valley [in Great Zimbabwe], *to offer it to our ancestors.*

J.F.: When was this, what year?

Chief: 1950? Or ...

VaMatambo: No, 1960 something ...

Aiden Nemanwa: No, it was 1960s, something.

[...]

VaMatambo: It was our father that finished the beer, with Muzenda, and he was chased by the police and the soldiers.

J.F.: oh! ... you were chased away?

Chief Nemanwa: When they went inside [Great Zimbabwe], *they were spotted by a plane that passed overhead, while they were kneeling down, appeasing the ancestors. Then the police of the museum, they came to my house, I was guarding inside the 'temple' they asked me if I had seen the people offering the beer, drinking and offering beer. And I said 'No I haven't seen them'. ... If I had said something, they would have been put in jail.*

J.F.: Ok, So Mr Muzenda came here to do what, exactly?

VaMatambo: VaMuzenda came here to ask for help to take the land, from the biggest spirits of the land, in Zimbabwe here. They came here to look for ideas for how take the land, from the biggest spirits of the land, for Rhodesia is ours, it is Zimbabwe.

Aiden Nemanwa: Wait there, I have something to add. They were requesting the midzimu yemuZimbabwe [ancestral spirits of Great Zimbabwe] *that used to speak, to help them in that war, to tackle the task of war.*

J.F.: So Great Zimbabwe, the mhepo [winds] *of Great Zimbabwe, are they the mhepo dzenyika yose* [winds of the whole country]?

Aiden Nemanwa: I have said that there are two types of spirits in Great Zimbabwe; that voice that was speaking in Zimbabwe, that was the very first big spirit, so everything that can be requested comes from there. Like Muzenda, who comes from Gutu, he knows that here is the biggest spirit of all of Zimbabwe, and that spirit is of Nemanwa, it talks to Nemanwa. So he came here to request from Nemanwa, to talk with the Mudzimu wemuZimbabwe to ask for help to fight the war. That is it that is how it happened.
(Group discussion with Nemanwa Elders, 18/7/2001)

For his part, Chief Mugabe told me a story that similarly illustrates the association between Great Zimbabwe, the nationalist struggle, and the authority of the ancestors, though his story related specifically to ethnic struggles within nationalism between a Shona dominated ZANU, and Ndebele dominated ZAPU, as he offered his explanation as to why it was ZANU, and not ZAPU that formed the first African government after independence.

During the chimurenga, you could get into Great Zimbabwe with a request, so I can still remember when I was in Bulawayo, all the big people, the authoritative people came, like Joshua Nkomo, Parerenyatwa, George Nyandoro, Mugabe was very young at that time, and Chikerema. The leader was Joshua Nkomo from Matabeleland. And the one they trusted among the Shona was Parerenyatwa. That is when they came there to Great Zimbabwe, and there is a cave there, to be found in the ground. They wanted to show who is the one who is allowed to rule this country.

So the first one to enter was Nkomo, because he was the eldest. When he got inside, he found a very big snake, which was all over, and was the thickness of this [indicates a speaker] it was round, with its mouth open. So Nkomo ran out and said 'ahh what I have seen in there is very tough!'

Then Parerenyatwa entered, and he found the snake just like that, and he gathered his courage and just grabbed the snake by the neck, and then the snake changed and became a staff. And that is the staff that is used right now for people to be successful in life.

When Parerenyatwa got it, and brought it out of the cave, then everybody knew that Parerenyatwa is the one that is wanted by the spirit for him to get hold of the country. Then as they stayed and stayed, Nkomo was the one who was the eldest, and he made plans to kill Parerenyatwa. And so Parerenyatwa was killed and that staff was taken by Nkomo.

As they stayed like that all the Shona people were not happy. So one day they had a meeting in Tanzania, they met all the leaders talking about the country. So when they arrived there, they paid a certain girl, who was supposed to take care of their clothes, where they were staying. So Nkomo put the staff with his clothes. So the staff was taken by that girl and she hid it. As they finished the meeting, Nkomo did not think that the staff was no longer there, and when they had all gone, that girl took the staff from where it was hidden, and gave it to some boys who were supposed to give it to the Shona people. And then that's when we hear, it was later given to Mugabe. That is where he got the powers to rule. So whoever has got power to rule, should have that particular staff. (Interview with Chief Mugabe, 22/11/2000)

Chief Mugabe went on to present his own particular slant on why the name Zimbabwe was chosen for the new nation.

The power of liberation of the country came from Great Zimbabwe, but long back they were not much interested in making the country have that particular name, Zimbabwe, because it was really supposed to be called Great Zimbabwe. But now the Queen was not very happy about it being called Great Zimbabwe, because there was going to be a clash of names, Great Zimbabwe, Great Britain. So it became just Zimbabwe. (Interview with Chief Mugabe, 22/11/2000)

In both cases the narrators have a stake in presenting Great Zimbabwe's national political significance as a sacred site, though each through the prism of his own local claims to the site. There probably is some truth in the widely held view that some nationalist leaders did indeed visit Great Zimbabwe and conduct a ceremony, or make an offering to the spirits at Great Zimbabwe. I have tried to follow up Muzenda's visit to Great Zimbabwe from other sources but have been only marginally successful. According to Dawson Munjeri,[3] former executive director of NMMZ, when he accompanied a group of foreign dignitaries around the Great Zimbabwe in early 1980s, Simon Muzenda explained how in the early 1960s he and other nationalists, including George Nyandoro, came to Great Zimbabwe specifically to ask for guidance and permission to fight the war of liberation, from the ancestral spirits. Dawson Munjeri suggested that I try to arrange an interview with Simon Muzenda, which I did; unfortunately my attempt was unsuccessful. Nevertheless, this does suggest that what I have so far painted as a 'western-educated' and 'modernist' nationalist elite, must have been aware of, and indeed had some sympathies with, the idea of Great Zimbabwe as a 'national' sacred site. Indeed we should be careful not to overly aggregate the 'nationalist elite'; some of course would have taken their own mythology about the role of the ancestors and *Mwari* more seriously than others.

In chapter 7 of his book on the Matopos hills, Ranger focuses on the role played by the *Mwari* shrines in the history of nationalism. He describes a very early visit to the Dula shrine by Joshua Nkomo in 1954 as a 'key moment in the cult's nationalist history', and suggests that 'narratives of Nkomo's visit to Dula, ... came to be widely distributed in Matabeleland, [and] bestowed a sacred legitimacy upon his leadership' (Ranger 1999: 216, 17). A decade later, after the name Great Zimbabwe had been widely adopted for the nationalist movement, a visit to that site may have been a conscious attempt on the part of individual nationalists from the Masvingo area to invoke upon themselves a similar kind of 'sacred legitimacy' as Nkomo evidently received in Matabeleland from his visit to the Dula shrine. If so then it worked. But I think it may have been less self-conscious than is implied above. There is a significant difference between the role of the *Mwari* shrines in the Matopos and Great Zimbabwe during this period. The *Mwari* cult shrines already existed and had a fairly well established, and large, geographical area of influence that stretched beyond Matabeleland, and well into Masvingo province. Great Zimbabwe did not and does not. Rather I am suggesting that through nationalism, the war of liberation and the close associations and collaborations of spirit mediums and guerrillas, Great Zimbabwe became increasingly viewed as a 'national' sacred site; but there is no evidence to suggest that before the 1960s it ever had the influence of the *Mwari* shrines, nor does it now, though many 'traditionalists' think it should. We could entertain the suggestion that visits by senior ZANU politicians to Great Zimbabwe in the 1960s were attempts to create a cult centre, separate from the Matopos shrines, that are so conspicuously located in Matabeleland, dominated then by ZAPU. However, since independence no efforts have been made to recognise Great Zimbabwe's 'national' sacredness, rather this has been quite literally repressed, as I will indicate below. Furthermore, the Matopos shrines are not 'Ndebele shrines' by any means, and have significant influence with many Shona clans in Masvingo province and even beyond.

Pondering the origins of the idea of Great Zimbabwe as a 'national' sacred site, I asked VaKanda whether, in his 'political education', a specific link had been emphasised between Great Zimbabwe and the ancestral heroes of the first *chimurenga*. His answer was informative.

JF: Was there ever a connection made between those masvikiro and Great Zimbabwe, explicitly, or was it implicit?

VaKanda: Ya it was a bit implicit. It was not explicit. Remember we were educated that once we liberate our country, it was going to be called Zimbabwe. And the connection was that this very woman, this very brave woman, this legend, was actually also fighting for the freedom of this country called Zimbabwe. So that was the connection. But whether, when they were fighting, they had this connection with Dzimba Dzemabwe, that we cannot know. That is beyond our knowledge. (Interview with VaKanda, VaMuchina, MaDiri, 16/3/2001)

The clear implication is that the idea of Great Zimbabwe as a sacred site must have emerged in the thoughts and discourses of 'traditionalists' and guerrillas fighting the war – who were experiencing, practising and using the guidance of the ancestors on a daily level – rather than the political nationalists far away from the battle field. Some nationalists, of course, would have also made the link, but it was not an explicit part of the political mythology of nationalism. Eddison Zvobgo stressed that Great Zimbabwe's 'sacredness' was considered fairly insignificant in relation to its 'usefulness' as a 'rallying point'.

Yes, I think it was clearly a place of worship, there was a religious element to it. And the nationalists did not perceive it in that way, but many masvikiro and so on did. I was told over and over again that the link between culture and religion is exemplified by the temple at Great Zimbabwe.

As practical politicians we did not worry whether it was linked to religion or not. We had found a rallying point, a very useful one, and everybody then accepted that. *Southern Rhodesia was named after Rhodes, and then that name* [Zimbabwe] *was born, and that was that.* (Interview with Dr Eddison Zvobgo, 18/8/2001)

Evidence for this notable lack of interest in Great Zimbabwe's role as a 'national' sacred site also emerges from the absence of a military policy towards the site. Surrounding areas, including both the neighbouring hotels, and several farms in the vicinity, were attacked by guerrillas on several occasions. In 1979 the monument was closed to the public, and NMMR staff were evacuated, but apart from the burning of the 'Karanga village' (Ucko 1994: 276), Great Zimbabwe seems to have seen relatively little action during the war. When I asked Comrade Nylon, who was a guerrilla commander in that area during the later stages of the war, whether they ever went into the area of the monuments themselves, he said it was too well guarded.

We couldn't go into Great Zimbabwe, the ruins, because it was well protected by the Rhodesian forces there was a battle camp. There was also Machonya [guerrilla term for the local army camp on Shepherds plot next to Great Zimbabwe, where the Lodge at the Ancient City Hotel stands today]. *But as you might have heard, there was a battle at that place, which came out on the radio, and we actually attacked and bombarded the place, and took away the guns.* (Interview with Comrade Nylon, 8/8/2001)

This stands in marked contrast to the explanations of many local people who frequently claimed that they had told the guerrillas that Great Zimbabwe was 'too sacred' for the guerrillas to fire their guns there. When I asked Comrade Nylon whether he had received orders in relation to Great Zimbabwe from his own commanders in Mozambique, it became clear that there had been no specific 'policy' about Great Zimbabwe at all.

J.F.: So were you told to go to Great Zimbabwe, from Mozambique?

Comrade Nylon: Ya, we can say that there were areas where each of us worked, so our area was Gaza province, and within Gaza province there were sectors so that was sector one of Gaza.

G.Mazarire: I think what he is asking is about whether you were told to attack the Great Zimbabwe hotel from Mozambique or whether it was just part of the front?

Comrade Nylon: No we thought about doing that when we were there, because it was within our sector, our operating zone and it was part of the strategic plan as far as that sector was concerned.

That was instructed by me as the commander of that detachment, any contact that was made was entirely sanctioned by me.

J.F.: I would like to know whether you had specific orders from your leaders, your comrades about Great Zimbabwe. Were you told specifically to look after Great Zimbabwe, or do this at Great Zimbabwe, it is an important area, or were you just told to look after the whole area without any specific instructions for Great Zimbabwe?

Comrade Nylon: No we didn't have specific orders apart from going into the area, to see what the situation was, I would sanction what was necessary for the situation. (Interview with Comrade Nylon, 8/8/2001)

This supports my argument that whilst Great Zimbabwe was used as a 'useful rallying point' for nationalism, representing African ingenuity and future aspirations, the perspective of Great Zimbabwe as a place of 'national' sacred importance was not wholly embraced by the nationalist leadership. Rather it emerged among 'traditionalists' and guerrillas on the ground, not just in Masvingo but far across Zimbabwe, who stretched the mythology of national ancestors to include Great Zimbabwe. Some nationalist politicians undoubtedly did embrace the 'theology' of nationalism – what I have argued constitutes Chatterjee's 'authentic' and 'original', 'spiritual domain' – while others saw the ancestral authority it provided as a useful means of politicising 'traditionalists' among the rural masses, and ensuring their co-operation. Certainly nationalist songs and poetry often invoked the names of Nehanda, Kaguvi and Chaminuka, and even *Mwari* himself (see Pongweni 1982). One poem written by Emmanuel Ngara in Lesotho in 1978 invoked 'the spirit of Mwari' at Great Zimbabwe, and deserves mention here to illustrate that how widespread the 'nationally sacred' perspective on Great Zimbabwe became.

Stirs in the Temple

The Temple lay in ruins for a hundred years
Mwari lay buried and groaning for a hundred years
But the soul of the Temple continued floating
And the ancestral emblem, the Fish Eagle,
never ceased to flap its wings
For I saw it flying over the dilapidated Temple
Circling the Temple and ministering to the spirit of the buried God
They chained our hands and minds and captured the Temple
They brought us the Vulture which sucked our souls
And swallowed our gold and drove our cattle to their kraals
They brought us the Dove
But a Dove begotten of the eggs of the Vulture
They concealed the Eagle and muffled its voice
Mutapa was chained, the Temple was a ruin
And Mwari breathed the breath of a dead god.

But now the soul of the Temple stirs
The spirit of Mwari wakes the sleeping stones
And Mutapa's ancestral emblem flutters wings of battle Rising from the rising
And reciting the prophecy of the Manifesto.
Listen, oh listen, the Temple Bird sings the war song!
Listen, oh listen, the spirit of Mwari proclaims the war dance!
Hark and hearken, the ancestral lion is possessed
And roars the prophetic roar
That shakes the heart of the frightened usurper
And shakes the Temple walls into walls of life! (Emmanuel Ngara 1992: 23–24)

Ngara's emphasis on the 'stirs in the temple', invoking 'the spirit of Mwari' that 'wakes the sleeping stones' represents Great Zimbabwe as a 'national' sacred site, which is an idea that grew out of (but separately from) the explicit nationalist mythology of the first *chimurenga*, and the imaginings of a nationalist struggle empowered by ancestral, and divine, legitimacy. Therefore Ngara's poem seems to embody the 'authentic' and 'original' aspect that Chatterjee referred to as anti-colonial nationalism's 'spiritual domain'. In contrast, the poem by M. Zimunya, *Zimbabwe (After the Ruins)* which opens Chapter 4, invokes a very different perspective on Great Zimbabwe.

The mind that dreamt this dream massively reading into time
And space the voice that commanded
the talent that wove the architecture:
Friezes of dentelle, herring bone, check patterns, chevron and
all the many hands that put all this silence together,
The forgotten festivals at the end of the effort:

All speak silence now – silence (M. Zimunya 1982: 99–100)

With its heavy emphasis on silence, it seems to embody the 'mimicking' or 'derivative' nationalist use of Great Zimbabwe as national heritage; a past African achievement that demonstrates the historical precedence of the nation. Of course, these two perspectives are not necessarily mutually exclusive; neither is it possible to definitively put individuals into either camps. Indeed, during the liberation struggle such differences were of little significance, as Fay Chung herself acknowledged. But once independence had been achieved, and especially in the very movement from liberation struggle to postcolonial state (as the reins of government changed hands) these differences came into sharp focus. This was exemplified at Great Zimbabwe by the events of 1980–81 when Ambuya Sophia Tsvatayi Muchini returned to the area having been released from prison shortly after the February 1980 elections. She claimed to be the spirit medium for the spirit of Ambuya Nehanda, heroine of the rebellions of 1897. Her story illustrates not only the disparity over the role of Great Zimbabwe within the ideology/mythology of the nationalist movement, but also some of the tensions between continuity and change that were played out in the immediate period following independence. Perhaps most of all her story illustrates Chatterjee's argument that ultimately 'the root of postcolonial misery' lies not 'in our inability to think out new forms of the modern community but in our surrender to the old forms of the modern state'(Partha Chatterjee 1993: 11; also 1996: 222).

THE STORY OF AMBUYA SOPHIA TSVATAYI MUCHINI,
ALIAS AMBUYA NEHANDA

In his 'thinly fictionalised' (Ranger 1999: 218) account of the experiences of guerrillas and spirit mediums in Masvingo province during the liberation struggle, Daneel (1995: 12–13) describes an event which happened at Great Zimbabwe shortly after the ceasefire at the end of 1979. It substantiates my argument that Great Zimbabwe came to be seen as a 'national' sacred site during the liberation struggle.

> That fateful day soon after the cease-fire … people trying to find their feet, only half believing that hostilities were really over. The new state was vulnerable, in the throes of birth. Far away from the conference tables a few spirit mediums still conveyed military directives from the spirit council-of-war to a sceptical peasant audience. A few of the most hated personifications of oppression, who had managed to get through the war unscathed, still had to be eliminated. Some guerrilla fighters who had hidden their weapons and failed to report to the UNO-monitored checkpoints were ready to obey the ancestral commands. In this last hour of retribution the voice of Nehanda was said to have ordered that the Great Zimbabwe area be cleared of foreign intruders. Hit squads converged on white farms near Lake Kyle, leaving behind the lifeless bodies of Abe and Magriet Roux and a few others.

> Weeds remembers the midnight ceremony at the ruins of Great Zimbabwe quite clearly. In the moonlight Nehanda the medium, in a simple black robe, stood out dramatically against the pale grey granite walls behind her. She was recounting the glories of the Mutapa rulers of centuries ago to her audience of ZANLA fighters. Then she spoke about the first chimurenga [the rebellions of 1897 against Rhodesian settlers] and the feats of the two spirit mediums representing the national ancestors Kaguvi and Nehanda. Their death defying stand against overwhelming odds remained an example to the guerrillas, strengthening their resolve not to lay down their arms.

> 'Who of you are prepared to follow in our footsteps? Who of you are prepared to perform brave deeds to cleanse the land – even if it means facing the gallows?' Nehanda's piercing eyes shifted from one guerrilla to the other, letting her challenge sink in. In her state of possession she was the epitome of proud black dynasties.

> Weeds saw in a flash the disfigured face of his father. His resolve hardened. Mazhindu the quick tempered, the white Gutu farmer responsible for that scar, still had to be punished. As if pushed by fate, he rose and walked slowly towards the medium. In front of her he sank to his knees, clasped the knobs of the two ebony staffs she held in her hands firmly in his own. Then he rested his forehead against his knuckles and repeated the vow he had made as a boy to avenge the humiliation of his father. Still kneeling, he heard the steady voice of Nehanda above him commending his plan of action to the guardian ancestors of Zimbabwe. With a sense of quiet elation he realised that his own destiny was being ritually tested against their oppressors in the name of freedom and justice. This was not make-believe. There was purpose in his stride when he left the gathering and walked off into the night. (Daneel 1995:12–13)

Ambuya Sophia Tsvatayi Muchini, alias Ambuya Nehanda, had returned to Great Zimbabwe having been released from prison shortly after the February 1980 elections that brought ZANU (PF) to power. She had first moved into the Great Zimbabwe area in 1974 claiming to be the spirit medium for the spirit of Ambuya Nehanda, heroine of the rebellions of 1897, and spent much of the period of the war there, apparently assisting guerrillas. According to Garlake,

> She recruited people for guerrilla training and was asked to become a guerrilla 'leader'. Harassed by the Rhodesian forces, she saw one of her sons shot dead in front of her and her young children. She was imprisoned for her activities for six weeks in 1978 and again from July 1979. (Garlake 1983:16)

When she returned, after her release in February 1981, the staff of the now renamed National Museums and Monuments of Zimbabwe (NMMZ) had already begun to re-establish themselves at Great Zimbabwe, having returned from their evacuation during the latter years of the liberation war. Immediately frictions developed between Ambuya Sophia and NMMZ staff. Only a week or so before the independence celebrations, on 10 April 1980, Cran Cooke, the Regional Director of NMMZ in Victoria Province, and in charge of Great Zimbabwe, wrote the following report to his superior the Executive Director, Des Jackson, about the activities of Ambuya Sophia.

> This woman who claims to be a spirit medium has been a considerable nuisance to the staff at Zimbabwe, and has also committed a number of offences.
>
> (a) Cutting trees and bamboos
> (b) Building and occupying huts on our land
> (c) Encouraging large numbers of young Africans to assemble at her huts and create noise during day and night.
> (d) Stating to Africans that only Europeans have to pay entrance fees
> (e) Disturbing the labour force by making them take their boots off in the ruins area.
> (f) Placing stones across roadways which interfered with our lorries, tractors etc. and also impeded the rabid jackal trapping exercise.
> (g) Telling at least one visitor (white) to take his boots off.
> (h) Refusing to move to a site which we offered her, near a small ruin outside the main ruins area but adjacent to a small ruin.
> (i) Her acolytes threatening our gate attendants
> (j) Keeping chickens within this area.
> (k) Continuing to build huts after a number of warnings by ourselves, police and district Commissioner.
> (l) Removing stones from Conical tower and placing offerings in the cavities thus formed. (10 April 1980 NMMZ file C5)

But she was not without support from higher up in the fledgling administration. A hand written letter dated 5 April 1980, from Deputy Minister of Health , Dr Mazorodze, addressed 'To whom it may concern', stated,

> We [were] informed that this spirit medium has been subject to considerable harassment lately because she has taken residence on privately owned property. May I request that she be left alone until the D.C., Fort Victoria and the local Tribal Leaders have had time to sort out her place of residence. (5 April 1980 NMMZ file C5)

Nevertheless, after further complaints from Cooke, he was able to report on 19 June 1980, that 'the police have destroyed the hut and removed all the possessions belonging to her'(19 June 1980 NMMZ file C5).

But the saga continued. Ambuya Sophia, refusing to leave despite a suspended sentence against her, took up residence on land outside the boundaries of Great

Zimbabwe, and went about her night time activities with renewed vigour, much to the distress of Cran Cooke and others in NMMZ. Apart from being a general 'nuisance to staff' and breaking various National Monument bye-laws, Cran Cooke's reports included some colourful descriptions of sacrifices carried out at various sites around the ruins. It is clear that from his perspective the activities of Ambuya Sophia threatened the tourist potential of the site.

> On Sunday night and Monday night sacrifices were performed at the Great Enclosure. On Sunday a goat was killed and blood splattered on the steps and walls of the Eastern entrance. Apparently the body was then dragged along the ground to the conical tower. A white chicken was killed by cutting it's throat and left at the tower. Beer (African) was poured on the lemon trees outside the curio shop. All signs have since been removed. The matter has been placed before the District Commissioner to see if he can persuade the P.F. to curtail these activities. The police are also investigating.

> Although there is no direct evidence to put the blame on Sophia, the staff at Zimbabwe are convinced that it is her doing.

> Whether an approach to the Minister of Tourism or the Minister of Health, who is supporting Ngangas, could be of any use I don't know, but if this goes on Tourism could well be affected by these sacrifices. (28 August 1980 C. Cooke to D. Jackson, NMMZ file C5)

As the episode continued through the rest of 1980 and into 1981, it was noted with increasing concern that she was being visited by armed ex-ZANLA guerrillas, and members of the newly formed Zimbabwean National Army. She also seemed to have close links with members of the provincial ZANU (PF) party, and even certain ministers of the new government. In particular it was noted that the Minister of Health, Dr Ushewokunze, visited her on several occasions.

One of the key worries of Cran Cooke and NMMZ at the beginning of 1981 concerned widespread rumours that plans were being made at the ZANU (PF) headquarters in Salisbury (Harare) for some kind of ceremony to be held at Great Zimbabwe over the independence weekend in 1981. During the forthcoming ceremony the remains of war dead from the Chimoio guerrilla camp in Mozambique would be buried at Great Zimbabwe. In a memo to the deputy executive director of NMMZ, Ted Mills, Cran Cooke made his concerns known and alleged that Ambuya Sophia 'may be mixed up in this'.

> I have had a talk with Chief Charumbira on the question of the proposed ceremonies at Zimbabwe Ruins from 17th–20th April.

> He tells me that a delegation from ZANU (PF) in Salisbury visited him some time ago. They indicated to him that they wished to hold a burial ceremony at the ruins of material obtained from Chimoio. He would have nothing to do with it and sent them to the local Zanu (PF).

> The ceremony is to include beer drinking, killing and eating of oxen and tribal dancing as far as he and John Thokozane have gathered. John has been asked to store 12 bags of rapoko for beer making, this he has refused.

> The local men Musike and Manwa are associated with the Department of Education, but the Regional Director does not know them by those names. They are both active members of Zanu (PF).

So far I have been unable to make any progress locally, but have asked John if he is approached again by either of these men to send them to me.

I think that before this matter gets out of hand an approach should be made to headquarters of Zanu (PF), if necessary through a Minister, to nip it in the bud.

Over a public Holiday, such ceremonies would cause havoc and give great annoyance to tourists. It would undoubtedly get out of hand with the amount of beer envisaged.

Under our Act it is an offence to consume intoxicating liquor within a national monument, a point which could be stressed at any meeting with the authorities concerned.

The so-called Nehanda/Sophia may be mixed up in this. However, I have heard that the real Nehanda was brought to Zimbabwe to give advice on the Independence Day Celebrations.

Comrade Ushewokunze was seen to take Sophia through the Zimbabwe Ruins Hotel. I have been unable to find out anything about this meeting.

What all this signifies I cannot think, but we obviously have to tread carefully. (6 January 1981 C. Cooke to Ted Mills, NMMZ file C5)

This particular memo is revealing in a number of ways. First of all it clearly indicates that after independence there were efforts by people at high levels within the ruling ZANU (PF) party, who recognised Great Zimbabwe's 'national sacred' role during the liberation struggle, to organise a large *bira* ceremony at Great Zimbabwe. It betrays Cran Cooke's own opposition to these plans, but also his uncertainty about how to proceed. His suggestion about the need to 'tread carefully' indicates that he was not sure who in the government was involved in these plans, and where support lay for NMMZ's opposition to both Ambuya Sophia's presence and the planned ceremonies.

The memo also reveals that while the elders of some local clans were heavily involved in these planned ceremonies, Chief Charumbira, for one, was not. Furthermore it illustrates that this Ambuya Nehanda was not undisputably considered to be *the* spirit medium for Nehanda at all; another one, the 'real Nehanda' was herself also implicated in these planned ceremonies. This last point was made very strongly to me by many local actors in the area, and even Comrade Nylon. Contrary to what Garlake and Ucko have written about her being a 'highly respected spirit medium' (Garlake 1983:16) who was 'of extreme importance to the revolutionary anti-imperialist movement' (Ucko 1994:274), most people I spoke to, including Aiden Nemanwa, were adamant that she is mad, or has been made mad by the troubling *mashave* that do possess her. Comrade Nylon suggested that she was working 'both sides' during the war,

But our opinion was that she was an impostor, that she was working on both sides. So we tried once or twice to attack the place and the third time we realised it was probably because of this woman's magic that we weren't succeeding. The bottom line was that we considered this Ambuya Nehanda as an imposter, she worked with us, as much as she worked with them. (Interview with Comrade Nylon, 8/8/2001)

It seems from a later report by Cran Cooke (16 March 1981 NMMZ file C5) that Ambuya Sophia had inadvertently became involved in the disputes over Great

Zimbabwe between the neighbouring Mugabe and Nemanwa clans. Whilst the latter saw her as impostor, and mad, she did apparently have the support of Mugabe's people, though I should emphasise that I was never told about this by any members of the Mugabe clan, and Chief Murinye and his council told me they also considered her mad. Nevertheless for Cran Cooke, her alliance with one side of this local dispute reinforced his opinion that 'not only is it undesirable to have any such celebrations at Great Zimbabwe, but the tribal implications make for a very dangerous situation'.[4]

These security fears were seconded by J.Whitelaw, the Assistant Police Commissioner in Victoria Province, who described Ambuya Sophia's large entourage of ex-ZANLA guerrillas and armed soldiers from the newly amalgamated Zimbabwe National Army. She also had the support of members of both local and national ZANU (PF) structures. In his words:

> Sofia has built up a considerable reputation, particularly as some form of healer and seems to be well known to the Minister of Health, Dr USHEWOKUNZE, his deputy, Dr MAZORODZE and probably other Ministers, Senators and Members of Parliament. She receives numerous visitors including a large number of ZANLA and Zimbabwe National Army members from many parts of the country. It is thought, among other things, that they come to be 'cleansed after the war'. They pose by their very presence and numbers, a threat to law and order and have already been responsible for several incidents of violence and malicious injury to property both at the Ruins and in Fort Victoria. (Letter from J. Whitelaw, Assistant Commissioner for Officer Commanding Police, Victoria Province, to The Commissioner, Police Headquarters, Salisbury, 12 March 1981. NMMZ file C5)

In this same letter Whitelaw claimed that despite Ambuya's 'considerable reputation' among some ministers, he had the support of Mr Geza, the Under-Secretary in the Ministry of Lands, Resettlement and Rural Development who apparently agreed that Ambuya Sophia's 'continued presence in the Ruins area is highly undesirable' and that she should be prosecuted for trespass. For their part, Des Jackson and Ted Mills in the Executive Directorate of NMMZ had made their appeals against this planned ceremony, and the continued presence of Ambuya Sophia at Great Zimbabwe, to the Ministry of Home Affairs, where apparently they found some sympathy and support.[5] It seems clear that there were significant divisions among the ranks of the new ZANU (PF) government over the issue of Ambuya Sophia, and the planned ceremonies with which she was involved. At Ambuya Sophia's trial in the High Court later that year, Dr Ushewokunze mentioned how he became involved with her, and the plans to organise a ceremony at Great Zimbabwe for the Independence Day celebrations of 1981. On 18 December 1981, the *Bulawayo Chronicle,* reported what he said.

> In December last year, said Dr Usehwokunze, a message channelled through ZANU (PF) summoned the Prime Minister, Mr Mugabe, the Army Commander Lieutenant-General Rex Nhongo, and himself to Miss Muchini's dwelling. As the Prime Minister and General Nhongo could not go, Dr Ushewokunze with two senior Army Officers visited the medium over the Christmas period. In a trance she told them that a libation ceremony should be staged to stave off impending unrest in Zimbabwe and to restore peace between races and tribal groups. ZANU (PF) decided to comply the suggestion and bought livestock and rapoko for the brewing of traditional beer with funds from the Ministry of Education and Culture. Unfortunately the ceremony did not take place said

the former Minister because the rapoko went bad and 'events over took it with the arrest of the accused'. On a second visit in February to check on arrangements for the ceremony, Dr Ushewokunze told the court he discussed with Miss Muchini impending racial strife and disunity and 'her security, vis a vis where she was staying' (*Bulawayo Chronicle,*18 December 1981)

The ceremony of reconciliation that Ambuya Sophia Muchini had been part of organising therefore never happened. On 27 March, after a gun battle at her home near Great Zimbabwe, during which several of the armed ex-guerrillas accompanying her were shot, she and the rest of her entourage including her children were arrested. The annual report of 1980/1 for NMMZ's Southern Region (NMMZ file O/3) stated that Ambuya Sophia's arrest was ordered by the Minister of Home Affairs after two double murders of white settler farmers by ex-guerrillas under her instruction. As indicated by reports in the *Bulawayo Chronicle* of her first trial in Fort Victoria, and her second later on that year at the High Court in Salisbury,[6] Ambuya Sophia was accused of ordering the murder of all the whites who lived in the vicinity of Great Zimbabwe. Some ex-guerrillas who had already been convicted of the murders, were promised leniency if they testified against her, which they did. Having suffered nine months being held in police custody, and threats of death (Garlake 1983:17), even her children testified against her. During the trial Dr Ushewokunze was himself heavily implicated. Several witnesses suggested that he had known about the killings, and had sent the guerrillas to guard Ambuya Sophia. He was sacked from his ministerial position in October, and himself gave evidence at the trial in December, denying his involvement in the murders. On 18 December 1981 Ambuya Sophia Muchini was found guilty and sentenced to death. Passing sentence Mr Justice Pitman stated that he found Dr Ushewokunze's denial 'bald and unconvincing', and believed that he had sent ex-guerrillas to guard Ambuya Sophia and had known about the killing of the whites.

When Ambuya Sophia herself gave evidence to her trial, the *Bulawayo Chronicle* (16 December 1981) reported that she claimed,

that as well as Nehanda's spirit she also represented the spirits of Monomutapa and Chaminuka. Making frequent biblical references, she told the court that she had lived in the region of Great Zimbabwe since 1974. She claimed to have been detained in 1979 and often harassed by the police to move after her return the following year. Denying complicity in the murders of Mr Abraham and Mrs Margaret Roux and those of Mrs Helena van As and her grandson Philip, she said that after her home was burnt down she had heard a message from God. He told her to organise a ceremony to bring an end to 'the great chaos all over the country'. He also said that there would be shooting in Bulawayo and in Gwelo. (*Bulawayo Chronicle*, 18 December 1981)

Since her conviction and sentencing to death, she has been pardoned and released. She returned to the Great Zimbabwe area in 1986, and now lives a few miles away under Chief Murinye. I have visited her on three occasions, first in 1997, and then again in December 2000, and July 2001. On each occasion she has appeared wild and intimidating, dressed in animal skins, her arms jangling with copper bangles, beads around her neck. She always seemed angry, unwilling to talk, and keen to rant and scold. For myself, and those who accompanied me on these visits, her agitated and animated performance was sometimes even frightening. For her, Great Zimbabwe is 'God's place', as she made clear during my last visit to her which I noted as follows.

> Everything, the soil, the rocks everything belongs to God, even us people, black and white we belong to God, but we are refusing to listen to God's rules.

> She is particularly shouting at me, asking me why I keep coming to see her, if I refused to follow God's rules.

> She talks about the fuel prices, saying how low they used to be; 15 cent to go to Masvingo; 5 kg of *upfu* [mealie meal] used to be 5 dollars, now its 74 dollars .... (Fieldnotes, 13 July 2001)

Later on during the same visit, she described the events that happened at Great Zimbabwe in the early 1980s, after independence.

> She says that she went to live within the ruins area in the 70s but was chased away by the soldiers that were guarding the place. When the soldiers had gone, she came back from Gokwe. Then the police surrounded her house with guns, and started shooting.

> Here she stops, she seems to be remembering bad things, she talks about her child, she talks of a shirt so dirty with blood, and then she returns sharply from a faraway gaze: I was arrested, she says.

> They put me into prison without food or water, and no clothes for twenty days. Since they released me they never said sorry, never apologised. Why do they torment God's people, killing them leaving only the children? she says pointing to her son, and Pardon [my research assistant].

> [...]

> Great Zimbabwe to her is 'God's hill' and she emphasises that now people have to pay to go in, so only white tourists can pay and go in, but black people don't go in, because they can't afford it. She laments the refusal to do rain making ceremonies at Great Zimbabwe ... . (Fieldnotes, 13 July 2001)

Towards the end of this, our last encounter, I offered her some *bute* snuff I had brought along as a conciliatory gesture, but she refused it, telling me that when she had seen her words 'in newspapers and books laid before me, then I will take your *bute*, not before'. I sensed that day that she had warmed up to me, and her words often come back to me. It seemed evident to me though that she must have been through quite an ordeal that has not quite left her. Yet despite her change in fortune, it is remarkable that her concerns about the importance of Great Zimbabwe, and how it should be managed, continue to be widely shared by war veterans and 'traditionalists'. Among 'traditional connoisseurs' of the local clans of Nemanwa, Mugabe and Charumbira, the continued 'silence of Great Zimbabwe' is attributed to the perceived failure of NMMZ, and government to follow *chikaranga* or to consult with local elders about the management of Great Zimbabwe. The related issue of the need to hold a national ceremony at Great Zimbabwe to thank the spirits for independence, and to settle the spirits of the war dead, has even wider, and increasing currency among 'traditionalists' across Zimbabwe today. Since Ambuya Sophia Muchini's arrest and trial, there have many calls for such ceremony at Great Zimbabwe, but to no avail. VaMhike made his sympathies with the concerns of the local 'traditional' leadership clear. Furthermore he expressed his view that the political opposition that the ruling party is now facing in the form of the MDC, is a direct result of the continued failure of government to have carried out this important ceremony at Great Zimbabwe.

> *In Nemanwa here, we have got the masvikiro, we have got Chief Murinye, Makangamwe, these are of the Moyo clan, the VaDuma. They actually have their svikiro Ambuya VaZarira whom they consult on what to do. They are still complaining of the situation whereby a ritual has not been made in order to cleanse the war veterans after the war, or to brew beer to tell the masvikiro in Great Zimbabwe that Zimbabwe is now liberated. That is the way forward. Our Government has not yet done that. It is one of the complaints that the masvikiro are ever mentioning or repeating, saying 'Please come and do this ritual otherwise we won't be in a position to experience this opposition that we are facing [ie MDC ?]' It is actually sort of a punishment. It is not an opposition that comes on earth, an opposition like in Britain, that is there are Conservatives or the Labour party. Ours is some sort of punishment, that we have this uprising from within the government, as part of the task of co-operation between the government and the masvikiro. That is the situation. It is not an opposition, because it has more or less western bias, and it is a situation nobody can understand.* (Interview with VaMhike, 26/6/2001)

I asked VaKanda why the government has failed to carry out the necessary rituals to acknowledge the role of the spirit and the *masvikiro* during the war. His reply reinforced what I have argued above about the difference in perspectives of guerrillas and the nationalist political elite.

> *VaKanda: That's a tricky one, because there used to be not that direct involvement in consulting the masvikiro at national level. It's a bit difficult now to discern whether there is that plan to consult the masvikiro. And we as war veterans we have actually started to initiate something like that by going to Chinoyi.*
>
> *JF: Where Ambuya Nehanda stays?*
>
> *VaKanda: Yes, by going to Chinoyi, and by trying to organise a bira at national level at Matonjeni and at Great Zimbabwe, you see? But we don't seem to get as much support from the national leaders as we would like. To tell you frankly, my opinion is that when we came to independence there were a lot of forces that wanted to dilute us, as it were, to dilute our revolution. You know from, especially, the West. Remember we had a lot of people who had been educated in the West, in America and Britain and so on, who didn't quite have a league with what we had; our experiences in the war. So that when they came into positions of power and authority, they didn't seem to realise that this was important. So they didn't take up this issue of linking up with the past. They didn't actually see the significance and the importance of reaching out into the past. So processes of assimilation, in certain instances actually happened, where people would say well this is not Christian, these are not Christian principles and that sort of thing. So it has not been very clear. It is not as we would have wanted it to be.* (Interview with VaKanda, VaMuchina, MaDiri, 16/3/2001)

I also asked Eddison Zvobgo why a big national ceremony that had been planned in the 1980s failed, and in his reply he also pointed to opposition from Christians. Betraying what I have argued to be a lack of interest from the political elite in Great Zimbabwe's 'nationally sacred importance' in his comment 'we never pushed it strongly', he also gave some reason for optimism in his recognition of the growing feeling among 'traditionalists' that the failure to have conducted a 'national' ceremony has been the cause of the nation's troubles.

> *Yes that was widely canvassed but it never took off. There was real opposition to that type of ceremony from the leading churches in the country, who argued that it was idolatry. We never pushed it strongly. But I think that will be done. Because the elders and chiefs think that many of the mistakes that have befallen us, come from the fact that we have not performed such a ceremony and of course the country has had more than its fair share of problems and troubles. Some day I can see that happening.* (Interview with Dr Eddison Zvobgo, 18/8/2001)

This optimism may be justified. I have already mentioned that NMMZ co-sponsored a local ceremony at Great Zimbabwe in July 2000. It passed off without any dispute emerging between the local clans, an issue which NMMZ has always used to justify deliberate prohibition of ceremonies at Great Zimbabwe. While there are only limited signs that NMMZ may be willing to yield to some of the demands of local traditionalists about the management of Great Zimbabwe – NMMZ would still prefer to persuade local communities to become involved into NMMZ management plans – there have recently been some indications that a 'national' ceremony involving the reburial of remains taken from mass graves in Mozambique and Zambia, is being planned. In 2001, both Chief Charumbira, who as an MP is on the Council of Chiefs in parliament, and Dawson Munjeri, mentioned that such plans were being made, but no further details were made available.

Such a ceremony would obviously fit well with the political requirements of ZANU PF's 'Patriotic history' (Ranger 2004), as indeed a number of recent national events held at Great Zimbabwe have done. These include several flamboyant 'Unity Galas', a large televised ceremony (June 2003) marking the return of the missing half of one of the Zimbabwe Birds from a Berlin museum (Ranger 2004: 226–7), and a smaller 'local' event marking its return to 'the Masvingo chiefs' in May 2004 (*The Herald* 6 May 04 and 8 May 04).[7] But despite the government's prominent use of such events as opportunities to solidify its own rural and 'traditionalist' support – especially in the face of political opposition from the MDC – as I develop in Chapter 10, for many elders in Masvingo these events have reinforced and strengthened, rather than dissipated, existing concerns about the management of Great Zimbabwe. Nor has the likelihood of permission being granted to other stakeholders, local, regional or otherwise, to carry out rituals at Great Zimbabwe been increased in any way by these events. In May 2003, just before the return of half-bird the following month, a *svikiro* from Zaka, called Dickson Marufu, was turned away by police and NMMZ who worried he 'may tamper with physical structures at the national shrine' (*Daily News* 14/5/03).[8] After all, as Munjeri's own response to Chief Mugabe's request to hold a joint ceremony with war veterans in October 2000 indicated, NMMZ has not only 'national' concerns to worry about, but also the requirements of the World Heritage Convention, and 'international prestige' to consider.

The failure of the post-independent government to engage with and act upon the view of Great Zimbabwe as national sacred site, from where the authority of ancestors was bestowed upon the liberation struggle, and which grew out of nationalism's own mythology of the first *chimurenga*, supports or exemplifies Chatterjee's argument that it was in the movement from anti-colonial nationalism to postcolonial state that there has been 'a surrender to the old forms of the modern state'. Aiden Nemanwa himself put it:

During the colonial regime, white people controlled Great Zimbabwe. After independence, people are not yet independent, they are still following the ways of the white people. As there is not yet independence, spiritual independence, I have knowledge that I cannot yet divulge until independence. When Great Zimbabwe is again ruled by the traditional custodians, then there will be independence. Great Zimbabwe and the other shrines are still ruled in the white man's strategic ways. (Interview notes, Interview with Aiden Nemanwa, 21/11/2000)

Notes

1. Perhaps the best overarching analysis and description of this lively discourse about the role of 'traditional religion' for the mobilisation of popular support for freedom fighters is provided by Maxwell (1999:120–148), but see also Alexander (1995).

2. More recently, Ranger (2004:215–6) has stated that 'my first two books about Zimbabwe – *Revolt in Southern Rhodesia* and *The African Voice in Southern Rhodesia* – had been "nationalist historiography" in the sense that they attempted to trace the roots of nationalism'.

3. Dawson Munjeri mentioned this to me after we had finished our formal 'interview' on 11/5/2001.

4. 'Report on the proposed Burial and other ceremonies to take place at Great Zimbabwe on Independence Day Week-end' 16/3/1981, NMMZ file C5.

5. Memos dated 3 and 24 February 1981, NMMZ file C5.

6. *The Bulawayo Chronicle* June 18, 19, 20; July 29; October 13, 14; December 8,9,11, 12, 15, 16, 18, 19. My thanks to Terence Ranger for providing me with his notes of the reports of this episode that appeared in the *Bulawayo Chronicle*.

7. My thanks to Sara Rich Dorman for passing these newspaper articles to me.

8. My thanks to Terence Ranger for passing to me several newspaper reports relating to these events.

CHAPTER 8

NMMZ AND GREAT ZIMBABWE: THE PROFESSIONALISATION OF HERITAGE MANAGEMENT

The management of Great Zimbabwe as a national and international heritage site has become increasingly 'professionalised' since independence in 1980. Mirroring the continued dominance of 'academic' representations of Great Zimbabwe's past, its management as 'heritage' did not witness a radical rupture in the movement from colonial to independent state. After a brief period of uncertainty that immediately followed independence, National Museums and Monuments of Zimbabwe (NMMZ) very rapidly established itself, with international assistance from UNESCO and other international institutions, as a thoroughly 'professional', and 'modern' heritage organisation. It now has considerable expertise in scientific and technical conservation, particularly though the continual monitoring, surveillance and preservation as well as restoration of dry stone walls. Similarly, the processes and 'technologies of power', that began during the colonial period, through which Great Zimbabwe became increasingly alienated and separated from the surrounding landscape, communities and their history-scapes, have continued. Fences are constantly replaced and rebuilt, entrance fees are charged, and police or security guards are employed to guard against illegal trespassing, grazing, wood cutting and poaching.

Great Zimbabwe's role as an international heritage and tourist site has become solidified through its status as a World Heritage Site, which has, I will argue, entrenched and consolidated NMMZ's authority over the management of the site. World Heritage status allows NMMZ to appeal to not only the *national* but also the *international* significance of the site to justify its policies and interventions. As international heritage discourse has become influenced by development discourses with an emphasis on community participation, NMMZ has increasingly recognised that heritage management involves more than just concern with the materiality of heritage, and the scientific conservation, restoration and presentation of remains of the past. Since the late-1990s, NMMZ has adopted the language and rhetoric of local community development. Members of NMMZ I spoke to re-formulated its *raison d'etre* as impartial mediation between all stakeholders. While NMMZ has made some serious attempts in recent years to involve local communities (highlighted by the *bira* held at the Chisikana spring in July 2000) it is very significant that these efforts only began in the late-1990s, once NMMZ's authority at Great Zimbabwe had been thoroughly and irrefutably established. Given the complexities of local communities and the competing attachments at stake the language, concepts and practices of 'local community participation' that heritage discourse generally, and NMMZ specifically, have adopted are not sufficiently sophisticated to allow effective or meaningful consultation. There is a huge disparity between the rhetoric of local community involvement that NMMZ uses and the actual extent to which members of local communities experience any sense of control or influence of the management process. The new Shona Village project (which NMMZ often cites as a prime example of its efforts to involve locals in the distribution of financial resources) illustrates how 'local participation' often appears more like co-optation.

NMMZ: CONTINUITY AND CHANGE AT GREAT ZIMBABWE

The immediate period after independence was characterised by both change and continuity for National Museums and Monuments of Zimbabwe. Apart from the organisation's name, which scarcely changed from its predecessor, the National Museums and Monuments of Rhodesia (NMMR), the organisational structure and its legislative authority, the National Museums and Monuments Act of Rhodesia [now 'of Zimbabwe'] 1972, was also directly inherited from the pre-independence period. Initially most of the NMMZ staff, both nationally and locally, remained the same. In a sense, independence actually provided the opportunity to complete the establishment of the National Museums and Monuments as the proper administrative authority at Great Zimbabwe. From 1951 to 1976, Great Zimbabwe had been managed under a peculiar dual arrangement between what had been the Historical Monuments Commission (set up by the 1936 Monuments and Relics Act) and the Department of National Parks (set up by the 1949 National Parks Act). The former was concerned with the preservation of the ruins, archaeological research and the site museum, while National Parks managed the whole estate as a national park, and ran tourist facilities such as the guest lodges, the camping ground and the caravan park. As Ndoro (2001: 42) has argued, during much of the 1960s 'the top priority for most curators in the country was archaeological research. No clear policy or management plan existed apart from attempts to satisfy research based archaeological questions'. This changed with the National Museums and Monuments Act of 1972, which brought about the 'amalgamation of the Monuments Commission and the various city museums in the country', and 'meant that for the first time all archaeological property (finds and sites) were under a single curatorial administration' (Ndoro 2001: 16). It was, as Ucko has put it, 'a typically European system, combining museums and site recording and preservation under a newly created Board of Trustees' (Ucko 1994: 238).

Four years after this Act, and, importantly, only four years before independence, responsibility for the Great Zimbabwe estate was transferred from the Department of National Parks to the NMMR, which became solely responsible for both the Great Zimbabwe ruins, and the estate around it. From the correspondence (National Archives, File H15/10/1/3,10,20) between the Director M. Raath and the Board of Trustees, it is clear that this transfer was envisaged in relation to development plans for Great Zimbabwe that had been drawn up by William Van Reit in 1973. These plans involved extending the revenue earning potential of Great Zimbabwe and were linked to the proposed development of a tourist complex north of the Kyle road. The introduction of entrance fees, a ring road around the ruins, and improvements to the site museum were also planned. With the transfer of tourist accommodation and facilities from National Parks to NMMR, curatorship, the preservation of remains of the past, and the generation of revenue came together under one mantle at Great Zimbabwe, which for NMMR (and by extension its successor NMMZ) signified a key moment in the development of 'heritage management'. The Van Reit Report also highlighted the need to close the golf course, move the entrance gate and build new staff accommodation at the proposed township to be built at Nemanwa Growth Point. Few of these plans were realised before independence due to increasing shortages of funds, a huge drop in tourism, and the deteriorating security situation which accompanied the intensification of the liberation struggle. Entrance fees were introduced to the consternation of some residents in Fort Victoria, and the golf course was closed in 1978 (NMMZ files O/3). Because of the 'serious security situation' (NMMZ file O/3) the site was eventually closed to the public on 30 June 1979, and NMMR staff were evacuated shortly thereafter.

In February 1980, after the ceasefire and the elections, the site was re-opened under the directorship of Cran Cooke, after the previous director, Peter Wright, resigned at the end of 1979. Cooke's preoccupation during the first year was clearing the site of vegetation and the re-establishment of staff and tourist facilities damaged during the war and from by a bush fire in 1979. He continued the pre-independence development plans, and began work on alterations to access routes and the displays within the site museum (NMMZ File O/3). He also had to contend with continuing fence cutting, poaching and cattle grazing on the estate. Considerable effort was spent repairing fences to keep people and cattle out. Clearly, in terms of both policy and practise, there was very little immediate change within National Museums and Monuments at independence. But there were some very significant changes in the social and political environment within which NMMZ was working; both from 'above', that is the new government, and 'below', from the local communities.

Despite the immediate continuation of pre-independence approaches to Great Zimbabwe's management, it does appear that for some in the new government, particularly in the new Division of Culture in the Ministry of Education (Ucko 1994: 244), the euphoria of independence was seen as a good opportunity to rethink national approaches to the management, presentation and educational use of 'heritage', museums and the past. Recognising that 'for the majority of Zimbabweans in 1981, museums ... seemed to have little point – many (see Ucko 1981) considered that the whole museum service should be disbanded, on the grounds that the concept of a museum was European and exaggerated the static nature of collections' (Ucko 1994: 239). In what could have amounted to a profound challenge to continuing archaeological priorities on the 'distant', 'remote' and 'dead' past, a concept of 'culture houses' – 'of a multifaceted and dynamic "past" safely housed under local control' (Ucko 1994: 237) – was widely mooted. But others, particularly it seems some of the NMMZ 'old school' (Ucko, pers. comm. 25/9/04), did not 'readily accept the desirability or necessity of change; nor indeed [could] they envisage a role for museums different from their colonial role' (Garlake 1982b: 32). A foreign consultant's report (Ucko1981) on 'culture houses' that advocated strongly for 'the establishment of a past meaningful to its local populations' (Ucko 1994: 247) was considered too provocative, and was subsequently buried (pers. comm. 25/9/04). Although one 'culture house' was eventually opened in Murewa in 1986,[1] the 'structural and bureaucratic divides within Zimbabwean bureaucracy' that were inherited in 1980, especially between the Ministries of Education and that of Home Affairs, meant that ten years after independence, policy changes to heritage management in its broadest terms were 'fewer and less striking than some may have expected, or even hoped for' (Ucko 1994: 241, 256). An opportunity for a radical change that might have reversed a growing gulf between an increasingly 'professionalised' NMMZ – informed by 'processual archaeology' (Ucko pers. comm. 25/9/04) – and the majority of Zimbabweans had not been taken up.

If there were some (albeit unsuccessful) challenges, in the early 1980s, to NMMZ's inherited approaches to heritage management at a national policy level, this was mirrored by events at a local level around Great Zimbabwe. In 1980 Ambuya Sophia Muchini returned to Great Zimbabwe claiming to be the spirit medium of the legendary Ambuya Nehanda. After a year and a half of dispute and 'conflict' with NMMZ, she was eventually arrested and convicted of murder in 1981. Her tale illustrates both the continuity of NMMZ's policies towards local community involvement at Great Zimbabwe, and the changing social and political environment which NMMZ faced immediately after independence. Ambuya Muchini was not the only spirit medium calling for a national ceremony at Great Zimbabwe (though her determination, and the circumstances of her removal were unique). In the early 1980s, there were numerous attempts to conduct both local and national ceremonies. Like the involvement of several ZANU PF officials with Ambuya Muchini, some of these attempts had 'official' support, which made NMMZ's position difficult. Apart from these calls for ceremonies at Great Zimbabwe, which have continued to this day, NMMZ also faced new problems with 'squatting' on the Great Zimbabwe estate, as locals took advantage of the political situation after the ceasefire in 1979. This occurred in the context of wider territorial disputes among clans, and the proposed designation of farms surrounding Lake Kyle (Mutiriki) as a National Park. Again demonstrating the uncertainty of NMMZ's position given the political and social milieu of that time, it was not until 1983 that these settlers were eventually removed from the estate, and only in 1985, after the elections of that year, that 'squatters' were removed from the neighbouring and newly created national park.

NMMZ staff were also in an uncertain position in relation to the new government. The ambiguity of the government's position in relation to ceremonies at Great Zimbabwe was obviously problematic. Attempts to re-appropriate Great Zimbabwe for the nation also caused some concern. The desire to avoid the word 'ruins', and to rename parts of the site was only begrudgingly accepted by Cran Cooke (letter from C. Cooke to T. Huffman, 11 June 1981, NMMZ file H2). The memorial plaque of the Allen Wilson Patrol, on site of the original grave, also caused problems. In a quarterly report at the end of 1980, Cran Cooke reported that

> The remains of the memorial grave of Allen Wilson and his men has also given offence. This was not erected on the site of the original burial shown by early photographs. Therefore the small plaque has been re-erected in an area nearby but away from the main pathway. The European built surrounding wall is gradually being

removed and the stones used for re-construction and display in the museum area. This action has been taken to avoid the defacement which has taken place on one Pioneer monument in the Fort Victoria area. (Quarterly report, 31 December 1980, NMMZ File C1a)

This plaque was removed entirely in 1983. The numerous unannounced visits by international and national state dignitaries illustrate the strained nature of communications between NMMZ and government ministries at this time. On one occasion President Julius Nyerere of Tanzania visited Great Zimbabwe with President Banana of Zimbabwe. Previous notice had been given and arrangements made, but security and crowd control was left to the local ZANU PF. When members of the press climbed over dangerously unstable walls, Cran Cooke made his feelings known in a memo to the Deputy Executive Director, Ted Mills.

> Enclosed cutting from the Herald of 4/12/80. Note the look of apprehension on the face of President Banana as he glances up towards a news hound who was sitting on one of the most dangerous structures flanking the narrow passage. It was just after this incident that the Minister of Tourism, the police and security and myself shouted at him to get off the wall. This sort of behaviour has got to be controlled in the future before some V.I.P. suffers severe injury from falling stones or pressmen. My report on the visit will follow shortly. (Memo from C.Cooke Regional Director, to Ted Mills, Executive Director, 8 December 1980. NMMZ file C6 Correspondence From Ministry)

Perhaps unsurprisingly, given the trouble he had had with Ambuya Muchini, and his general mistrust of new government, Cran Cooke tendered his resignation at the end of 1981 and was transferred to Bulawayo to be the regional director of the western region in January 1982. The possibility of an African Director at Great Zimbabwe emerged for the first time. 'Africanisation' was a widespread policy of the new government, and this included NMMZ. As Dr Mahachi, then Deputy Executive Director of NMMZ put it,

> At that time, from about 1980/81 onwards, there was this drive by government to try and introduce the local people into various arms of government, and NMMZ was part of that. But I think that in the case of NMMZ there was also that recognition that, you know, Great Zimbabwe for instance is quite important in the struggle for liberation. Immediately after independence quite a number of things happened at Great Zimbabwe, which clearly brought it to the fore in terms of political thinking. So there was a feeling that we cannot continue to have such monuments continue to be run and interpreted by Rhodesians, as it were. (Interview with Dr G. Mahachi, 15/5/2001)

Having graduated in history from the University of Zimbabwe in 1982, Dr Mahachi himself joined NMMZ at the Queen Victoria Museum under these circumstances in 1984, as a 'cadet curator of archaeology'.

> It wasn't easy, but I wouldn't say that there was hostility, but there was uneasiness about such developments. I think that when there is action that is directed at introducing certain members of the community into a system, those that are already within that system obviously feel a little threatened. And so there would be that kind of uneasiness, but I personally must admit that I worked quite well with the archaeologists who were in the system at that time. Those are the people who actually introduced me to archaeology. (Interview with Dr G. Mahachi, 15/5/2001)

A shortage of archaeologists, and especially African archaeologists was a big problem for most of the early 1980s. Great Zimbabwe did not have a resident archaeologist until

1984, and the first appointment of an African archaeologist at Great Zimbabwe was only in 1987.

In the context of this wider 'Africanisation', Cran Cooke's resignation in 1981 was therefore an opportunity to re-appropriate Great Zimbabwe. Dr Ken Mufuka's appointment as regional director in May 1982 was almost certainly made in this context. Given the lack of trained African archaeologists at the time, his position as associate professor of African and Western civilisation at Lander College in South Carolina, USA certainly gave some authority to his appointment. He set about his task with great zeal, and it was in the midst of 'post-independence euphoria', as his co-author James Nemerai has put it, that his *DZIMBAHWE: Life and Politics in the Golden Age* was published in 1983. As I have already discussed, this work was not well-received by academic historians and archaeologists, who criticised its 'romanticism' and unreliable use of oral traditions. It was, however, very popular among the 'lay' audience, and should perhaps be seen as part of a wider effort on Mufuka's part and that of NMMZ generally to widen the popular interest in Great Zimbabwe. As Mufuka himself put it in the 1982/83 annual report, 'Our main emphasis in the second year of national transformation was to fix in the conscience of black Zimbabweans the historical significance of Great Zimbabwe' (Annual Report 1982/83, NMMZ File O/3). In September 1982, James Nemerai's appointment as the first Education Officer heralded an increased emphasis on the educational role of NMMZ and Great Zimbabwe. School visits have continued to be regular feature at the site. He has suggested that his appointment may have been an attempt by NMMZ to control the numbers of visiting school children,

> One thing is that there were quite a number of schools coming to Great Zimbabwe, now the people who were working here did not know how to work with the schools when they arrived here. So NMMZ as an organisation wanted to employ a teacher. So I came here as a teacher, my work was to work with the school children, inform them about the history of Great Zimbabwe, and the history of Zimbabwe in general. Maybe the reason why they wanted someone here, was for disciplinary reasons, to make sure that the schools would behave. They had problems with the schools when they came here. Now I know about the rules and regulations from the Ministry of Education, and we inform the teachers that if they go against the rules of NMMZ, I would report them to the Ministry of Education. So we wanted to keep order, so that the schools who were visiting here would share our heritage, with visitors who are also coming here. (Interview with James Nemerai, 20/11/01)

Apart from Mufuka's own attempt to provide a new African interpretation of Great Zimbabwe's past, and the increased emphasis on the site's educational role, there were other significant developments during his period as Regional Director. Tourism plans for Victoria (Masvingo) province continued to be developed around 'the belief that Zimbabwe – Kyle Dam complex will become the major centre of tourism. We are in the enviable position to set the tone of development in the region' (Annual Report 1981–1982, NMMZ File O/3). There was a general optimism that Great Zimbabwe would indeed be a major source of revenue, and a central part of the development of the entire region. In line with these expectations, the building of Nemanwa Growth point was begun to provide accommodation for NMMZ and hotel staff.

Related to his desire to create an African interpretation of Great Zimbabwe's past was Mufuka's apparent enthusiasm for holding a national festival for *n'angas* ('traditional' healers). He became quite involved with the arrangements for this festival, and with the Masvingo Publicity Association who were its organisers. This later led to a 'conflict of interest' with his position as Regional Director and he found himself strongly criticised by the executive directorate of NMMZ. The festival was eventually cancelled from 'above' due to 'political reasons' (Meeting of local board of trustees, 9 November 1983, NMMZ file C1a).

While Mufuka may have empathised with local communities, he nevertheless seemed pleased to announce in his annual report of 1982/83 that 'squatters' on the estate had at last been removed. In particular he described as 'ingenious' a plan by Dr Matipano to persuade the settlers to move away.

> Just before independence, eleven peasant families had moved onto the land belonging to National Museums and Monuments under the impression that they were doing an heroic thing. The damage done to the environment, by cutting firewood, overgrazing and burning grass between 1979–1983 became an eyesore to us. Various efforts to remove them were unsuccessful. Dr Matipano, Chief Executive for National Museums had an ingenious plan which brought immediate results. He brought down from Harare the National Board of Trustees and representatives of seven ministries involved, including those from the secret services and the police. When the peasants were required to help the police with certain enquiries regarding their occupation of museum property, they left in peace. (Annual report 1982/83, NMMZ file O/3 Annual Reports)

But 'squatters' remained in areas designated a national park to form a new buffer zone for Lake Mutiriki, and problems of trespassing, grazing, poaching, fire wood cutting, and fence cutting continued at Great Zimbabwe until their final eviction after the 1985 elections. Such problems with local communities have never been satisfactorily solved, and in recent times have re-emerged in the context of war veteran-led land reform and deteriorating economic conditions across Zimbabwe as a whole (Fieldnotes 'Walk around Great Zimbabwe with E. Matenga' Regional Director, 15 July 2001). Given the deep historical ties that different local communities claim with the land, these issues are unlikely to be resolved. They represent the continued alienation of local communities from both Great Zimbabwe, and land surrounding Lake Mutirikwi which has been designated a 'recreational park' (Fontein 2005). NMMZ policies regarding the integrity of the estate boundaries represent one of the most profound continuities in the management of the site since independence. Despite Mufuka's own enthusiasm for different interpretations of its past, in this respect he offered no alternative approach.

During Mufuka's tenure NMMZ began to look to UNESCO both for help with the preservation of Great Zimbabwe and concerning the possibilities for extending its international recognition. The Executive Directorate invited UNESCO to send a consultant to advise on the preservation of Great Zimbabwe; Hamo Sassoon visited Zimbabwe for two months in May/June 1982. In his first draft of his first annual report in July 1982, Mufuka described the UNESCO consultant as follows.

> Mr Sassoon is a character from the United Nations (UNESCO) sent to advise us on the preservation and possible recommendation that the Great Zimbabwe be placed on the World Heritage List. We still await his recommendations. (Annual Report 1981/82, NMMZ File O/3)

In a memo to Mufuka, the Executive Director of NMMZ, Mr H.D. Jackson requested that certain sections of the report be re-written.

> Mr Sassoon's investigation is directly related to the preservation of Great Zimbabwe and could well be married to your remarks under that head. It is precisely because of our concern for the permanent preservation of Great Zimbabwe that we invited UNESCO to send a consultant to advise us on remedial measures that could be undertaken. Your remark that he 'is a character from the United Nations (UNESCO) sent to advise us …'. may well be misunderstood in many quarters. (Memo from Executive Director to Regional Director, 26 July 1982, NMMZ file O/3)

Sassoon's report (September 1982) dealt with the general management of Great Zimbabwe, and 'stressed the desperate condition in which the monument was, particularly the lack of any maintenance strategy' (Ndoro 2001: 48). It also provided a draft plan of action for particular parts of the site that needed attention. Sassoon suggested that 'where collapse has begun, preservation must involve rebuilding' (Sassoon 1982: 10), though he also emphasised that 'I do not think any of the work which I have suggested should be done without close archaeological supervision'. He added that 'I find it very disturbing that there is no trained archaeologist on site at Great Zimbabwe, nor apparently is the site subject to frequent and regular inspection by any archaeologist' (1982: 20–21). Sassoon highlighted the need to focus on other archaeological remains on the site, most specifically the remains of *dhaka* structures in the Western Enclosure on the Hill Complex and in the Great Enclosure.

With the wide nature of his brief, Sassoon also considered the future development plans. While he acknowledged that visitor numbers were likely to increase, and this would put pressure on the site, it seems he doubted whether the figure of a million visitors a year which 'was mentioned' would be fulfilled (Sassoon 1982: 17–18). He looked in detail at the proposed 'ring road development plan' which involved building a 'a 6.5 km ring road on which electric trolleys would circulate carrying visitors through the ruins area' (Sassoon 1982: 17). He raised two objections, one concerning the prohibitive cost of the project at 3.36 million Zimbabwe dollars, expected to be borne by the Zimbabwean government; and second, that 'there is not the slightest doubt that the ring road and its vehicles would constitute a serious violation of the environment' (1982: 19). Summarising his disapproval, Sassoon raised the stakes by appealing to world opinion about the value of the site.

> I would therefore summarise views on the ring road development plan by saying that it is an extremely expensive way of spoiling one of Zimbabwe's most beautiful assets; and it is unnecessary. If the government implements this plan, it will earn the disapproval of thinking people throughout the world. Great Zimbabwe is not just a local asset; it is a world-famous site and the world is interested in what happens to it. (Sassoon 1982: 20)

This appeal to the international value of the site is characteristic of the means by which the 'World Heritage system' (Fontein 2000) operates. In 1985, the ICOMOS recommendation to the World Heritage Committee that Great Zimbabwe's nomination to the *World Heritage List* should be upheld, specifically referred to the Sassoon report, and mentioned the 'ring road project' stating that,

> It would appear necessary to postpone the installation of tourist facilities which are expensive and dangerous (like the project to build a road around the site) in order to

better investigate, conserve and manage one of the most important archaeological sites on the continent of Africa. (ICOMOS Report to World Heritage Committee on the Nomination of Great Zimbabwe to the World Heritage List, 25 June 1985)

The Sassoon report therefore led to an important change in NMMZ management policy. It was the first step that led to the site's nomination to the *World Heritage List* in 1986, and formed the basis of management plans that were subsequently developed, after further consultancy missions from UNESCO in 1987 (Rodrigues and Mauelshagen 1987) and 1990 (Walker and Dickens 1992). The extensive tourist development plans were dropped or scaled down dramatically. The main emphasis and resources of the organisation became focused on the scientific preservation and conservation of the site, according to the standards and guidelines offered by UNESCO's *World Heritage Convention* (1972), and other international heritage bodies such as ICOMOS.

But these changes in policy and practice took time. At a meeting of the Local Board of Trustees in November 1983, the Executive Director, Ted Mills, announced that Sassoon's report 'had been accepted in principle it was only the delay in appointing archaeologists which delayed implementation' (Minutes of meeting, NMMZ file C1a). However, in the immediate period after Sassoon's report, Great Zimbabwe had already witnessed more unsystematic, 'Wallace-style', re-constructions in the Eastern Enclosure, by Ken Mufuka (Ndoro 2001: 46), despite the absence of archaeological supervision or photographic surveys. It seems that Mufuka may have been inspired by both the Sassoon report, and his own visit to Great Britain in 1983.

> Great Zimbabwe and the outlying ruins are in a state of disrepair. The United Nations Report by Hamo Sassoon (1982) emphasised the need for urgent attention of the ruins. My own visit to Great Britain in April–May 1983 shocked me into new consciousness. In every place I went, whether it be a Roman palace at Fishbourne or a cathedral at Winchester, the British were busy restoring their national heritage. I later found out at the Commonwealth Institute that in Asia, the rebuilding of national monuments either falls under the Ministry of the Interior or the Ministry of Defence in order to emphasise their importance. (Annual Report 1982/83, NMMZ File O/3)

At the meeting of the Local Board in November 1983, the Deputy Executive director, Dr Matipano, reported that they were awaiting a team of archaeologists from Germany, and

> in the meanwhile all reconstruction work must cease. ... Mr Mills reported that he had received a report from our archaeologists expressing alarm at reconstruction work that they had seen. This was incorrect and no photographic records had been taken. Dr Mufuka denied any work had been done. (Minutes, NMMZ File C1a)

It was at this time that Mufuka was beginning to find himself in a variety of disputes with the executive directorate of NMMZ over issues ranging from his role as secretary of the Masvingo Publicity Association to his disregard for rules and regulations concerning the publication of a tourist brochure, and a series of bizarre accusations of mealie-meal theft against the executive directorate. Added to the controversy between Ken Mufuka and Stan Mudenge that surrounded the visit of Prince Charles in 1984, and the very mixed reception that his *DZIMBAHWE* had received, it appears that Ken

Mufuka's term as Regional Director was far more controversial than had been anticipated. His departure from Great Zimbabwe was similarly shrouded in controversy, when he resigned, apparently without warning, the following August. It later transpired that he had been 'required to work outside the United States of America for a two year period in order to gain a life professorship at his college in the USA' (Minutes of Meeting of Local Board of Trustees, 9 May 1984, NMMZ File C1a). This having been completed he simply resigned and returned to the USA. Perhaps the minor controversies that surrounded Mufuka's term embody in a small way the larger state of flux that NMMZ had been in since independence in 1980. It was during the period after Mufuka's departure, when Dawson Munjeri took over as the Regional Director at Great Zimbabwe, that NMMZ really began to solidify its policy, practice and experience in relation to the management and preservation of Great Zimbabwe. As Dawson Munjeri put it

> *I worked with Ken Mufuka, ya I was his deputy for the whole of 1984. ... And then after he had left I became the substantive regional director for Great Zimbabwe. And it was during that time that we were able to get resident archaeologists for Great Zimbabwe. It was the first time that we came up with, really, a proper approach to the preservation of Great Zimbabwe ... .* (Interview with Dawson Munjeri, Executive Director NMMZ, 11/5/01)

And the development of this 'proper' approach was intricately linked to NMMZ efforts to prepare the nominations of Great Zimbabwe and Khami for the *World Heritage List*, in which Dawson Munjeri was himself deeply involved. This experience seemed to have been the basis of his promotion to head office, shortly after Great Zimbabwe's successful nomination to the *World Heritage List* at the end of 1986.

> *So those were my experiences from 1983 to 1987, I was the resident director and it was also during that time that we fought very hard to compile the nomination dossiers to have Great Zimbabwe on the World Heritage List, together with Khami. I was responsible for that exercise. And after that, I was shot upstairs, when I became the deputy executive director here at Head office, 1988 to 1993. And from 1993, I became the substantive Executive Director for the National Museums and Monuments as a whole, up to now.* (Interview with Dawson Munjeri, Executive Director NMMZ, 11/5/01)

One of the first issues that arose after Mufuka's departure was what to do about his re-constructions.

The Regional Director reported that the then Minister of Home Affairs Cde Simbi Mubako had physically inspected the reconstruction made by Dr Mufuka at the East Ruins.

Dr Matipano gave the background leading to the Minister's visit viz:　that the Consultative Committee had inspected the area and had recommended that action be taken. But since the nature of such action was likely to create publicity, it was felt necessary to inform the Minister before proceeding with work. Cde Mubako consequently toured the area but was transferred to another ministry before he had made known his final decision.

Mr Makasi informed the meeting that the Minister had made it known to him that he and Mr Munjeri were of the opinion that if the demolition of Mufuka's work was done, there was a danger of destroying both Mufuka's reconstruction and the original. He had not yet made up his mind.

Mr Munjeri also informed the meeting that he had covered the area with the stonemason involved in the reconstruction and the latter was reluctant to reveal the exact areas that had been reconstructed. He also read a Report made by Miss Caroline Thorpe and even at that stage it was becoming impossible to know the extent of the reconstruction. In light of that it was considered unsafe to dismantle the Mufuka work.

Mr Mills stressed the need to have a full photographic record before anything was done to the structures. He also agreed that there was a danger that more harm than good could be caused. Mufuka's work was no better or worse than that of Wallace and the rest. The meeting agreed that the reconstruction should be left as it is – at least for the foreseeable future. There was work of a more urgent nature than 'De-Mufukaisation'. It also urged any urgent photographing programmer for Great Zimbabwe. (Minutes of meeting of local Board of Trustees, 2 August 1985, NMMZ file C1a)

While it was decided that Mufuka's reconstructions should not be dismantled, it was clear that from then on any structural interventions would have to take the issue of 'authenticity' much more seriously. As Webber Ndoro (2001: 62) has pointed out, all preservation work that has been carried out at Great Zimbabwe since Mufuka has been based on principles of 'authenticity' that derived from the Venice Charter of 1964 and were institutionalised through UNESCO's *World Heritage Convention* of 1972. Central to this concern with 'authenticity' was the envisaged role of both archaeologists and specialist 'conservation officers' trained in a variety of architectural surveying, monitoring and conservation techniques. In November 1985 another UNESCO consultant, Mr Bulenzi, visited Great Zimbabwe and emphasised that consultants in photogammetry and architectural conservation were needed. He recommended that part of the financial assistance NMMZ was to receive from UNESCO should be spent on training a Zimbabwean photogammetrist (Annual Report 1985/86, NMMZ file O/3). It became obvious that NMMZ did not have the capacity, in terms of both equipment and expertise, that this new 'internationally-derived', and 'science-dominated' approach required. According to Ndoro (2001: 48) there was not even an accurate map of the Great Zimbabwe estate. NMMZ had to turn outside for help. Indeed, NMMZ now considered the preservation of Great Zimbabwe to be a task that it could only handle with extensive technical and financial assistance from both national and international institutions, as the following quote illustrates.

> Co-operation between the National Museums and Monuments Organisation, the relevant UZ departments, and the Surveyor General's office and UNESCO/UNDP is a sine-qua non for the fulfilment of the preservation programme. … For their part the government of Zimbabwe and the UNDP have provided invaluable financial assistance. Upon proper interaction of these factors lies the future of the programme. It is an exercise which calls for material and human resources and it is hoped none of these elements would be found wanting. The prize preservation of a prime heritage calls for total commitment from all; the alternative is too ghastly to contemplate. (D. Munjeri, Annual Report 1985/86, NMMZ File O/3)

In 1987, two more UNESCO consultants a geologist (Rodrigues) and a photogammetrist (Mauelshagen), were commissioned with the specific brief of evaluating the state of preservation of stone structures at Great Zimbabwe. Like Sassoon before them, they emphasised the perilous state of the stone walls (Rodrigues and Mauelshagen 1987; Ndoro 2001: 48) and recommended that a intervention team, including trained stone masons, be set up at Great Zimbabwe and essential equipment be acquired. Mauelshagen in particular, recommended 'the adoption of photogammetry to monitor the movement of wall structures' (Ndoro 2001: 48). It was also suggested that further research programmes be established on documentation and identification of intervention priorities, and to evaluate different preservation techniques.

In 1988 the Swedish Agency for Research Co-operation (SAREC) provided funding for archaeological research at Great Zimbabwe. This funding contributed to the training of archaeologists, and, importantly, artefact conservation technicians. In line with the requirements of the new management approach to the conservation of Great Zimbabwe, the SAREC project 'provided field training for archaeologists on alternative, conservation friendly methods of archaeological research' involving 'remote sensing techniques such as magnetometer and phosphate analysis' (Ndoro 2001: 48; Sinclair et al., 1993) But the role for archaeologists at Great Zimbabwe was changing. NMMZ now 'prioritised the preservation of the monument rather than academic archaeology' (Ndoro 2001: 49). Therefore, while archaeologists have increasingly dominated NMMZ, and the management of Great Zimbabwe – as witnessed by the growing relationship between, and flow of people from, the archaeology section of the History Department at the University of Zimbabwe to NMMZ – archaeology itself became more involved with the preservation and management of archaeological remains, and less with research for its own sake. Excavations at Great Zimbabwe have therefore tended to take the form of 'rescue excavations' in the context of the reconstruction of collapsed walls. Given the amount of as yet unsorted, unexamined material these excavations have produced, there are no current plans for further excavations at the site (Fieldnotes 'Walk around Great Zimbabwe with E. Matenga, Regional Director, 15 July 2001).

At the beginning of the 1990s there was a further UNESCO consultancy mission which overlapped with an ODA funded joint project between the Universities of Zimbabwe and Loughborough (UK) (Walker and Dickens 1992). Focusing particularly on different methods of monitoring the movement of dry stone walls, as well as identifying causes of wall collapse, this project introduced the use of strain gauges for monitoring movement in stone structures. This is a much cheaper method of monitoring than photogammetry, and is more accurate than using glass wires to detect wall movement, a technique which had been introduced at Great Zimbabwe several years before (Ndoro 2001: 48, 62). A combination of strain gauges and glass wires remains the main method of monitoring wall movement at Great Zimbabwe today.

The ODA project was funded for two years, after which NMMZ was unable to continue the same due to lack of laboratory facilities and equipment. As Ndoro (2001: 49) has pointed out, this exposed 'the limitations of dependence on donor funds'. NMMZ began to focus on establishing its own facilities. In 1992 the Conservation Centre was built at Great Zimbabwe, which meant that for the first time all NMMZ operations for Great Zimbabwe, as well as for the entire region, were based at the site itself. NMMZ also began to develop its own approach to preservation and reconstruction that took on board limitations of finance and expertise. While many international experts, including Sassoon (1982) and the engineers from Loughborough had suggested the use of 'consolidates and geogrids' to improve the stability of the walls, NMMZ chose instead to focus on training 'traditional' stone masons (Ndoro 2001: 49). With extensive and continuous identification, monitoring and documentation of problematic sections of walling, restoration and reconstruction have again taken central place in the preservation of stone structures at Great Zimbabwe. Thus NMMZ has been able to combine the 'traditional' skill of stone masonry with much more 'scientific' and 'technical' monitoring and documentation methods, to establish its own approach which meets the criteria of 'authenticity' espoused by international organisations and conventions such as UNESCO, ICOMOS, and the *World Heritage Convention* (1972).

Several restorations and reconstructions have taken place at different parts of Great Zimbabwe since the late 1980s. The growing confidence of the conservation team at Great Zimbabwe was perhaps best exemplified by the reconstruction of the Western entrance of the Great Enclosure in 1995, which had been inaccurately rebuilt by Wallace in 1912. This is by far the most ambitious and high-profile reconstruction project that NMMZ has yet undertaken. It provided a unique opportunity for NMMZ to demonstrate, both nationally and internationally, its technical expertise and experience, establishing its authority as 'professional' heritage managers. Under the headline 'Great Zimbabwe's western entrance to get a facelift' the national newspaper, *The Herald* printed an article on the proposed reconstruction on 17 March 1994. Describing the background to the proposed reconstruction, the correspondent articulated the reasons for the work and, importantly, emphasised NMMZ's concern with 'authenticity'.

> The entrance is said to have been originally different from its present state. During the early part of this century the monuments were reconstructed by a curator St Claire Wallace.
>
> Mr Wallace rebuilt the monuments based on those found in Saudi Arabia because at the time it was believed the ruins were not built by Africans. The reconstruction was not based on scientific studies as rock conservation was then limited.
>
> The entrances are an open door system contrary to the open door system found in sketches of the monuments by Carl Mauch and others.
>
> [...]
>
> At least one entrance to the hill complex has its original structure. The NMMZ executive director, Cde. Dawson Munjeri yesterday said the west-end entrance was already showing signs of imminent collapse, which leaves them with no option but to rebuild it.
>
> [...]
>
> Cde. Munjeri said the existence of the original entrance at the hill complex to the present day shows that the original structure had a longer life span than that adopted by Wallace.
>
> [...]
>
> It is interesting to note that the collapsing section is the very section that was reconstructed by Wallace ... the sections that were [not] rebuilt are intact, he said. (*The Herald*, 17 March 1994)

Emphasising that contrary to the lack of 'scientific studies' behind Wallace's work, this reconstruction was to be based on thorough research − 'to make the changes as

Figure 8.1: Reconstructing a wall along the 'modern ascent' to the Hill Complex, at Great Zimbabwe. (Author 2001)

authentic as possible' – the correspondent for *The Herald* ended by describing NMMZ's capacity for the job.

> Over the years the NMMZ has developed technology, some of which is indigenous, to help undertake projects such as this one. Reconstruction of the entrance will be done using pictorial materials from Mauch to assume the original structure. (*The Herald,* 17 March 1994)

It seems obvious that the correspondent had the co-operation of NMMZ to produce this article, and it therefore demonstrates how NMMZ uses appeals to 'research' and its own 'technology' – a combination of 'indigenous' and 'scientific' techniques – as well as by contrasting its efforts to earlier 'inauthentic' reconstructions, to claim authority and legitimacy for its interventions. The importance of technology is also often framed in terms of the need to comply with international standards of heritage management. This was demonstrated by an article in the organisation's *Conservation News*, in which O. Nehowa described the use of computers and surveying to exemplify how 'the Great Zimbabwe National Monuments Conservation team is trying to keep up with world standards in monuments conservation' through the application of technologies (Nehowa 1993: 4).

During the early 1990s a comprehensive, long-term management plan for the preservation of the site was produced. Illustrating the role that archaeologists have come to play in heritage management, it was an archaeologist, David Collett, who drafted the *Master Plan for the Conservation and Resource Development of the Archaeological Heritage* (Collett 1992). With a focus much wider than Great Zimbabwe alone, this ambitious document subsequently became the backdrop of NMMZ management policy for heritage sites across Zimbabwe. It was presented at an international Donors Conference which was organised, with the help of UNESCO, to raise funds for the extensive list of projects that were proposed. Looking beyond just the preservation of archaeological resources, this plan, and the conference, took a wider view of heritage management that emphasised the 'potential economic development which may arise from a better management of the archaeological resources' (Ndoro 2001: 50). Despite its wide focus on all heritage sites in Zimbabwe, Great Zimbabwe was to become NMMZ's 'flagship', both in terms of prestige and as a major source of income. In particular it was hoped that by increasing public awareness of the monument, income from tourism would increase, thereby providing the funding for its continued conservation and other NMMZ projects. NMMZ was to be 'self-sustaining', even if this led 'to conflicts with other values associated with the heritage e.g. educational and cultural values' (Collet 1992: 7, cited in Ucko 1994: 263).

In a sense, the *Master Plan* was indeed the 'first comprehensive document relating to heritage management in Zimbabwe' (Ndoro 2001: 50), because conservation, management and revenue-earning potential were firmly interlinked, unlike in the early 1980s when they had seemed almost in opposition to each other. Despite some problems with the 'regional administrative politics of NMMZ'[2] (Ndoro 2001: 50), Great Zimbabwe should certainly be seen as NMMZ's 'flagship'. It is the focus of most of NMMZ's resources – it has the largest staff of any of the regions, and is the only site which receives direct government funding for its preservation. It is also NMMZ's biggest money earner. For this reason it is sometimes suggested that the regional director at Great Zimbabwe has relatively more influence in the general administration of NMMZ than other regional directors (Ndoro 2001: 50), and often the post of regional director at Great Zimbabwe looks like a 'stepping stone' to the executive directorate, as both the current and the former executive directors are former regional directors at Great Zimbabwe.[3]

The aim that Great Zimbabwe should 'pay for itself' has not yet been realised. Recent economic, social and political problems facing Zimbabwe have had their toll and deteriorating visitor numbers mean it is unlikely to do so in the near future. Nevertheless, it is clear that NMMZ's management policy and practice at Great Zimbabwe has been driven more by financial concerns and the needs of international heritage organisations than by the concerns of local communities. As one observer noted in 1994, NMMZ was 'apparently committed to a course of opting out of any concept of a living past, and of archaeology in the present' (Ucko 1994: 261). Elders and 'traditionalists' among local clans complain that Great Zimbabwe is 'treated like a business', and 'ruled in white man's strategic ways'. As Ndoro (2001: 69) has put it, 'what appears to matter to the heritage organisation is the feels [feelings] of the tourist and international organisations such as ICOMOS and WHC'. Taken together therefore, the emphasis on technological, 'science informed' approaches to heritage preservation – what Ndoro (2001: 69) has called '*technofixes*' – and the need to generate financial returns with which to fund such conservation, have contributed to the continuation of local communities' alienation from the management of Great Zimbabwe. Ndoro has summarised this very effectively.

> The adherence to the catholicity of the preservation movement as espoused by the Venice charter also guarantees that the local community cannot contribute meaningfully to the preservation or presentation of its heritage. Given the role Great Zimbabwe played as a rallying point for African Nationalism the exclusion of local community participation is surprising. This also indicates a lack of significance value assessment in the preservation of this monument. It appears the people who determine the value of the site are tourists and UNESCO through its charters. This position is hardly surprising given the tone of the Master Plan for Resource development (Collet 1992) and The Strategic Plan (1998) emphasis on income generation. National Museums and Monuments commissioned both these documents. (Ndoro 2001: 69)

But is not simply the embracing of international heritage preservation standards, and the 'science dominated' approach that it involves, or the requirements of international tourism that have led to the continued alienation of local communities from Great Zimbabwe. It is important to consider the processes of power that allow certain people or organisations, such as 'tourists' or 'UNESCO' to 'determine the value of the site', and exclude other ways of perceiving landscape, the past, and what to do with it. It is obvious that the National Museums and Monuments Act of 1972 charges and empowers NMMZ to preserve Great Zimbabwe for the 'nation'. Similarly, Great Zimbabwe's status as a World Heritage Site charges and empowers NMMZ to preserve and manage the site for the whole 'international community of nations' represented by UNESCO. This dual mandate is clear in a passage in the minutes of a meeting of the Local Board of Trustees in August 1985, when the issue of a 'District Heroes Acre' was discussed.

> The regional director [Dawson Munjeri] told the meeting that a request had been received for the alienation of 15 hectares of the National Monument estate for the purpose of converting it into a District Heroes Acre. He went on to state that in view of the fact that Great Zimbabwe is being considered for the World Heritage List, it was strongly felt that such a move would not be in keeping with such ideas. The Ministry of Home Affairs had therefore been advised accordingly and had in turn ruled that a District Heroes Acre could not be part of a World Heritage [site]. He reported that the matter had been resolved amicably. (Minutes of Meeting of Local Board of Trustees, 4 August 1985, NMMZ File C1a)

A piece of land on a nearby escarpment overlooking both Nemanwa Growth Point and Great Zimbabwe was later provided for a District Heroes Acre by Morgenster Mission. While this situation may have been 'resolved amicably', the issue of local ceremonies at Great Zimbabwe has never been entirely resolved, despite recent efforts by NMMZ. Until the NMMZ sponsored 'Chisikana ceremony' in July 2000, all ceremonies were prohibited. Since that date all requests are still channelled to head office where the final decision is made. Given both the widespread belief in Great Zimbabwe's 'sacred' role during the *Chimurenga*, and the early enthusiasm of some NMMZ employees, such as Mufuka, for conducting ceremonies at the site, it does seem odd that NMMZ came to reject all requests for ceremonies at Great Zimbabwe for such a long period after independence. The explanation lies in an incident that occurred in 1984.

Soon after Mufuka's departure, a request was received by NMMZ from Chiefs Nemanwa and Mugabe, to 'return artefacts to Great Zimbabwe in a ritual ceremony' (Minutes from meeting of Local Board, 8/8/84 NMMZ File C1a). A 'fracas' occurred during this ceremony when the representatives of the local clans became involved in a heated dispute. It was after this event it became an unwritten policy or, in Matenga's words, '*modus vivendi*' that ceremonies would not be allowed at Great Zimbabwe (2000: 15). This informal injunction against such ceremonies was based on a desire to avoid any negative publicity, and to remain to be seen as 'impartial' in local disputes and the 'fear' of being dragged into 'petty local politics' (Ndoro 2001: 60). In the context of Great Zimbabwe's elevation to the status of a World Heritage Site, this 'fear' was multiplied, and NMMZ was provided with a very powerful means of justifying its decision. This was clearly indicated in Dawson Munjeri's reply to Chief Mugabe's request to hold a ceremony with war veterans at Great Zimbabwe in October 2000, in which he emphasised his anxiety that 'the element of disrepute may creep in and Zimbabwe will be sanctioned by the World Heritage Committee' which was meeting the following month (Letter from D. Munjeri to Chief Mugabe, Ref G/1: EM/wcm, NMMZ File G/1).

As I will show in the next chapter, the status of 'world heritage' acts to extend the authorisation and legitimacy of the state, in this case NMMZ, in the management of heritage, through appeals to 'outstanding universal value', and preservation for 'humanity', which relay the responsibility of management policies to the international community. As Ucko has effectively put it in relation to Stonehenge, a site in many ways comparable to Great Zimbabwe,

> Part of the complexity of the politics of the past lies in the fact that the apparent congruence of interests of those agencies which deal with events at a pan-level can swamp all other interests. It is easy to ignore agonising local dilemmas of principle and action by invoking a putative 'world' identity and interest. Who could wish, for example, to be embroiled in disputes about the access of hippies, tourists and druids to Stonehenge if an alternative were to declare it a World Heritage site and to deny close access to everyone by insisting on protection? Part of the complexity of the whole situation becomes clear when one realises how much easier it must be for legislators and politicians to recognise the significance of a specific archaeological or historical site than to come to terms with claims for the sanctity of whole *areas* of land. Let alone for the sanctity of the earth itself. (Ucko 1990: xvii)

Notes

1. As Ucko has carefully discussed (Ucko 1994: 244), although the original 'discussion paper' proposing 'culture houses' in Zimbabwe suggested that one be built for every district in the country, and research (Ucko 1981) suggested 'overwhelming' support in all but one province, ultimately only one was ever built; the Murewa Culture House. Like the 'shona village' at Great Zimbabwe (see Chapter 9), which is based on an early initiative that Ucko (1994: 276) described as a 'virtual culture house', the Murewa Culture House too has suffered from being more imposed from above, than generated from below. This is particularly apparent in its choice of location; within easy reach of Harare for visits from important guests of the

government. Despite a lack of local consultation, however, there was a moment during the late 1980s, when 'there appeared to be evidence of community culture as well as significant educational role for this culture house' (Ucko 1994: 250). But this initial local success was short lived as a series of unpopular, imposed decisions – preventing the sale of alcohol and the exclusion of spirit mediums and *n'angas* – alienated and distanced local communities. The example illustrates effectively that however well thought-out any attempt at promoting 'living cultural' sites may be – including the proviso that inevitable local disputes be regarded as an essential part of their dynamic and vibrant character – their success is ultimately dependent on the very specifics of how implemented, and the 'degree of significance the people attributed to them, how they functioned in practice, and the position they would attain in practice' (Ucko 1994: 247).

2. W.Ndoro (2001: 50) has described how 'national interests were compromised', because different NMMZ regions wanted their development projects to be brought ahead of the schedule in the *Master Plan*. As a result several plans for Great Zimbabwe had to be postponed as more 'politically conspicuous projects' such as the Old Bulawayo project in Matabeleland were brought forward.

3. Both Dawson Munjeri (former executive director) and Godfrey Mahachi (current executive director) are former regional directors at Great Zimbabwe.

CHAPTER 9

UNESCO AND THE POWER OF WORLD HERITAGE

As I have argued at length elsewhere (Fontein 2000), the *World Heritage Convention* (1972) can be usefully viewed as a 'policy document' that gives 'institutional authority to one or a number of overlapping discourses' (Shore and Wright 1997: 18), namely those of 'heritage' and 'internationalism'. 'Heritage' as a concept can be traced back to a linear, progressive view of time and the past that arose from the European enlightenment, which became embodied in the disciplines of history and archaeology. They in turn appropriated authority over different experiences of the past, through a claim to objectivity. These disciplines and the concept of 'heritage' were closely associated with the rise of the idea of the 'nation-state', by providing legitimacy for national ideologies. Central to the idea of 'heritage', therefore, is that it should be preserved for the nation. Through UNESCO's *World Heritage Convention* (UNESCO 1972), this 'heritage discourse' – the idea that certain bits of evidence of the past should be preserved in a fast changing, indeed 'progressive' world – was taken up as part of not just the 'national' project but also the 'international' one. Thus 'heritage' discourses became aligned with discourses of 'internationalism' to constitute what I call the 'world heritage discourse'.

But 'internationalism' is more than a discourse. It exists, or can be *imagined* to exist, through international institutions such as UNESCO, in a similar way to that in which Abrams (1988: 58–89) has argued that the 'state' only exists as an idea, which is experienced through the 'state-system', which consists of the institutional manifestations of the 'state' such as the police, schools, museums, and so on. Such institutions act to reify, legitimise and indeed naturalise the discourses upon which they are founded. Therefore the *World Heritage Convention* not only established a world heritage 'discourse', but also the world heritage 'system', through which the idea of 'world heritage' is reified and naturalised, and can be *imagined* to exist. This 'system' comprises of an 'executive', the World Heritage Committee and the Bureau of the Committee, a permanent secretariat, since 1992 unified in UNESCO's World Heritage Centre (WHC), as well as the *World Heritage List*, the *List of World Heritage in Danger*, the World Heritage Fund, and the statutory texts and directives, such as the *Operational Guidelines for the implementation of the World Heritage Convention*, and the *Format for the nomination of cultural and natural properties for inscription on the World Heritage List*, through which the 'system' operates (UNESCO WHC).

But if the concept of 'heritage' is bound up with the idea of nationhood, does not the idea of 'world heritage' contain a contradiction or tension between 'national' and 'international' claims to any particular 'heritage', and, importantly, what to do with it? Indeed does this not reflect a contradiction between the discourses of 'nationalism' and 'internationalism'? There is only a contradiction, however, if we consider 'internationalism' as something opposed to 'nationalism', and as Gupta has noted, the 'nation' has always had to rely on something outside of itself.

> The nation is continually represented in state institutions such as courts, bureaucracies, and museums, which employ the icons and symbols of the nation. ... But, very important, the nation is also constituted by a state's external dealings with other states who recognise these practises as belonging to an entity of the same kind, thereby validating the ideology of nationalism. (Gupta 1992: 72–3)

Applying Gupta's work to the context of a Hutu refugee camp in Tanzania, Maalki suggested that 'internationalism' is 'fruitfully explored as a transnational cultural form for imagining and ordering difference among people – and that one of the underpinnings of dominant discourses of internationalism is the ritualised and institutionalised evocation of a common humanity' (Malkki 1994: 41). 'Internationalism' therefore reifies the idea of the nation through conceiving of the world as a 'family of nations' (Malkki 1994: 49) – 'globality is understood to be constituted by interrelations among discrete nations' (1994: 41). Therefore,

> Internationalism does not contradict or subvert nationalism; on the contrary, it reinforces, legitimates and naturalises it. In the process, the national order of things becomes a natural one – a moral taxonomy so commonsensical that it is sometimes impossible to see. (Malkki 1994: 62)

So on a discursive level there is no contradiction within the notion of 'world heritage', because it carries the assumption that any 'heritage' of 'outstanding universal value'[1] must also, almost by necessity, be of national value. Nevertheless, a tension does exist within the world heritage 'system'. It is a tension which in fact permeates UNESCO, and the 'UN family' and may be best framed as a tension between the sovereignty of member states (known as State Parties), and the influence of the 'imagined' international community, the 'family of nations' represented by UNESCO. This tension is embodied within the text of the *Convention* itself. Focusing on 'parts of the cultural and natural heritage [that] are of outstanding interest and therefore need to be preserved as part of the world heritage of mankind as a whole' (*Convention* UNESCO 1972: 1), it assigns responsibility for the protection of heritage to both individual, 'sovereign' State Parties, and to the 'international community' as a whole, as the following articles illustrate.

> Each State Party to this Convention recognises that the duty of ensuring the identification, protection, conservation and transmission to future generations of the cultural and natural heritage ... situated on its territory, belongs primarily to that State. It will do all to this end, to the utmost of its own resources and, where appropriate, with any international assistance and co-operation, in particular, financial, artistic, scientific and technical, which it might obtain. (Article 4, *Convention*, UNESCO 1972)

> the State Parties to this Convention recognise that such heritage constitutes a world heritage for whose protection it is the duty of the international community as a whole to co-operate. (Article 6, *Convention*, UNESCO 1972)

Beyond the text of the *Convention* this tension manifests itself in the way in which the world heritage 'system' operates, and is central to what I call 'the power of world heritage'. It is perhaps most visible in the nomination process by which a site is proposed for the *World Heritage List*. While the 'system' works at the level of the

'State Parties' – the 'state' is the level of agency, and only it can make a nomination for a site on its territory – at the same time, the criteria for successful inscription to the *List* are set by the 'system'. Indeed, as Leon Pressouyre noted (1996 (1993): 48), the influence of the world heritage 'system' is most robust and persuasive through the nomination process.[2] The final decision on whether a nomination is successful is made by the Committee, with the consultation of two advisory bodies, ICOMOS and IUCN. Sites are listed under particular criteria, as laid down in the *Convention*, and more specifically, the *Operational Guidelines*, which are announced upon inscription through a 'Statement of Significance'. State Parties are then required to ensure that the particular characteristics for which a site is nominated, its 'world heritage values', are maintained and preserved. Herman Van Hoff made this point during an interview at WHC in Paris, in June 1999, as my notes testify.

> *He responded by saying that the world heritage 'system' highlights a particular set of values associated with a site – world heritage values – which are never all the values associated with a site. This is always the case and can be a problem. The World Heritage Committee is concerned with the 'world heritage values' of a site, and looks specifically for these in nominations and periodic reports made by States. Furthermore it advocates that these 'values' or features should be maintained, at the cost of other values associated with a site if necessary. In this light, Herman Van Hoff mentioned the required periodic reporting by States, which refers very explicitly to the world heritage 'values' for which a property was listed, the 'statement of significance', and emphasises that these values in particular must be maintained.*
> (Interview notes, Interview with Herman Van Hoff, WHC, UNESCO, Paris, 28/6/1999)

This illustrates the *prescriptive* 'power of world heritage' – the fact that the world heritage 'system' prescribes, and attempts to impose a particular view on, or 'values' of 'heritage'. Furthermore, this *prescriptive* power relates not only to the values associated with a site or how it should be viewed, but also what to do with a site; how it should be managed. This was demonstrated by the ICOMOS report to the World Heritage Committee on its examination of Great Zimbabwe's nomination to the *World Heritage List* in 1985, which led to the abandonment of the proposed 'ring road project'. One glance at the *Nomination Format*, with its insistence upon detailed cartographic, graphic and photographic documentation, as well as historical accounts, legal requirements and detailed management plans, reveals that it is indeed a 'technology of power' which encourages, if not imposes, particular ways of seeing and managing 'heritage'. This effect is heightened if we consider that up until relatively recently, the concept of 'heritage' embodied by the world heritage 'system' was based on peculiarly European notions – most obviously the nature/culture division, which continues to be central to the structure of the world heritage 'system' despite recent changes. In relation to 'cultural heritage' specifically, Henry Cleere, then World Heritage Co-ordinator at ICOMOS, commented, 'heritage is definitely seen in a western, historical, art historical and archaeological (in the loosest sense) way' (Interview notes, 30/6/1999, ICOMOS building, Paris). Similarly, Galia Saouma-Forero, then senior programme specialist at the WHC, has noted (Interview notes, 30/6/1999, WHC) that the requirements of the *Nomination Format*, were also based on peculiarly European notions of, for example, legal ownership and management policies. In this light it seems obvious that the gradual 'professionalisation' of NMMZ's approach to heritage management since independence – particularly the increasingly 'scientific' and 'technical' approach to the preservation of Great Zimbabwe, and the concern with 'authenticity' – was closely related to the site's newly acquired 'world heritage site' status, and more generally, NMMZ's commitment to adhere to the principles of heritage management espoused by UNESCO's *World Heritage Convention*.

And yet to argue that the 'power of world heritage' is merely *prescriptive*, enforcing particular world heritage 'values', and particular ways of managing 'heritage', is to miss half the point. As far as UNESCO, and the world heritage 'system' is concerned, 'the state' is the level of agency. Only State Parties can make nominations to the *World Heritage List*, and ultimately they carry the responsibility for a site's management. The world heritage 'system' can impose very little without the co-operation of the State Parties. It has to use a 'stick and carrot' approach, as Mechtild Rössler suggested (Interview notes, 28/6/1999, WHC), but the carrot seems stronger than the stick. Inscription upon the *List of World Heritage in Danger*, often perceived as the 'Dock of Dishonour' (Pressouyre 1996(1993): 56), may be as far as the stick can reach, and it is rarely invoked without the consent of the State Party involved. A recent addition to the world heritage 'system's' 'arsenal' is the requirement of periodic reports every six years,[3] outlining the application of the *Convention* and the state of conservation of any World Heritage Sites, by State Parties. Again it is the State Party that compiles such reports, reinforcing my argument about the weakness of the world heritage 'stick'. In comparison, the incentives of financial, technical and capacity-building assistance offered by the World Heritage Fund seem much more substantial than the punitive alternatives, especially for less well-off countries such as Zimbabwe, though these funds are by no means infinite.[4] Perhaps it is most important of all to recognise that because the bedrock of UNESCO and indeed the 'UN family' is a discourse of 'internationalism' which, to paraphrase Maalki, 'reinforces, legitimates and naturalises the national order of things', there is an inevitable and fundamental commitment to 'the sovereign right of each state party'.[5] This, combined with the appeal to the 'outstanding universal value' of heritage, and the need to preserve it for all humanity, acts to 'reinforce, legitimate and naturalise' the role of the state, in this case NMMZ, in the management of what has been labelled 'world heritage', and indeed any 'heritage' at all.

Thus NMMZ are able 'to ignore agonising local dilemmas of principle and action' (Ucko1990: xvii) at Great Zimbabwe, and operate a '*Modus vivendi*' against local, 'traditional' ceremonies 'by invoking a putative "world" identity and interest' (Ucko 1990: xvii), as demonstrated by Dawson Munjeri's rejection of Chief Mugabe's request for a national ceremony at Great Zimbabwe, on the grounds that such an event 'may bring disrepute to the World Heritage Convention' (Ref G/1:EM/wcm, NMMZ file G/1). Following Ferguson's lead, presented in his work on development projects in Lesotho (Ferguson 1990), we could call this the 'anti-politics of world heritage'. According to Ferguson,

> In this perspective, the 'development' apparatus in Lesotho is not a machine for eliminating poverty that is incidentally involved with state bureaucracy: it is a machine for reinforcing and expanding the exercise of bureaucratic state power, which incidentally takes 'poverty' as its point of entry – launching an intervention that may have no effect on poverty but does in fact have other concrete effects. Such a result may be no part of planners' intentions – indeed it almost never is – but the resultant systems have an intelligibility of their own. (Ferguson 1990: 255–6)

He goes on to establish that crucial to this process is the de-politicisation of poverty, which means that a

> 'development' project can end up performing extremely sensitive operations involving the entrenchment and expansion of institutional state power almost invisibly, under cover of a neutral, technical mission to which no one can object. (Ferguson 1990: 256)

Under the rubric and guise of a seemingly de-politicised claim to 'universal value', and the need to preserve 'heritage' for humanity, the world heritage 'system' therefore reinforces and entrenches the authority of state bureaucracies, such as NMMZ. In this sense, the 'professionalised', 'scientific' and 'technical' emphasis of NMMZ's approach to the preservation of Great Zimbabwe represents not only the 'prescriptive' power of 'world heritage', but also the 'de-politicisation' that is essential for the 'anti-politics' effect of 'world heritage'. And both of these contribute to the continued alienation of Great Zimbabwe from local communities, and the marginalisation of other perspectives on its value, and appropriate management. As Herman Van Hoff (Interview notes, 28/6/1999, WHC) put it, 'the system does not [and perhaps *cannot*] concede to the need to listen to locals and NGOs'. If we turn now to consider the widespread reforms that have recently been made to the world heritage 'concept', and 'system', embodied by extensive modifications in the *Operational Guidelines* and the *Nomination Format*, it will become apparent that these reforms have, in a sense, attempted to address the *prescriptive* 'power of world heritage', whilst the inevitable insistence upon the 'sovereignty' of State Parties remains, as does the 'anti-politics' effect of 'world heritage'.

In recent years the world heritage 'system' has undergone substantial changes that have broadened not only the concept of 'world heritage', and the criteria embodied by the *Operational Guidelines*, but also, importantly, the management requirements of the *Nomination Format*. These changes came about because of a gradual realisation that the concept of 'world heritage' embodied by the *Convention*, and how it should be managed, was highly restrictive and indeed eurocentric, and that this accounted for the unbalanced nature of the *World Heritage List*, which had (and still has) a massive over-representation of European cultural sites. As Léon Pressouyre noted in 1996,

> If the concern of the authors of the Convention was to achieve an equitable distribution of the World Heritage Sites in the various regions of the world, as well as a numerical balance between natural and cultural sites, their hopes were dashed: 48% of the properties inscribed since the first session of the World Heritage Committee are cultural properties situated in Europe. The cultural heritage (and primarily that of developed countries with 'great monuments', witnesses to secular or even ancient history) appears to have been widely favoured by a procedure which was intended initially to be more ecumenical. (Pressourye 1996: 57)

These imbalances were perceived as a threat to the credibility of the *World Heritage List*, and by extension, the very *raison d'etre* of the *Convention* and even UNESCO, because they undermined the 'universalism' upon which the 'system' is built. They precipitated the adoption of the *Global Strategy for a representative and credible World Heritage List* by the Committee in 1994, at the same time as a second round of major changes to the *Operational Guidelines*. This was a 'framework and operational methodology for implementing the World Heritage Convention' (Saouma-Forero 1999: 1–2), which took a regional and thematic approach in order to identify certain types of 'heritage' and regions which are under-represented on the *List*.

The African continent is particularly under-represented, especially in terms of 'cultural heritage', as Pressouyre highlighted in 1996 when 'the African cultural properties inscribed on the World Heritage List only account for 5 per cent of all properties on the list' (Pressourye 1996: 60). Furthermore, illustrating the problematic nature of the definitions of 'cultural heritage' embodied by the *Convention* and the 'system', the number of 'natural' sites in Africa that are on the *List* is almost double that of 'cultural' sites (Saouma-Forera 1999: 2). This reverses the general trend of the *List*, which as a whole has far more 'cultural' than 'natural' sites, and seems to re-establish an old primitivist and 'naturalist' myth of Africa as an 'untamed', 'uncultured' and 'virgin' continent. Great Zimbabwe and the Khami ruins are among a minority of sites in Africa that do meet the pro-'ancient', and 'monumentalist' biases of the original criteria for world heritage contained in the *Convention*. As part of the *Global Strategy for African Heritage*, there have been a series of 'Expert Meetings' (Harare 1995; Addis Ababa 1996; Porto Novo 1998; Nairobi 1999) on the continent which have attempted to look into the application of these criteria in an African context. As the Senior Programme Specialist for the Global Strategy and African Heritage, Galia Saouma-Forero, emphasised the importance of allowing 'African experts' to define their own problems and possible solutions.

> *What I have been trying to do since 1994 is to relate to the African Experts and to inform them of the World Heritage Convention, and to let them express themselves on their difficulties. They have also defined their needs; for example the need to increase the amount of preparatory assistance they receive, designing training programmes concerned with site management, language courses, and school programmes. These have been designed by the African experts themselves. One of the main problems is to identify the problems that are faced, so I attempt to let African experts speak to African experts about these problems, so these African experts have been defining their own problems. Once the problems are identified, the solutions can be discussed.* (Interview notes, interview with Galia-Forero, WHC, Paris, 30/6/1999)

Apart from allowing 'African experts' to discuss their the needs in terms of 'institutional capacity building' and financial and technical assistance for the implementation of the *Convention*, these expert meetings also identified particular categories of 'cultural heritage' appropriate to African contexts. Recognising, as Laurent Lévis Strauss put it at the 1996 Addis Ababa meeting, the 'total interpenetration and inseparablity of nature and culture in African Societies' (Lévis Strauss 1996: 38) categories such as 'living cultures', 'spiritual heritage', and 'routes and itineries' were identified, which relate directly to the relatively 'new' and innovative concept of 'cultural landscapes' which was adopted in the *Operational Guidelines* in 1992. As Henry Cleere put it (Interview notes, 30/6/1999), the idea of 'cultural landscapes' put 'another dimension into the heritage concept' because it challenged the fundamental distinction between 'cultural' and 'natural' heritage, which is the 'question that has, little by little, become the Convention's stumbling block' (Pressourye 1996(1993): 28).

The distinction between natural and cultural heritage is one of the most striking aspects of the *Convention*, which continues to be reflected in the very structure of the 'system'. According to Pressouyre, the *Convention* was itself

> born of the encounter of two currents of thought: one, directly stemming from the Conference of Athens organised in 1931 under the aegis of the Society of Nations,

concerned the conservation of the cultural heritage and widely referred to the classical notions of 'masterpieces' or 'wonders of the world'. The other found its source in the First International Conference on the Protection of Nature, held in Berlin in 1913. This concept gained new force at the conference of Bremen in 1947 and led to the foundation of the IUCN in 1948. Delegates to the conference wished to transmit to future generations a number of 'virgin' natural sites unspoiled by mankind. (Pressouyre 1996(1993): 56–7)

Articles One and Two of the *Convention* make this fundamental distinction between 'cultural' and 'natural' heritage, and this is still maintained in the *Operational Guidelines*, which, despite recent attempts to merge the categories,[6] continues to divide the criteria for 'world heritage' into six for 'cultural heritage' and four for 'natural heritage'. The former are required to fulfil a 'test of authenticity' and the latter a 'test of integrity', and the main advisory bodies, ICOMOS and the IUCN deal separately with 'cultural' and 'natural' heritage respectively. Given that the nature/culture dichotomy has been thoroughly recognised and criticised in anthropological discourse as a product of the European enlightenment (e.g. Bloch and Bloch 1980; MacCormack 1980; Descola and Palsson 1996), the manifestation of this problematic distinction in the *World Heritage Convention* and 'system' shows the extent to which it was constructed upon very eurocentric discourses of heritage.

In 1992 the World Heritage Committee defined and adopted three categories of 'cultural landscape', including the 'clearly defined landscape designed and created by man', 'the organically evolved landscape' and the 'associative cultural landscape' (paragraph 39, *Operational Guidelines*: 9–10). The basis of the 'cultural landscapes' concept being to emphasise the inherent interdependence between 'natural' and 'cultural' aspects of heritage, these three categories reflect different degrees of this interdependence. The first focuses on landscapes deliberately created by humans, such as parklands and gardens. The second relates to landscapes that have 'evolved' or 'developed' over time, from an initial human 'imperative' in association with the natural environment. It breaks down into two sub-categories; 'relic' or 'fossil' landscapes where the 'evolutionary' process has stopped, but the material form is still visible; and 'continuing landscapes' which maintain an active 'social role' in society, and where the 'evolutionary process is still in progress'. The third category is the 'associative cultural landscape'. This is the most innovative because it acknowledges that a landscape can be 'cultural' as well as 'natural' purely through the 'cultural' values that people associate with it. Unlike the former two, this category does not therefore require any material manifestation of its 'cultural' features, and thereby allows for the recognition of the intangible values of a site, and the possibility of, for example, 'spiritual heritage' as identified by Dawson Munjeri, then executive director of NMMZ, at the first Global Strategy meeting for African Heritage in Harare in 1995 (Munjeri 1995: 54). Combined with the fact that the cultural landscapes concept also makes a strong case for what could be termed 'living heritage', this reveals that the heritage idea may no longer retain the same strong hold on a notion of the past as 'foreign' and 'distant' (e.g. Lowenthal 1985; Walsh 1992), as it once did.

The recognition of the intangible aspects of 'heritage', and the idea of associative landscapes and living heritage has raised problems with the 'notion of authenticity', as the difficulty of defining 'authentic culture', and the associated danger of 'freezing' traditions and preventing change, have been recognised. One of the recommendations of the Expert Meeting in Amsterdam in 1998 was that were the 'natural' and 'cultural' criteria of world heritage to be combined then the 'notion of authenticity' would have to be dropped, and, instead, the 'notion of integrity' currently applied to 'natural' properties would apply across the unified criteria. While these recommendations have not yet been endorsed, some of these problems with 'authenticity' have been addressed through other changes to the 'system', which included 'an emphasis upon the idea that traditional and customary law must be acknowledged as having equal weight as statutory law' (Henry Cleere, Interview Notes, 30/6/1999). Changes in the *criteria* for world heritage therefore coincided with changes in the legal and management requirements of the *Operational Guidelines* and *Nomination Format*, whereby 'customary law' and 'traditional management' practises are now acceptable. Added to this is the growing realisation of the important role of local communities in the management of heritage, which has filtered in from wider development discourses. Indeed the *Operational Guidelines* and the *Nomination Format* both now contain clauses that emphasise the importance of 'local community participation' in the management of heritage. For example, the 1998 edition of the *Operational Guidelines* noted that in terms of management and conservation,

> Informed awareness on the part of the population concerned without whose active participation any conservation scheme would be impractical, is also essential. (Article 34 *Operational Guidelines*, February 1998, UNESCO document WHC-97/WS/1 Rev)

And in relation to the nomination process, it stated that

> The nominations should be prepared in collaboration with and the full approval of local communities. (Article 41, *Operational Guidelines*, February 1998, UNESCO document WHC-97/WS/1 Rev)

While these far-reaching changes have had relatively little impact so far upon the imbalances of the *World Heritage List*, they have precipitated some tremendous new nominations. In particular the Rice Terraces of the Philippines Corderillas are often cited by those within the 'system' as a model example of a cultural landscape of 'outstanding universal value' (Rössler 1998: 19). With the adoption of the cultural landscapes concept, Australia and New Zealand sought to re-nominate sites that had already been inscribed under 'natural' criteria in the 1980s, so as to formally recognise the 'cultural' values associated with these properties. Together, the Australian sites of Uluru-Kata Tjuta National Park (formerly Ayres Rock–Mount Olga) and Kakadu National Park, along with Tongariro in New Zealand represent landmark nominations in that the management policies by which they are run were formulated with the consultation and indeed full participation of indigenous peoples for whom the landscapes are sacred. As a result, no visitors are allowed to Tongariro and access to parts of Kakadu and Uluru are similarly restricted. The inscription of the Sukur Cultural Landscape in Nigeria in 1999, which is almost entirely managed according to 'customary law' and 'traditional management practises' represents a recent example from Africa of the application of the new world heritage concept of 'cultural landscapes', and 'local community' and 'tradition-friendly' approaches to heritage management. An even more recent example, from Zimbabwe, was the inscription of the Matopos Hills as a cultural landscape in July 2003 (Ranger 2004: 228), 20 years after an initial nomination was submitted on the basis of, exclusively, their 'natural' values.

The changes to concepts of 'world heritage' and approaches to management that these recent nominations exemplify seem to represent an increasing 'anthropological awareness' in the world heritage 'discourse' and 'system', as was often suggested to me during my period as an intern at the World Heritage Centre in Paris in 1999. It also seems apparent that these changes have directly challenged what I have described as the *prescriptive* 'power of world heritage', by widening the 'world heritage' idea, reducing its eurocentric biases, and allowing for other ways of perceiving and, importantly, managing it. It could even be suggested that the increasingly pervasive concern with 'local community participation' across the 'world heritage' system, undermines what I called the 'anti-politics of world heritage' through which the role of state bureaucracy in the management of heritage is reinforced, legitimated and naturalised.

There may be some validity in the suggestion that the *prescriptive* 'power of world heritage' has been reduced, and yet, on the other hand, this is not conclusive because there is a sense in which the 'world heritage' system continues to be a 'standard setter' to use Mechtild Rössler's words (Interview notes, 28/6/1999), defining, and occasionally re-defining, the 'world heritage' criteria and management requirements. Some would not agree with this at all. When I asked Dawson Munjeri whether NMMZ's apparent, recent shift in policy towards the notion of 'community participation' was a result of a change in UNESCO's approach, he argued the opposite at length, suggesting these changes reflect bottom-up feedback from 'local' to the 'global' level.

I would put it the other way around, and say to what extent has UNESCO changed from a perspective where it did not see the intangible values of a property, to the point that it is now realising that. I think we have never changed, none of us has ever changed, we have always realised that, not withstanding what legislation we might have inherited from the past, the essence of it is that in practise we could not operate without taking the reality of the situation on the ground. As people responsible for the preservation of this heritage, it can only survive on the basis of the respect that it is given by the indigenous populations. So the very survival of heritage is dependent upon the appreciation of that heritage by the local communities, and the populace at large. So that's our understanding. Prior to that, the strategy was to have as many policemen around as possible, security guards, guarding this and that. But you cannot be everywhere, and we are not just talking of Great Zimbabwe, we are talking of an immense heritage spread throughout the country. Some of it we have never been there; some of it, we are scantily there, we cannot be all over. But it has survived where it has survived by virtue of it being respected.

And I think UNESCO too, having been much involved in programmes of a cultural dimension and the like, must have come to the same scenario. As you know it has been funding, it has been sending out experts, it has been giving out all the technical advice, but still heritage is at risk, if I can use that term, and I think it has been that realisation, that unless you link the dimension of people's involvement and people's appreciation, then everything is an exercise in futility. So that is the realisation that we had at the local level, which cumulatively must have got up to UNESCO levels. And therefore obviously that level with all this experience both at local and at global levels, UNESCO then took the definitive policy decisions to try and see it from that position.

[...]

To answer your question more specifically, ... the movement within Zimbabwe – after the realisation that without a total appreciation of heritage in its totality, as widely defined, both intangible and tangible aspects, cannot survive, without the total involvement of the people, the generators of that culture – it has reached higher levels, and those higher levels have responded accordingly, by formulating policies. So it really has been a question of inputs up, and outputs down, and I think … cumulatively, globally, this has been the part elsewhere. I have been involved in quite a lot of work in Nigeria, Botswana, Namibia, and other countries, and you look at it, and its almost the same thing, simultaneous kind of realisation all over, that's what I would say. (Interview with Dawson Munjeri, NMMZ, 11/5/2001)

Galia Saouma-Forera's emphasis on the role 'African Experts' have played in *Global Strategy* meetings seems to concur with Munjeri's view. My hesitation here relates to the extent to which these 'African Experts' can claim to represent people whose experience of what is being labelled 'African heritage' may be very different from their own. In terms of NMMZ, I have already argued that since independence NMMZ has been dominated by archaeologists, and in particular, as Peter Ucko emphasised (pers. comm. 25/9/2004), a specific, 'modern' and 'objectivist' strain of the discipline known as 'processual' archaeology. One wonders, therefore, to what extent they are able to represent the views and perspectives of 'traditionalists' and elders among local communities. In this light, the changes to the 'world heritage' system may have been a response to the needs and experiences of heritage practitioners within State Parties.

Another way of looking at this issue relates to the motivation behind the *Global Strategy* which I suggested was based on a perceived need to re-address the imbalances of the *List*, which were considered a threat to the credibility of the *Convention*. In other words, to improve and validate the 'universalism' upon which the 'system' is built and relies, it has had to acknowledge the differential nature of what it wishes to 'universalise'; that is, 'heritage'. Yet by changing the notion to make it less eurocentric – by introducing an idea such as 'cultural landscapes' for example – it is still supplying the terms or perhaps 'tools' by which this is done, and also the requirements that go with it. To put it very simply, the scope of the concept of 'world heritage' may change, but the label 'world heritage' remains, as does the claim to 'outstanding universal value' and the corresponding need for its 'preservation for humanity'. In this sense, the world heritage 'system' is inherently *prescriptive*, however 'ethnographically aware' its concept of 'heritage' becomes.

As for the 'anti-politics' effect of 'world heritage', the commitment to the 'universalism' that is the bedrock of UNESCO as a whole means that however pervasive the concern with 'local participation' becomes across the world heritage 'system', the insistence upon the sovereignty of State Parties remains. Representing the level of agency within the 'system', it is up to 'state parties' to determine what nominations are made and how, and importantly, what 'local community participation' might actually amount to, and the circumstances in which it can be applied. In the case of Great Zimbabwe, this allows NMMZ to use the complexity of local, regional and national disputes and claims over the 'sacred' values of the site, as well as the site's national and international 'heritage' and 'tourist' importance, to frame its own role as 'impartial intermediary', and justify its control over the site's management, whilst espousing the rhetoric of 'local community involvement'. This was exemplified by Munjeri's lengthy response when I asked about NMMZ's apparent reluctance to acknowledge Great Zimbabwe's sacred importance, as witnessed by the delay of almost 20 years between independence and the Chisikana ceremony of July 2000.

The delay, I could say … Let me put it this way, from my own experience at Great Zimbabwe, initially there was actually a potential for chaotic situations.

I received people from Plum Tree, from Mt Darwin, from Chimanimani, from Chinoyi. … spirit mediums coming there, all saying 'Well, I'm now possessed by the spirit that controlled this area, I am the spirit', not withstanding the fact that you were talking of Nemanwas, and then the Mugabes. Let me say that there are actually quite a number of other stakeholders not necessarily resident there, who used to come. I still remember one time there was a lady who, and it was about 4 am, she knocked at our door, and she came from Plum Tree and she said – she was speaking in languages that I could not understand. She had come all the way on foot, to come now to take over the site which truly belonged to the spirit that was possessing her. And we had a whole day of trying to cool her down, and so forth, and if we had allowed that situation, you can imagine the reaction it might have had on those who were in the immediate premises, and so forth. That is one scenario.

It has never died out, we are still involved in some of these serious situations, and we have always tried to say, 'No, there is a role for everyone.' To the extent that we have played that intermediary role, I think we have been able to keep the peace which exists. To the extent that we are now talking about a stakeholder group that we work with, and that works with us, means that we are bringing order to the whole scenario. It was definitely not that kind of scenario at that time, and it never could have been. And I can well imagine that we could have been caught up in a very tricky situation, 'oh yes, you are siding with so and so, … .' And that kind of thing. This was a scenario that we could not allow.

And as for the fact that you are saying that NMMZ has not handed over to the spiritual leadership … . Spiritual leadership does not need to be in control for it to be in charge. Spiritual leadership recognises the temporal arm. I'll give you an example which we draw from our own traditions. The chiefs are the temporal arm, anointed and blessed and also given the go ahead by the spiritual leadership. The two are not inseparable, but the two definitely know each others' rights. The fact that the chief is responsible for the land that is allocated under the traditional system, does not in itself deny that the basis of that land is spiritual leadership.

So what we can say, in actual fact, our physical presence, is for a purpose, and we have never pretended to take the spiritual dimension. Nor can we say that the spiritual dimension has purported to take our role as the physical custodian of a site … the way we want to see it is to realise that we each have a role to play.

I know of sites in other countries that move under their own momentum, and I can give you a very excellent example that we now have on the World Heritage List, Sukur Cultural Landscape in Nigeria … There you have this symbiotic relationship of the Hedi, who is the spiritual leader, resident on top, and the temporal arm, resident at the bottom as the chief. The two have delineated responsibilities but they all come together to give stability to the site. So the bottom line of what I'm saying is that even where a site or sites have actually been looked after without even … . an official presence – you do not even have a Monuments Commission of Nigeria on that side, because I think they have not yet set up. Anyway it's not necessary, but that relationship still exists. We do not have, in the case of Great Zimbabwe, that kind of scenario.

Why? It comes back to what I am saying. If for any reason, the experiment that we are having works, and we have the Mugabes, and the Nemanwas integrated, or working as a unit, then we can foresee a time when that role can be played – assuming there are not those who are not necessarily physically there, who lay a claim on the site.

You know, you should never, never, forget that dimension. In a site like Sukur … it's all close knit, it's known who is the traditional leader of that place from time immemorial. That kind of thing is not as clear when you consider Great Zimbabwe.

You are talking here of a site […] where you can talk of a local significance, and a national significance. Where you have a local significance – in fact most of the sites that fall under NMMZ – we would only pay a courtesy visit, just to make sure it's okay. We have no presence there at all because they are of local significance, and there is no conflict and so on. But when you are talking of a national site, it means exactly that. There are many players, and stakeholders, you can have them in structures, you see. Stakeholders at national level, and now you have stakeholders at international level, because it's a World Heritage Site. So it has to meet certain obligations and so forth. So you have international, national, regional and local, and then on the other hand you've got spiritual, temporal and so forth. (Interview with Dawson Munjeri, NMMZ, 11/5/2001)

This response sums up NMMZ's current position on its own role in the management of Great Zimbabwe as an 'intermediary' within a complex web of stakeholders that cross local, national and international interests and perspectives. It also, for our purposes here, illustrates how the world heritage 'system' leaves it to 'state parties' to determine what 'local community participation' might actually amount to, and in what circumstances it should be applied.

It may also be constructive to consider how the idea of 'cultural landscapes' could, in certain circumstances, actually increase the 'anti-politics' effect of world heritage by allowing the label 'heritage' (and indeed 'world heritage') to apply to new and different types of property, thereby encouraging the 'entrenchment and expansion of institutional state power' (Ferguson 1990: 256) into new areas. This can be illustrated with reference to the Matopos Hills in south-western Zimbabwe, where the various *Mwari* cult shrines are located (Daneel 1970, 1998; Ranger 1999, 2004b; Nyathi 2003). The complex history of these hills has been extensively described in Ranger's *Voices from the Rocks* (1999), the purpose of which was, as it states on the back cover, to 'reinstate culture into nature'. Ranger outlined with remarkable lucidity how these picturesque hills have been at the centre of a multiplicity of historical disputes, rivalries and conflicts, at the root of which lie different ways of 'seeing' the Matopos. As he put it in another paper, while Europeans variously appealed

> to the heroic memories of the campaigns in the Matopos in 1896, or to the sacred significance of the hills as the burial place of Rhodes, or to the romantic idea of preserving a specimen of 'the old Africa', or to the scientific idea of preventing soil erosion and silting, or to the conservationist idea of preserving species of fauna and flora. ... Africans variously appealed to their own heroic histories and to their ideals of utopian nature or to their own ideologies of conservation. To the Banyubi, the hills were an ancestral home; to the Ndebele they held the grave of Mzilikazi. To African cultivators, they offered not a wild but a domesticated environment in which their stock had opened up grazing, and where they could achieve good crops of vleis and sponges. These traditions of agriculture ran directly counter to European scientific conservationism. The hills were the site of the central shrines of the cult of Mwali, or Mlimo, with its own elaborated set of ecological observations and prescriptions. (Ranger 1989: 218)

Despite this complex, multiplicity of ways of 'seeing' the Matopos, in 1962 it was the conservationist appeals of white Rhodesians that were acted upon, and a specified area within the hills became a National Park, and the inhabitants evicted. Revealing a similar continuity in the movement from colonial to post-colonial state as I have described in relation to Great Zimbabwe, in 1983 an unsuccessful nomination for the Matopos Hills was prepared for inscription on the *World Heritage List* which again focused extensively and exclusively upon the 'natural' values of the area. This reflected the continued dominance of a conservationist view of the hills as 'natural heritage'.

With all the changes that had occurred in the world heritage 'system' and within NMMZ, by the late 1990s a new nomination for the Matopos Hills as a 'cultural landscape' was being prepared. In 1995, working groups at the *First Global Strategy Meeting for African Heritage* (held in Harare) recommended the Matopos as both a natural and cultural landscape, as 'living heritage', and as a religious site 'associated with rain-making ceremonies' (*Report of the First Global Strategy Meeting* 1995: 104–6). The final nomination dossier (for a summary see Nyathi 2003: 8, 11) was finally approved by the World Heritage Committee on 3 July 2003 (Ranger 2004: 228). But while it could be said that 'culture' has been 'reinstated into nature' – as more of the complex and multiple values associated with the Matopos have been formally recognised – at the same time, labelling the area as a 'world heritage' site may also amount, in effect, to a further appropriation of the hills by the state, to encompass not only the 'natural' aspects of the landscape, but also those other, frequently contested, features contained within the area, like the graves of Cecil Rhodes and the Ndebele king Mzilikazi, and, of course, the various *Mwari* cult shrines.[7]

This point is made all the more poignant given that the *Mwari* shrines in the Matopos – whose spiritual importance stretches far beyond the hills themselves, reaching into Masvingo province where I conducted my fieldwork, but also as far as Botswana and the Transvaal (Werbner 1989; Nthoi 1995) – have been largely beyond the control of national politicians. This is well illustrated by Ranger's account of what happened during the 'troubles' in Matabeleland in the late 1980s (Ranger 1999: 253–2). A quote cited by Ranger from a certain Mrs Lesabe also makes this point clear.

> The solution, my dear, is that the government should keep away from meddling with the shrine. All politicians out of the shrine matters. Chiefs, all other chiefs except those that are known to be connected with the shrine. There are certain families who are directly involved, from their great grandfathers or grandmothers, who have continued to do the tradition. But because government, politicians, chiefs are interfering too much these people have decided to fold their hands. (Mrs Lesabe, cited in Ranger 1999: 261)

Yet the possibility of increased 'state interference' was directly raised by the then Minister for Home Affairs, Dumiso Dabengwa, during an interview with Jocelyn Alexander and Joanne MacGregor in 1996, when he suggested that a keeper may be paid by NMMZ to look after shrines

> We've been talking to chiefs Bango and Malaba … asking them to submit suggestions about Njelele and Dula. We asked if they could provide us with a keeper who will be paid by the department of National Shrines and Monuments. It was a national monument and we want to keep it as such. It was once deproclaimed and we've since had it reproclaimed. (Cited in Ranger 1999: 289)

Given that 'the Matopos – and Njelele in particular – … are living culture in the sense that they are various, constantly changing, contradictory, often challenging or confrontational to states' (Ranger 1999: 286), Ranger may be right to express his cynicism and state that 'it is hard to imagine the priest at Njelele as a paid state functionary' (1999: 290). Nevertheless the prospect does illustrate how despite the increasing 'anthropological awareness' of the world heritage 'system', and particularly the 'world heritage' criteria and management requirements, it still

> can end up performing extremely sensitive operations involving the entrenchment and expansion of institutional state power almost invisibly, under cover of a neutral, technical mission to which no one can object. (Ferguson 1990: 256)

LOCAL PARTICIPATION OR CO-OPTATION AT
GREAT ZIMBABWE?

Since the mid-1990s the issue of local community participation has been taken increasingly seriously by NMMZ.[8] Does this represent the *prescriptive* 'power of world heritage' in action, influencing ideas and approaches to heritage management at state level, or, alternatively, that the world heritage 'system' was itself influenced by the experiences of State Parties? It is clear that this change came about within a wider context of increasing international recognition of the importance of local community involvement in any 'top-down' intervention, be it 'development' or 'heritage management'. In a memo to the Board of Trustees, Dawson Munjeri appealed to the 'global' nature of the issue in his request for contributions towards a NMMZ policy on the issue.

> The board has already accepted in principle the idea of local community participation ... in fact so global now is the issue that during the last week of October and beginning of November 1996 the International Council of Museums and Ethnography (ICME) is holding a special conference in Bhopaal (India) to address this issue. (Memo from executive director Dawson Munjeri, to the Board of Trustees, 22 Oct 1996. ref AD.A6/DM/sn NMMZ File N1(a))

This policy does not mean that NMMZ will in the near future 'hand over' control to local communities, even if they were to resolve their disputes. As Matenga put it (1998: 12), 'returning custody of Great Zimbabwe to either Nemanwa or Mugabe is out of the question'. Rather NMMZ increasingly sees itself as an intermediary within the complex web of local, national and international stakeholders. There is a sense in which NMMZ is attempting to broaden the recognition of the values associated with the site, to include the 'local sacred values', alongside the more established national and international 'heritage' values – as a monument to past African achievement for example – for which the site gave its name to the new country, and for which it was originally inscribed upon the World Heritage List. As Munjeri put it, it is not about 'displacement' but rather 'synchronisation'.

> *Yes that's right. I think it is in the interest of all. This is the most important thing of all ... erm it also brings a better understanding of the values, that because the Nemanwas have their own values that they attach, and the Mugabes have got their values that they attach ... and those values are very valuable for the site itself, it only reinforces those values. And to the extent that we can benefit, and everyone can benefit from combined values, I think this is important. It should never be a situation of displacement, but a situation of synchronising.* (Interview with Dawson Munjeri, 11/5/2001)

Recognising Great Zimbabwe as a 'shrine' or 'sacred site' increases its 'heritage value', especially in the context of the World Heritage 'system's' adoption of concepts such as 'living' and 'spiritual heritage'. Of course NMMZ are not able by themselves to run Great Zimbabwe as a 'shrine'; for that they need local communities. And this has been at the centre of NMMZ's 'dilemma' as Matenga put it.

> *If the local people are not involved ... its only the local people who can effectively run the place as a shrine, but if they are coming here to fight, then it's not necessary at all. You see? So this is our dilemma. You see the point? That okay, it adds value to the site if it is regarded, both in theory and in practice, as a shrine, but now if this means that people are coming here to fight, then there is a problem.*

Now the shift is coming because, what I've done is, this is what I thought, my initiative, ... I've always wanted to encourage the local community to work, to co-operate together, and use the site as a shrine, and therefore to add ... to give it its deserved value, but now if they are coming here to fight, it means we have to put a stop. (Interview with Edward Matenga, 31/10/2000)

Given the situation that arose with Ambuya Sophia Muchini residing at Great Zimbabwe after independence, and the dispute or 'fracas' that flared up between the Nemanwa and Mugabe clans at a ceremony in 1984, NMMZ needed to first establish, irrefutably, its own authority as the 'temporal' custodian of Great Zimbabwe, before 'officially' recognising it as a sacred site and turning again to local communities, once the disputes had 'cooled down' as Munjeri put it (Interview with D. Munjeri, 11/5/2001). NMMZ have therefore succeeded in 'bringing order to the whole scenario'.

To the extent that we have played an intermediary role, I think we have been able to keep the peace which exists. To the extent that we are now talking of a stakeholder group that we work with, and that works with us means we are bringing order to the whole scenario. It was definitely not that kind of scenario at that time. It could never have been. And I can well imagine we would have been caught up in a very tricky situation. (Interview with Dawson Munjeri, 11/5/2001)

Thus, having established their authority, NMMZ are now able to take a 'stakeholder approach' and occasionally entertain ideas of ceremonies at Great Zimbabwe, which they could not have done in the early 1980s. However, the Chisikana ceremony of July 2000 was unique. It was the first officially sanctioned ceremony held at Great Zimbabwe since the early 1980s, and involved the participation of both Nemanwa and Mugabe. It was also initiated and organised by NMMZ itself. Beyond merely tolerating and permitting ceremonies, NMMZ has emerged as a potential sponsor for ceremonies at the site, whilst it also maintains its veto on ceremonies of which it may disapprove, as demonstrated by its rejection of Chief Mugabe's request to hold a ceremony with war veterans in October 2000. Coupled with the indications that a 'national' ceremony is being considered, involving the reburial of remains taken from mass graves in Mozambique and Zambia,[9] alongside the two well publicised state events (May 2003 and May 2004) marking the re-unification and 'return' of one Zimbabwe Birds, there is therefore a strong impression that NMMZ is becoming increasingly open to the possibilities of holding both 'local' and 'national' ceremonies at Great Zimbabwe. This is acceptable as long as it remains in full control of such occasions and the management of the site in general.

But if NMMZ's concern with involving local communities in the management of Great Zimbabwe springs from a desire for the site's 'sacred' and 'spiritual' values to be recognised, represented, and maintained, along with the interests of all the other stakeholders whose 'values' must also be upheld, then does 'local participation' not ultimately amount to the 'co-optation' of locals into the larger 'heritage project' of NMMZ, or indeed that of the World Heritage 'system'? As Chief Mugabe himself put it, 'they are just respecting us in terms of when they want some ceremony to be done, but all in all it is now becoming their own particular place' (Interview with Chief Mugabe, 20/11/2000). This point about 'co-optation' was further illustrated by Edward Matenga's comment, made as we passed some re-constructed walling during our walk around the site one Sunday in July 2001, that 'from the NMMZ point of view, it is important that we can persuade people so that our proposals or plans are accepted, and indeed happen' (Field notes, 15 July 2001). Then he could 'envisage ... a short bira being held before such a project, and again ... when it is finished, as an opening'.

On 20 September 2000, after a long process dating back to an initial project proposal in 1996 (NMMZ file H2), the new display at the Great Zimbabwe site museum was officially opened by the guest of honour, the Minister of Home Affairs, John Nkomo. There was an extensive guest list of local, provincial and national dignitaries, which included Chiefs Charumbira, Nemanwa and Mugabe. In his speech to introduce the guest of honour, the chairman of the local committee of the Board of Trustees, Mr Makonese stated:

> Today I wish to stress the excellent relationship between National Museums and Monuments and the local communities; we feel that these local communities are vital stakeholders in the exhibition and presentation of our history and culture and their active participation is central to the success of Great Zimbabwe. This close relationship was recently demonstrated in the collaboration between staff at Great Zimbabwe and the Nemanwa, Mugabe and Charumbira Communities at the opening of the sacred spring which has great historic and symbolic value. This spring is close by to the south of the museum; we are proud of our 'living history' site and it is brought alive by the local communities. This ownership and involvement of the local communities is also illustrated through the commitment by the local Nemanwa Primary School to care for and look after the Great Zimbabwe site.

> The exhibitions that you see here today have been born out of a process of extensive consultation between staff of Great Zimbabwe and the local communities. It is their history and they have played a key role in its interpretation; it is a presentation by them, about them, and for them. (Mr Makonese, Chairman of the Local Committee of the Board of Trustee, NMMZ. Speech at Opening of the Great Zimbabwe Site museum, 20 September 2000, NMMZ file J20)

The opening of the new display was a good opportunity to make a strong public statement about NMMZ's commitment to local communities. It is therefore startling to note that while there were some substantial changes in the new display – most importantly the emphasis upon Great Zimbabwe's continued occupation since the fifteenth century (rather than a stress on its abandonment in 1450), and some limited references to its sacred value for the Nemanwa and Mugabe clans, as well as for the 'nation' as a whole – there is no evidence to suggest that the new displays were in fact 'born out of a process of extensive consultation' with local communities. Apart from brief references to periods of Nemanwa and Mugabe occupation of the site, the oral traditions from these and other local clans are entirely absent.[10] There is no mention of the sounds or the *Voice* that used to be heard amongst the walls, or the importance of the spring, or the underground caves and *nhare*. Nor are the history-scapes of these local clans discussed; there is no mention of the Nemanwa/Mugabe battles, or the origin of their *zvidawo*, or how Charumbira came into the area as an elephant hunter. The displays and the text follow a much more conventional historical and archaeological pattern, outlining the chronology of the site, focusing on the archaeological remains and the technological achievements of the people who inhabited the site between 1150 and 1450, rather than the attachments and values of local clans to the site. Great Zimbabwe is presented, to paraphrase Peter Ucko (2000: 80), as an entirely 'static' and 'dead' site. Even *Pfuko ya Kuvanji*, the mysterious walking pot, is displayed, thoroughly de-contextualised. No mention is made of the rich stories that have been constructed around it. In sum, as Ndoro has put it,

> The presentation and interpretation in the museum displays and the guidebook is derived from archaeological sources. No mention or reference is made to 'myths, legends, oral histories or folklore' related to the monument. Yet … it is clear that there are many legends and oral traditions pertaining to the monument of Great Zimbabwe. The official presentation and interpretation mainly focuses on the site as a relic with no relevance to today's socio-economic or cultural environment. The monument is presented as a bygone era. (Ndoro 2001: 112)

Rather than having been 'born out of a process of extensive consultation' with local communities, as Mr Makonese claimed, the displays and the texts were put together by three NMMZ archaeologists: the curators, Dr Caroline Thorpe and Ashton Sinamai, and the Chief Monuments Inspector K.T. Chipunza, who was the project manager (NMMZ file H2). On 30 June 1999 a seminar was held at Great Zimbabwe 'to critically examine the text that will form the basis of the captions, labels and explanatory notes for the exhibits' (Matenga, Memo, 28 June 1999, NMMZ File H2). The motivation for this meeting stemmed from the realisation 'that Great Zimbabwe has a complex historiography and our desire is to consult widely on this important responsibility to present the site to the public' (Matenga, Memo, 28 June 1999, NMMZ File H2). But this 'desire to consult widely' did not encompass any elders or *masvikiro* from local clans; instead it brought together more 'professionals' – archaeologists from the University of Zimbabwe, members of the Conservation Department at Great Zimbabwe, several NMMZ display artists – to compliment the three NMMZ archaeologists working on the project. Contrary therefore to Mr Makonese's very public announcement that 'it is a presentation by them, about them, and for them', the museum display was not created by them, about them or even for them. All the texts are in English, which, as Ndoro pointed out (2001: 115), means 'that 70% of the African community [across the country] is alienated from the site'.

In the case of the new museum display therefore, NMMZ's rhetoric about 'local community participation' does seem particularly empty. In defence of NMMZ, I should note that during our interview Munjeri acknowledged that the display in the museum was limited to a representation of Great Zimbabwe as an archaeological site – a 'freeze situation' – and that this was not entirely satisfactory. He argued that it is perhaps an exaggeration, given its size, to call it a museum, and he suggested that 'site interpretative centre' would be more appropriate.

> *The limitations currently, as a site interpretative centre, you are perfectly right, it is trying to make the visitor understand the historical processes that led to Great Zimbabwe, OK. I don't think there had been much scope, within the confines of that museum, for anything more.* (Interview with Dawson Munjeri, 11/5/2001)

For the future, he assured me, a visitor centre is envisaged which would allow a more comprehensive representation of the multiplicity of perspectives and interpretations of Great Zimbabwe's past and present importance.

But apart from the limitations of space, there is also the continued dominance of a 'professional' archaeological view of Great Zimbabwe, which reflects the dominance of

Figure 9.1: The view from the old Shona village at Great Zimbabwe, looking west over the valley ruins. (Author 2001)

that discipline in NMMZ itself. It seems 'the untenable dichotomy between a remote archaeological past and a more accessible non-archaeological one' has continued 'to be fostered by the NMMZ', maintaining archaeology as an isolated 'practice of interest and relevance to only a small elite' (Ucko 1994: 274–5). To reverse this trend, and for the aspirations of genuine and meaningful consultation and participation by local communities to be achieved, seems to require NMMZ to 'de-professionalise', or alternatively, for what Robert Chambers (1997) has called a 'new professionalism' (1997: 228). An emerging body of literature[11] which explores what an 'engaged' anthropology, and indeed archaeology, might look like, suggests that fruitful, collaborative 'heritage' projects between anthropologists, archaeologists and local, 'indigenous' peoples are possible projects which 'feature a vision of reciprocity in academic research' (Clifford 2004: 21) even as they avoid the dangers of ethnic and religious politicisation (e.g. Ratnagar 2004), and of 'free floating' commercialism (Ucko 2000). And in this respect, the widely canvassed but ultimately, abandoned idea of Zimbabwe 'culture houses' which circulated in the early 1980s, was an early opportunity for effective engagement with communities in the representation and management of 'tradition', heritage and the past. While, with the exception of the Murewa Culture house (see Ucko 1994: 244–56), this idea has since been largely neglected across Zimbabwe, at Great Zimbabwe something like a 'culture house' has existed for a long time.

One of NMMZ's most profound, and long-running efforts to involve local communities at Great Zimbabwe is the Shona Village – a 'virtual culture house' (Ucko 1994: 276) – the origin of which can be traced back to the efforts of Lillian Hodges, who was director at the site between 1959 and 1973. At that time it was called the 'Karanga village' and was located near to the site museum, at the centre of the estate. Ucko (1994: 276) describes how, despite being 'embedded in expatriate intiative and toursim', this village 'became significant to the local population since, when the tourists were not present, the two "employees" lived in traditional ways, and local people began to use the village, carrying out certain rituals, and, before entering, making the right obeisances'.[12] During the war it was burnt down, and eventually, in 1988, a new village was built at a larger site on the eastern side of the central ruins area (see figure 9.1) to depict 'the way of life of a traditional Shona family unit' of the nineteenth century (NMMZ pamphlet 'The Shona Village', NMMZ file H4). It involved far more activities than the previous one (which was smaller, and consisted mainly of static exhibits), including demonstrations of traditional pottery making, a *n'anga* (traditional healer), a blacksmith, carvers and two groups of performing musicians and dancers. Visitors were to learn about and participate in these activities.

It has proven to be very popular with visitors. When it was closed to the public in February 1997 due to the advanced state of deterioration of the buildings and the exhibits, there was

> a public outcry and a petition to have the decision reversed was made through agents of the mainstream political party ZANU PF. Eventually NMMZ obliged and opened the village in January 1998. (E. Matenga, undated 'Management of the Shona Village' NMMZ file H4)

The Shona Village has also become a pillar of NMMZ's efforts to involve local communities at Great Zimbabwe. Nearly all the activities are run by members of local communities, who are technically tenants of the site.[13] While it provides them an opportunity to generate income from tourists, they are also required to pay NMMZ a combined monthly rent,[14] and half the gate takings. It is often stressed that 'although a curator has been assigned responsibility for the village, this appears to be largely advisory rather than managerial' and 'there is minimal involvement of the NMMZ management' (Pwiti and Ndoro 2000: 1) in the village. However, NMMZ actually keeps fairly tight control on what takes place in the village. All the village performers and occupants are required to sign a contract with NMMZ, which enforces a strict code of conduct, allows only very limited leave, and only one day of notice in event of discharge. During my period of fieldwork, I became closely acquainted with many occupants of the village, and there is no sense in which they experience the kind of control of the village that NMMZ often ascribes to them in their literature. I was told on various occasions that they had been forbidden to discuss the history of Great Zimbabwe with visitors (presumably because of a fear that they would mis-represent it).

In November 1994 Makwachata, the resident *n'anga* in the village dating back to Lillian Hodges's time, was discharged following a claim, attributed to him, that he could cure AIDS, which appeared in a local newspaper. NMMZ would not tolerate such an 'unauthorised press release'.

> Attached please find a copy of the leading news item of the Masvingo Mirror of October 28, 1994. The article is a statement attributed to Mr B. Makwachata, the n'anga presently 'practising' at the shona village. This is the second unauthorised press release by Mr Makwachata, the first one having been published on May 27 1990 by the Sunday News. I have in the mean time requested Mr Makwachata to leave Great Zimbabwe. (Memo from Regional Director to Executive Director NMMZ, ref H$:GMwcm, 1 November 1994, NMMZ files H4)

Similarly, the dismissal of Simon Nyanda (a wood carver/sculptor) in 1995 related to the nature of his sculptures and art work which were causing offence to visitors, as was relayed in a memo from the administrative officer for tourism to the regional director in April 1995.

> You will remember that sometime in January this year we spoke to Mr Simon Nyanda the sculptor in the shona village about the type of art he is currently doing. The art, so to speak, is out of context with the objectives of the shona village and the reasons for which the sculptor was contracted to work there. That discussion resulted in some of his work being removed from the village to the conservation centre for storage.

> Mr Nyanda promised us that he will begin to work on items that conform to our agreement. However I notice with disappointment that since then the sculptor has continued with his 'compositions' much to the chagrin from some of our visitors to the village who find his art and explanations confusing and embarrassing.

> It is on the basis of the above observations and complaints from our visitors that I recommend that Mr Nyanda's contract be terminated immediately. He should remove all his material from the village by 13/4/95. The other material at the Conservation centre can be removed after the holiday. National Museums and Monuments of Zimbabwe will look for another sculptor to replace him. (Memo from Administrative Officer for Tourism to the Regional Director, Southern, 12 April 1995, NMMZ file H4)

Clearly NMMZ does keep a fairly close eye on what happens at the village, which undermines their own claim that it is self-managed, and, by extension, their use of it as an example of 'community participation'. The Traditional Village Re-location Project, which was being implemented towards the end of 2001, was framed even more explicitly in terms of 'local community participation', but my own experience indicates that the beneficiaries, the village performers and occupants themselves, have not experienced any significant sense of empowerment from this project so far.

There was a gathering momentum from 1995 for the re-location of the Shona village to a new site within the vicinity of Great Zimbabwe (see e.g. Ndoro and Pwiti 1997). The motivation for this was partly drawn from the increasingly deteriorating condition of the village and its exhibits. However, the main reason was a belief that its position so close to the central part of the ruins was inappropriate, and the cause of a fundamental mis-interpretation amongst visitors, many of whom thought it represented a village at the time of Great Zimbabwe's original construction and occupation, rather than a nineteenth-century village (Ndoro and Pwiti 1997: 6, Pwiti and Ndoro 2000: 5, see also Ucko 2000: 79–80). It was decided that the re-located village should be called the 'Traditional Village', 'to avoid unnecessary freezing of The Village to a specific time period and thus place restrictions on the activities' (Pwiti and Ndoro 2000: 4). It is hoped that it will become a tourist attraction in its own right, to some degree independent from Great Zimbabwe, and that it might then fulfil more of its potential as a generator of income. It was also taken as an opportunity to 'strengthen the partnership' with the local community (Pwiti and Ndoro 2000: 4), and it was felt that 'the community must assume more responsibility in terms of looking after the place and increasing the activities' (Pwiti and Ndoro 2000: 4). In December 2001 (when I finished the main period of fieldwork for this book), it was already nearing the final stages of completion.

While it is, perhaps, still too early to judge the success of this project, it is clear that despite NMMZ's intentions that the project should 'strengthen the partnership' with local communities, some doubts remain about the extent to which the planning process for the project itself included, to any substantial degree, any meaningful consultation with the occupants themselves. Instead of consulting directly with local communities and the village occupants, NMMZ contracted out the development of the project to two archaeological consultants, Dr Pwiti and Dr Ndoro, from the University of Zimbabwe (Pwiti and Ndoro 2000). They conducted an 'ethnographic survey' among local communities in order to gather information for the design of the village, and a very limited 'community survey' using questionnaires among selected rural and urban areas in Masvingo district to find out the ' "communities" attitudes to the proposed re-location of the existing village' (Pwiti and Ndoro 2000: 33). They also had various meetings with the Shona Village occupants and performers themselves.

The decision had already been taken to move the village to a new site. Apart from concerns about some aspects of the plan for the new village – such as the suggestion that the village occupants should offer overnight accommodation for visitors, and on those nights they should stay at the village rather than to return to their own families – I got the sense that most of the occupants had no problems with the designs for the new village. Indeed once the actual building work began, which they themselves were contracted to do, more enthusiasm about the project was generated. There was approval for the 'traditional' techniques and materials that were to be used. However, most of the Shona village performers and tenants that I spoke to were not in favour of re-locating the village at all. They liked the old site, amongst the ruins, where they could see visitors coming from afar (see figure 9.1), and feared that the re-location of the village to a site much further away from the centre of the ruins would cause the number of visitors to the village to dwindle. The two groups of performers who used to draw into the village visitors they had spotted walking amongst the ruins, through the sounds of their singing and dancing were very unhappy. Aware that they would not be able to persuade NMMZ to keep the current site of the village, the present tenants submitted a request for an alternative site, near the main entrance road into the estate. It was turned down, and the new village was sited far away from the main area of the monument, behind the hill, alongside the Kyle Dam Road, and close to the Conservation Centre. It is interesting to note that the very reasons they preferred the site near the entrance gate into the monument – it is highly visible and likely to attract the attention of visitors – were the same reasons why for NMMZ this site was unacceptable.

> A problem with this location is that the site is so close to the main point of entry into the monument by road that it will be the first thing to catch the eye of the visitor and thus is likely to generate even worse confusion than is the case presently. The second point is that it is too close to staff houses for the comfort of its residents. (Memo from E. Matenga to executve director, 31March 2000, ref H/4:EM/wcm, NMMZ file H4)

As the consultants themselves put it,

> … it is the visibility which creates problems. It is right in front of the Great Zimbabwe Monuments entrance. It is our view that should this site be preferred, then NMMZ might as well maintain the present site. This site does not address the reasons for the need to relocate in the first place. The village at this location will be the first attraction any visitor coming to Great Zimbabwe would encounter. It would present new problems of presenting the monument itself. For these reasons we do not recommend this location. (Pwiti and Ndoro 2000: 9)

While NMMZ desired to separate the village from the monument, so as to clarify and solidify their interpretation of Great Zimbabwe in the minds of visitors (and thereby maintaining 'the untenable dichotomy between a remote archaeological past and a more accessible non-archaeological one' – Ucko 1994: 274–5) for the occupants of the village the high visibility of the old village, and its close proximity to the monument was key to its success. Aware of their dependence upon NMMZ for their positions, and the income which it generates, there was a sense in 2001 that they had resigned themselves to the wishes of NMMZ, and inevitably co-operated with the re-location of the village to the site dictated by the consultants from the University of Zimbabwe. It remains to be seen whether this project has been a success, and whether the village occupants have grown to appreciate their new village. Given the circumstances of the planning process, however, it seems unlikely that they will have any greater sense of inclusion, ownership or 'partnership' with NMMZ , than they had with the old village. In a sense, therefore, they were successfully 'co-opted' into a NMMZ project rather than meaningfully consulted.

During a return visit to the area in July 2004, it was apparent that some of the performers' fears about dwindling visitors had been realised, although this was undoubtedly also related to the effects of wider economic, social and political problems.[15] Whatever the cause, the result was that one group of musicians and dancers, the Nezvigaro *Ngoma* Group (see figure 9.2), no longer performs at the new village, having decided that the rewards were not worth the effort.[16] The other performers, the Great Zimbabwe *Mbira* group (see figure 9.3), only returned after a long, deliberate absence, because NMMZ insisted they turn up, despite the lack of visitors, or forfeit their right to play at the village (Fieldnotes, 21 July 2004). This is hardly an encouraging sign of a 'strengthened partnership' with the local communities. Furthermore, for a project that was partly intended to increase the depth of 'local community participation' in the interests of emphasising the 'living' nature of the site, it is ironic that it has caused the removal to the periphery of the estate those very aspects that did create a 'living atmosphere' amongst the stone walls: the sounds of drums, ululation, singing and

Figure 9.2: The Nezvigaro *Ngoma* group, performing at the old Shona Village. (Author 2001)

dancing which used to reverberate out across ruins from the old Shona village, enticing visitors to draw near. During 2000–01 it often occurred to me, as I strolled amongst the stone walls, that at least the silence of Great Zimbabwe – the absence of the *Voice*, and the sounds that local 'traditionalists' spoke of – was being filled by the 'traditional sounds' of *mbira*, drums and singing from the performers in the Shona village. In contrast, when I returned in July 2004 'the rocks and towering walls seemed to have succumbed to an all prevailing and almost oppressive silence' (Fieldnotes, 21 July 2004). Indeed it appears that the silence of Great Zimbabwe may now be completed, and the landscape finally 'frozen' 'as a palimpsest of past activity' (Bender 1998: 26).

To be fair to NMMZ, it is still, perhaps, early days since the shift in NMMZ's approach in the mid-1990s. As the deputy executive director, Dr Mahachi said in relation to the Chisikana ceremony of July 2000,

> *You know, it would have been unthinkable only two years ago, for NMMZ to actually pay for a beast, for NMMZ to commission women to brew beer, and invite the local community to do this at Great Zimbabwe, but that happened.* (Interview with Dr G. Mahachi, 15/5/2001)

Figure 9.3: The Great Zimbabwe *Mbira* group, performing at the old Shona Village. (Author 2001)

And this has clearly contributed to a growing sense of optimism, particularly among some members of the Nemanwa clan, that things are changing for the better. This is partly a result of the efforts of Edward Matenga, under whose directorship the changing policy was put into practise. His own approach seems to have been crucial in changing the relationship between local communities and NMMZ. During our interview Dr Mahachi confirmed this.

> *J.F.: I was wondering that myself; how much has this change been partly a result of Matenga himself, or would it have happened anyway?*
>
> *Dr Mahachi: Erm, no, I wouldn't say it would have happened anyway, but there was one requirement: to have that appreciation. That appreciation of the need to work with the local people, it's quite important. For instance if I am director at Great Zimbabwe, and I just moved from my home, to my office, from my office into town, I drink in the Chevron hotel, whatever, you know there is really no interface between the office and the local community, such a thing would be difficult to do.*
>
> *But if you are the type of person who goes to attend funerals in the local community, who talks with people, you break down some of these boundaries, you begin to talk about Great Zimbabwe itself, informally.*
>
> *J.F.: I suppose if you interact personally, you are no longer 'the director', you are Edward Matenga …*
>
> *Dr. Mahachi: That's the critical thing, I believe, and I think he has done that quite well.* (Interview with Dr G. Mahachi, 15/5/2001)

This 'breaking down of boundaries' is the note of optimism with which I want to leave the reader. There is no doubt in my mind that many members of NMMZ do take the issue of 'local community participation' seriously; the desire for meaningful consultation with all stakeholders is, in my opinion, genuine. But for this genuine desire to be reflected in the experience of local stakeholders, and not just NMMZ's rhetoric, it is important for NMMZ to accept that it may have to loosen its control over, and even 'deprofessionalise' its own approach to the management and representation of Great Zimbabwe, to allow space for the effective inclusion of other perspectives on its past and its management. As Pikirayi commented (my notes, 8/7/2004) in the 'Final Plenary Session' of a recent conference on heritage in southern and eastern Africa,[17] heritage managers need to take a more proactive approach to engagement with local communities. In the next, and final chapter, I will explore to what extent this is possible in the context of the recent emergence in Zimbabwe of a new historical discourse of 'authoritarian nationalism' (Raftopoulos 2003) which Ranger (2004) has called 'patriotic history'.

Notes

1. This a fundamental criterion for inscription to the *World Heritage List* – see articles 1 and 2 of the *Convention* (UNESCO 1972) – which itself exemplifies an 'institutionalised evocation of a common humanity' (Maalki 1994: 41).

2. As he put it 'the Convention's effectiveness is never as strong as it is during the preliminary investigation of the nominations submitted by the States with the intention of including a property in the World Heritage List' (Léon Pressouyre 1996 (1993): 48).

3. This was approved by the World Heritage Committee at its 22nd meeting in December 1998.

4. The World Heritage Fund established by the *Convention* in 1972, mainly receives its income from the compulsory contributions of State Parties, which amounts to about 1 per cent of their UNESCO dues. The funds available are relatively small, in comparison to other UN bodies. In 1998 the Committee approved a budget of US$5,426,000 for 1999 to be spent on international assistance for the conservation of heritage sites, preparing nominations, technical and emergency assistance, training programmes, periodic reporting and monitoring, as well as the operation of world heritage programmes such as the Global Strategy. Requests for financial assistance are considered and granted by either the Chairman of the Committee, or the Bureau, or the Committee as a whole, depending on the amount involved.

5. Resolution 24, *Implementation of the 1972 Convention for the protection of the World Cultural and Natural Heritage*, Records of Twenty-ninth Session of the General Conference of UNESCO, Paris 21 October–12 November 1997.

6. The World Heritage Global Strategy Natural and Cultural Heritage Expert Meeting, 25–29 March 1998, in Amsterdam, discussed whether the natural and cultural criteria in the *Operational Guidelines* should be unified into one set of criteria (UNESCO Document WHC-98/conf.203/inf.7). Such discussions continue, but no recommendations have yet been endorsed by the Committee. Herman Van Hoff (Interview notes, 28/6/1999) suggested that such innovations are often too fast for the Committee, who do not always understand some of the issues involved. This illustrates how change often takes longer in the 'system' than in the wider 'discourse'.

7. Indeed, for some this is seen as imperative. In a recent paper, Jackson Ndlovu (2003) of the National Museum in Bulawayo, made a powerful and spirited call for

the value and meaning of Entumbane – the grave of Mzilikazi – to be reflected upon, and interpreted for the nation and indeed the world, so that Mzilikazi, the 'father of the Ndebele nation', can take his due place as a 'Zimbabwean hero'. In his view, a distinction must be made between Mzilikazi's importance to his Khumalo clan, who were responsible for carrying out the original family rituals and the *umbuyiso* ceremony, and national observances and commemoration for which, in his opinion, NMMZ is 'the best placed institution' (2003: 4). I am grateful to Terence Ranger for providing me with a copy of this paper.

8. This is reflected in both an emerging academic discourse in Zimbabwe on the subject (Mahachi and Ndoro 1996; Ndoro 1996; Pwiti 1996; Pwiti and Ndoro 1999; Ndoro 2001; Ndoro and Pwiti 2001), as well as NMMZ efforts at other sites such as Domboshava (Pwiti and Mvenge 1996) and the Matopos (Munjeri et al. 1995, Ranger 1999 and 2004: 228).

9. This is an issue that has become increasingly salient across Zimbabwe in recent years, and probably relates to the growing dissent of war veterans in the late 1990s – which provoked increasing government concessions from 1997 onwards (see Kriger 2003; also Alexander 2003) – and also, more recently, to the emergence of what Ranger has called 'patriotic history' (2004). The issue has also taken on a regional dimension. During an International Conference on Heritage Management in Southern and Eastern Africa, convened in Livingston, Zambia in July 2004, the former executive director of the National Heritage Conservation Commission, N.M. Katanekwa delivered a paper that discussed SADC-wide efforts to co-manage 'joint freedom war heritage'; particularly the ongoing efforts of Zimbabwe and Zambia (Katanekwa 2004). He later confirmed (pers. comm.) that official delegations of chiefs, spirit mediums and others had come to *chimurenga* sites in Zambia to carry out certain rituals to appease and settle the spirits of the war dead. It was also a topic of conversation among spirit mediums in Masvingo Province during 2000–01, several of whom, like Ambuya VaZarira, expressed concern about the ritual procedures employed, and in particular, the presence of Christian priests (Interview, 16/8/2001). More recently, a 'cultural gala' was held at Chimioi in Mozambique to commemorate the Zimbabwean guerrilla fighters who were killed there during the liberation struggle (*Mail and Guardian*, 25/10/2004).

10. Only on one display panel are oral traditions and genealogies referred to; they are not those of local clans but of the ancestor Mutota of the Munhumutapa period from northern Zimbabwe. This reinforces the conventional view that local communities have no historical attachment to Great Zimbabwe. During our walk around the site, Matenga explained that this panel had been at the centre of some minor disagreements within NMMZ. Some felt that it was not appropriate at Great Zimbabwe and should really be placed in a museum in the north of Zimbabwe, as it relates to oral traditions of that region.

11. For some good recent examples relating to heritage projects in Alaska, see Clifford (2004) and Fienup-Riordan (1996 and 2000).

12. Illustrating the often fragile nature of relationships between heritage managers and local communities, but also how the village was thought of as a 'a good place to keep old treasured things', Ucko (1994: 261–2) mentions that after 'the

departure of a trusted curator', people requested the return of previously donated objects and family heirlooms.

13.   In the previous, much smaller, Karanga village, the *n'anga* and one female potter were paid full time salaries (Ucko 1994: 276).

14.   In 2000–01 it was Z$500, a nominal rent in relation to the cost of its upkeep, although for the Shona Village occupants this can amount to a great deal, in times when there has been a decline in visitors. In June 2000, they requested and received a 50 per cent reduction in the rent due because of the significant decline in visitor numbers related to international press coverage concerning the political situation in Zimbabwe. As Matenga put it (memo to Executive director, NMMZ, 6 June 2000, NMMZ file H4), 'we consider a reduction as a gesture of goodwill to the community'.

15.   Since 2000, declining wages, huge inflation, fuel shortages, combined with fears about political violence seem to have conspired, with drastic effect, on both internal and international tourism across the country.

16.   The members of Nezvigaro group live much further away (near Mutevedzi under Charumbira) from Great Zimbabwe than the *mbira* group, so it is no surprise they were first to stop. The new location probably put an additional half an hour of walking time on their journey to the village – an extra hour of walking everyday. They have, apparently, been replaced by another group, who also perform *ngoma* music, and from Charumbira (Field notes, 21 July 2004).

17.   Conference entitled 'Heritage in Southern and Eastern Africa: Imaginng and Marketing Public Culture and History', convened by the National Heritage Conservation Commission, and the National Museums Board of Zambia, the British Institute in Eastern Africa and the *Journal of Southern African Studies*, in Livingston, Zambia, 5–8 July 2004.

CHAPTER 10

BEYOND 2000: PATRIOTIC HISTORY AND GREAT ZIMBABWE

The stories of Great Zimbabwe that have been traced in this book illustrate how place and landscape are 'always in movement and always becoming' (Bender 2001: 4). The meanings and significance attributed to place and landscape are constantly being re-shaped, re-defined and re-negotiated within the changing spatial and historical context of wider discourses, struggles and contestations. Pre-colonial Great Zimbabwe was a sacred site that lay at the centre of disputes among local clans over superiority and land ownership. Colonialism saw Great Zimbabwe alienated from these local clans, and appropriated to provide historical and moral justification for the imperial project. It became the centre of a different contest, known as the *Zimbabwe Controversy*, between 'professional' archaeologists and 'amateur' Rhodesian 'antiquarians' over the authority to investigate and represent the past. Meanwhile the site itself acquired new meaning as a national monument and a tourist destination, which made not only the appropriation of Great Zimbabwe but also its desecration almost inevitable. Great Zimbabwe became silent; local people's responses to its past and its value were not heard, and the *Voice* and other mysterious sounds disappeared.

The rise of nationalism gave Great Zimbabwe new meaning as evidence of past African achievement, and future aspirations for the imagination of a new nation. It became a 'useful rallying point' for some, and emboldened as a national sacred site for others. Riding a wave of post-independence optimism and euphoria came calls for Great Zimbabwe to be re-claimed for the newly (re)created nation of Zimbabwe. New pasts emerged and the site itself came alive with the efforts of 'traditionalists', such as Ambuya Sophia Muchini, to assert Great Zimbabwe's national sacredness. In this same context local clans revived their claims to the custodianship of the site, disputing not only amongst themselves, but also with other 'traditionalists' from across the country. Amid this plethora of claims and interests, archaeologists re-asserted their authority to represent Great Zimbabwe's past by dismissing the new histories as 'romantic', and based on unreliable evidence. National Museums and Monuments of Zimbabwe sought to establish its own authority, with international assistance from UNESCO, as a thoroughly 'professional' and 'modern' heritage organisation. Becoming a World Heritage Site, Great Zimbabwe was invested with new meaning and significance, which both elevated its value to a new level, and contributed to the continued alienation of the site from both local communities and 'traditionalists' across the country.

In this ongoing 'movement' and 'becoming' of Great Zimbabwe, the Chisikana ceremony of July 2000 can be recognised as a key moment in the growing international and national recognition of the site's value as 'living' and 'spiritual' heritage, and in the development of new approaches to its management. Although I have sought to qualify the degree to which that and subsequent efforts by NMMZ amount to a move towards meaningful consultation with local stakeholders, its significance in terms of the 'breaking down of boundaries' between NMMZ and surrounding communities should not understated. The Chisikana ceremony was, without doubt an important, albeit long overdue, step for NMMZ to take, and it should be acknowledged as such.

In a much earlier version of the ending for this book, I sought to leave the reader with a sense of the optimism created by this event in 2000; with the possibility that there may be a chance that the *silence* of Great Zimbabwe, in all its manifestations, could be successfully filled (Fontein 2003a). However, the year 2000 also witnessed profound changes and developments across Zimbabwe's broader political, social and economic landscapes.[1] These included war veteran-led land reform;[2] an increasingly 'extreme and violent political intolerance' (Alexander 2003: 99) of any perceived opposition;[3] continuing economic decline, and in the drought of 2001–02, food shortages; all of which has been punctuated by several controversial, and hard-fought elections, and ever more restrictive legislation on citizenship, independent broadcasting, the media, and non-government organisations.[4] Some of these events formed the backdrop of my main period of fieldwork in 2000–01, but most have unfolded since that time. As a result, a second version of the manuscript for this book (Fontein 2003b) ended with a much more gloomy vision of the possibilities of successfully 'filling' Great Zimbabwe's multiple silences.

While these previous renderings have oscillated between optimistic and pessimistic forecasts, in this the last (and necessarily incomplete) ending I have a more modest aim. I want to avoid either extreme and rather focus on the continuing salience of the past, place and landscape as reflected in these ongoing events in Zimbabwe. In this chapter, I want to consider how Great Zimbabwe has featured in the emerging 'authoritarian project of the state' (Raftopoulos 2003: 217) and in particular, in its accompanying brand of 'patriotic history' (Ranger 2004), and to examine, as far as is possible, the nature of NMMZ's complicit (whether deliberate or not) involvement within it.

Ranger explains how he first became aware of a 'new variety of historiography' during the presidential election campaign of 2002. It appeared in the innumerable newspaper reports and television documentaries that 'insisted on an increasingly simple and monolithic history' (Ranger 2002: 60 cited in Ranger 2004: 218) which focused almost exclusively on the colonial wrongs of the past, and ZANU PF's heroic struggle for liberation. He depicts this patriotic history as 'explicitly antagonistic to academic historiography' (Ranger 2004: 218), having been born out of the perception among the ruling party that universities had been, as one ZANU PF member put it, 'turned into anti-government mentality factories'.[5] Ranger also highlights its complexity, which is most immediately apparent in the variety of its versions, authors and the contexts of its dissemination. These range from war veterans teaching history in youth militia camps, to journalists writing in the state-controlled press or broadcasting on state television and radio; and from the new history textbooks of the now compulsory secondary school history syllabus to the campaign speeches of President Robert Mugabe himself.

In the writings of one of its key protagonists – Professor Tafataona Mahoso, chairman of the media commission – President Mugabe himself is portrayed as representing 'deep ancestral memory', as 'the reclaimer of African space' and 'the African power of remembering the African legacy and African heritage which slavery, apartheid and imperialism thought they had dismembered for good'. This is contrasted with the 'bogus universalism' of the Zimbabwean opposition movement – the MDC – and 'their British, European and North American sponsors'.[6] Thus, as Ranger points out, Mugabe and ZANU PF are made to appear as 'custodians of history' while the MDC represent 'the end of history' and 'an a-historicised, globalised morality' (Ranger 2004: 223). Beyond merely over-emphasising a simplistic and monolithic version of the past, in which the president and the ruling party are endowed with unrivalled liberation credentials, this particular strand of patriotic history seems to bear some resemblance to the efforts of nationalistic writers such as Mufuka and Chigwedere immediately after independence in 1980, to respond to the calls for a new history. Perhaps it is not surprising then that another key author of patriotic history identified by Ranger is Aeneas Chigwedere himself, who in his current role as Minister of Education 'has done most to advance patriotic history and to combat "bogus universalism" ' (Ranger 2004: 225). Indeed it seems that Chigwedere's fortunes have seen a revival of a sort, and his contributions no longer 'remain in largely muted form in public discourse', as I suggested in Chapter 6. To a certain extent, this patriotic history can therefore be seen as a revitalisation of the attempts of the early 1980s to not only rewrite African history, but also the very means by which it is written. It raises the possibility that this emerging discourse of patriotic history – politicised, narrow and monolithic as it may be – is another attempt to create, in Chatterjee's terms, an 'authentic' and 'original' national history, wrestled from the hands of 'derivative' academic historians in universities and colleges. Certainly we can imagine that contributors like Professor Mahoso and Aeneas Chigwedere would like to think of themselves as 'authentic' narrators of the African past.

In the emerging literature on the 'Zimbabwe crisis' there has been a tendency to see current events either in terms of a radical rupture with, or in terms of a logical continuation of the past. A good example is Norma Kriger's argument that the contemporary prominence of war veterans in Zimbabwean politics 'recalls the early post independence years' (2003: 191). In contrast, Ranger (2004) has sought to highlight how current ZANU (PF) rhetoric differs from that of the early 1980s in its inclusion of ZIPRA's war record, the absence of a modernising, reconstructing and welfare agenda and the 'creation of a new history in which Zimbabwe was a colony until 1980' (White 2003: 97). Similarly on the land issue, or rather *issues* (see Hammer and Raftopoulos 2003: 18), Moyo (2001: 11) has argued that the 'essence' of the land occupations has remained largely the same throughout the independence period, while Alexander (2003: 99) has emphasised some of the differences in the ideology of the land occupations of the 1980s, from those that began in 2000. Other commentators, especially 'western journalists' (Worby 2003: 67), have tended to see recent events 'more simply and ahistorically', as evidence of a 'retreat' from, or 'end of modernity' in Zimbabwe; a sentiment that is also repeated in metaphors of 'exhaustion' and 'plunging' that have been employed by some academic observers (for example, Bond and Manyanya 2003; Campbell 2003). Worby, however, demonstrates that the current, predominant focus on issues of sovereignty represents not so much a 'retreat' from 'modernity', but rather a redefinition of the nation/state. In his view, the modern state involves a tension between sovereignty and development, and currently in Zimbabwe 'the see-saw of political modernity has tipped to one side – the side of sovereignty' (Worby 2003: 68).

Worby's argument about the centrality of sovereignty alongside, or in tension with, development in the idea of the modern nation state, has strong echoes with Chatterjee's argument about the co-existing 'authentic/spiritual' and 'derived/ material' domains of anti-colonial nationalisms (1993: 6, also 1996: 217). In these terms, Mahoso's emphasis on 'deep African memory' and 'reclaiming an African legacy' relates directly to the increasing urgency of sovereignty at the expense of development, which Worby identifies. Furthermore, both can be taken to reflect a recognition that 'the root of postcolonial misery', as Chatterjee would have it, lies in the 'surrender to old forms of the modern state' (1993: 11; also 1996: 222). Thus, the emergence of patriotic history in Zimbabwe might be more correctly regarded as reflecting not an 'exhausted' or 'plunging' nationalism but rather a revitalised and rejuvenated nationalist endeavour, however undeniably and 'indefensibly narrow' (Ranger 2004: 223), exclusive and alienating it has also often been.[7]

This perspective may enable a better understanding of some of the complexities, and apparent contradictions of patriotic history which Ranger draws attention to. One example here is the huge variety of authors who contribute to patriotic history, which defies any simplistic dichotomy between 'authentic' and 'derived' versions of national history. These authors include the Minister of Foreign Affairs, Stan Mudenge, whose well-received book, *A Political History of Munhumutapa* (1988), I have argued (see Chapter 6) represented the moment when efforts to establish 'an authentic national history' (Mudenge 1998: vii) by Zimbabweans working from within the sphere of the established academia took over from those working on its periphery. Ranger shows how Mudenge has contributed to the discourse of patriotic history, and 'drawn upon his historical data to paradoxical effect' to dismiss concern about Zimbabwe's suspension from the Commonwealth, and deliver a warning to the MDC about the dangers of depending upon 'foreign influence' (Ranger 2004: 226).

It is not just the inclusion of respected academics in the construction of patriotic history that illustrates how that discourse defies any simple opposition between 'authentic' and 'derived' means of producing history. Despite the appeals to 'deep African memory', and the rhetorical derision of the 'bogus universalism' of academic history by some of its most outspoken contributors, patriotic history does, in some of its more sophisticated incarnations, allow room for constructive engagement with the agendas of archaeologists at NMMZ and international institutions such as UNESCO and ICOMOS. This was perhaps most clearly demonstrated by the use that was made of the recent nomination of the Matopos Hills to the *World Heritage List*, which was finally announced by the World Heritage Committee on 3 July 2003. At a subsequent meeting of ICOMOS held at Victoria Falls on 29 October, President Mugabe himself made clear his vision of the relationship between patriotic history and world heritage.

> Zimbabwe was committed to preserving its heritage ... Zimbabweans had, through the agrarian reform programme, found joy because their greatest heritage – land – had been returned to them. 'Now that land has returned to the people, they were able once more,

to enjoy the physical and spiritual communion that was once theirs. For it must be borne in mind that the non-physical or intangible heritage is an equally strong expression of a people, manifesting itself through oral traditions, language, social practises and traditional craftsman-ship'. The objectives of ICOMOS were synonymous with Zimbabwe's philosophy. Comrade Mugabe said Zimbabwe valued Heritage so much that even the graves of the country's colonialists such as Cecil Rhodes were being preserved. 'We accept history as a reality'. (*The Herald*, 30 October 2003, cited in Ranger 2004: 228)

One striking aspect of this address is the means by which President Mugabe made reference to, and in a sense co-opted UNESCO's terminology of 'non-physical', 'intangible' heritage, and directly related it to his own land reform agenda. Of course, a convergence between the agendas of ZANU PF and the world heritage 'system' is not altogether surprising given both the ruling party's renewed emphasis on sovereignty (Worby 2003) and UNESCO's own pre-occupation with the 'sovereignty of state parties', which I discussed in Chapter 9. Alongside other examples,[8] this convergence illustrates how patriotic history may be part of a renewed and concerted effort to redefine the postcolonial state in a way that enables the co-existence of both the 'autonomous/spiritual/sovereign' and the 'derived/material/developmental' spheres of nationalism, thereby reversing 'the surrender to old forms of the modern state' that Chatterjee argued occurs in the movement from anti-colonial nationalism to independent state (1996: 222).

But if its complex rapport with other national and international schemes exposes patriotic history as a revitalised and rejuvenated nationalist endeavour, it also becomes crucial to consider its relationship with other diverse, multiple and often marginalised or silenced history-scapes, such as those that this book has focused on around Great Zimbabwe. What does patriotic history and a rejuvenated nationalist historiography offer such alternative voices? Can it fill Great Zimbabwe's multiple silences? So far most indicators suggest that although in its sophisticated forms patriotic history reconciles with versions of the past and 'heritage' embodied by organisations such as NMMZ, ICOMOS and UNESCO, it does not allow much room for radical alternatives, and different interpretations of the past or what to do with its remains. Ranger highlights a campaign recently launched by war veterans in Matabeleland for the removal of Cecil Rhodes's grave from the Matopos (2004: 227). Clearly on that issue, the overlapping of the world heritage 'systems' interest and that of President Mugabe's own version of 'patriotic history' seems to preclude the possibility of this particular 'local, radical war veterans' agenda' from being materialised. As Ranger puts it 'the listing does indeed mean that Rhodes's grave is safe' (2004: 228). Both these points – the compatibility between some of the interests of heritage professionals (UNESCO, ICOMOS, NMMZ), and those of ZANU PF; and the continued marginalisation of other interests, other versions of the past, and what to do with its remains – bring me to a consideration of recent events at Great Zimbabwe itself.

Sinamai has argued that Zimbabwe sites have always been 'used in making power visible to the "common person" ' (2003: 1) and in this respect patriotic history has been no different. Since the Chisikana ceremony in July 2000, patriotic history and the revitalised project of authoritarian nationalism have converged upon Great Zimbabwe in interesting ways. Two of a series of Unity Day 'cultural galas', which have been a feature of a populist strand of patriotic history promoted by the former Minister of Information, Jonathon Moyo, were held at Great Zimbabwe in December 2001 and 2002, and broadcast to the nation. As if anticipating the wider controversy these events have more recently become embroiled in,[9] the galas held at Great Zimbabwe caused a great deal of local alarm and disquiet. Chiefs around Great Zimbabwe 'vehemently opposed the Unity Galas as they feel that a sacred site is being used for entertainment purposes' (Sinamai 2003: 2). In an interview broadcast on ZBC in May 2002, Chief Mugabe 'did not mince his words', as *The Standard* (16 June 2002) later reported:

> I am aware that some people have gained mileage out of it, but this is unacceptable. You can not come and hold celebrations at this sacred place without consulting the traditional leadership, he said. However, Masvingo governor, Josaya Hungwe dismissed the chief's criticism saying the government did not need anyone's permission to hold national celebrations. (*The Standard* 16 June 02)

Some whom I spoke to in July 2004 blamed the galas for inadequate rains and resulting poor harvests, as well as the devastating effects of the HIV AIDS pandemic. As VaHaruzvivishe put it,

> Like when they held the Unity Gala at Great Zimbabwe. They held the Unity Gala there at the most important mountain. And the next day you could find so many condoms there, one or two kg of condoms in that sacred place!!!! That is why these things are happening. (Discussion with VaHaruzvivishe, Fieldnotes, 21 July 2004)

VaHaruzvivishe also echoed the startling claims made by other 'traditionalists' and chiefs from the Masvingo area, that a series of horrific road accidents which left 48 people dead in two days in June 2002, were 'the manifestation of the anger of the spirits over the organisation of a ZANU PF unity gala at Great Zimbabwe' (*The Standard* 16 June 2002).[10] Deriding the lack of prior consultation with the 'traditional leadership', Chief Murinye was quoted as saying

> We knew this debacle would come one day. How could the government allow a gala to be held at the Great Zimbabwe monument? You can't organise a function at a sacred place and have drunken youths and promiscuous elders coming to engage in sexual activities and to defecate the shrine.

> Has anything like that happened before in our history? he said adding, they even allowed musicians to come and play their guitars. These instruments have not been played at the shrine since the monuments were constructed. I don't want to frighten people, but mark my words. A lot of disasters will occur in the province unless Zanu PF swallows its pride and rectifies its stupid mistake. (*The Standard* 16 June 2002)

Neither the events surrounding these 'galas', nor the local responses to them, or indeed Governor Hungwe's dismissal of local concerns, bode well for a patriotic history that is responsive to alternative approaches to the past, place or landscape. The only saving grace is that, not surprisingly (and probably with the hand of NMMZ), no more cultural galas have been held at Great Zimbabwe since 2002.[11]

One spirit medium whose warning that the situation would 'worsen if the political leadership refuses to sort out its mess' was recorded in *The Standard*'s article of 16 June 2002, and made national news again the following year. This time, however, he seemed to have garnered more local political support. On Saturday 10 May 2003, Dickson Marufu, a *svikiro* from Zaka in Masvingo Province, turned up at Great Zimbabwe accompanied by various chiefs, government officials and security agents (*The Herald* 10 May 2003). His intention was 'to "unlock the mystery" surrounding the historic monument' (*The Herald* 10 May 2003). Claiming to be a medium for an ancestor called Sekuru Chinhope, but also referred to as Manhope, Manunure or Nunuratshe, he intended to 'perform a ritual ceremony' (*Daily News* 13 May 2003) to 'open up the underground' to prove his claims that 'under the monument ... lie some hidden treasures, medicines and a golden head statue of the architecture of Great Zimbabwe, which could salvage the country from its current economic crisis' (*The Herald* 14 June 2003). Evidently very tenacious in his efforts, Sekuru Chinhope had 'over the past three years gone to most Government offices and even Vice President Muzenda trying to get permission to prove his case' (*The Herald* 10 May 2003). That *The Herald* reported the story on the very day that the visit to Great Zimbabwe was made suggests that Sekuru Chinhope's intentions had already received some publicity. Along with the apparent alliance with local government officials,[12] this could be taken to indicate a moment of possible concurrence between the agenda of one spirit medium and that of the purveyors of patriotic history. This was further substantiated by the closing paragraph of *The Herald*'s initial article, which voiced the view of Great Zimbabwe's sacred national role during the liberation struggle.

> Some historians say the people who developed the medieval site [Great Zimbabwe] were the guiding spirits behind the war of liberation that brought about Zimbabwe's independence in 1980. (*The Herald* 10 May 2003)

This event clearly had the potential to represent a moment when the perspective of Great Zimbabwe as a 'national sacred site' – which I have argued developed in the discourses of guerrillas and 'traditionalists' during the liberation struggle, and has since been marginalised in preference for a 'national heritage' view – could be aligned with the demands of ZANU PF's patriotic history. But this was not to be. As the *Daily News* reported three days later,

> Things took a different turn when Marufu, accompanied by his chief, Reuben Ndanga, were denied permission by the police and officials from the Department of Museums and Monuments. Marufu and Ndanga, in the company of officers from the Department of Information and Publicity in the President's Office, went to the monument but came back empty-handed after a misunderstanding with the police. Sources said police and Department of National Museums and Monuments felt that the self-proclaimed spirit medium would tamper with physical structures at the national shrine.
>
> [...]
>
> Police yesterday confirmed that they barred Marufu from conducting the ritual ceremony. (*Daily News* 14 May 2003)

Given NMMZ's long experience of dealing with both the random arrival of spirit mediums at Great Zimbabwe and the long-running disputes between local clans, it is not surprising that the regional director's response went beyond merely stating the risk to the physical structure of the ruins, to emphasise that

> As far as we are concerned no function of that sort had been sanctioned by us or the local community and all the stakeholders. In the interest of protecting the shrine, any function has to be authorised first. (*Daily News* 14 May 2003)

When it comes to locally initiated, 'ritual' events, NMMZ has obviously managed to maintain its authority at Great Zimbabwe, and continues to frame its own role as 'intermediary' within a complex web of stakeholders. One might also have expected NMMZ's justification for preventing the ceremony to have included a reference to Great Zimbabwe's world heritage status, especially in the light of President Mugabe's subsequently stated view that 'the objectives of ICOMOS [or in this case UNESCO] were synonymous with Zimbabwe's philosophy' (*The Herald* 30 October 2003). The overlap between NMMZ's 'heritage' perspective, and the demands of patriotic history were even more evident during a huge nationally televised event held at Great Zimbabwe only a few weeks later, to re-unify two halves of one of the Zimbabwe birds.

Matenga described in his book on the Zimbabwe birds (1998: 38), how the upper half of one of the Zimbabwe birds had been removed from the ruins by Hall in 1902. While it remained in the country, the lower half was sold to Cecil Rhodes, and was 'eventually deposited in the Museum für Völkerkunde in Berlin by a missionary called Axenfeld in 1906' (Matenga 1998: 38). During the fall of Berlin in 1945, it was then taken to Leningrad by Russians until it was eventually recovered by the Museum für Völkerkunde after the collapse of the Soviet Union. It was during the organisation of an international exhibition entitled 'Legacies of Stone: Zimbabwe Past and Present' in Tevuren, Belgium (November 1997–April 1998), that the recovery of the missing half-bird by the Berlin museum was revealed, along with the discovery that it 'fitted' the upper half held by NMMZ (Matenga 1998: 61–2). The temporary re-unification of the two halves at Tervuren generated the incentive for NMMZ to pursue negotiations for the missing half-bird's return to Zimbabwe. NMMZ resolved, as Matenga put it, 'to recover the specimen in Berlin and allow it to return home to roost!' (1998: 62).

The ceremony held in May 2003 at Great Zimbabwe represented the culmination of NMMZ's efforts to retrieve the missing lower part of that Zimbabwe bird from Germany. It also represented the next stage[13] in a process begun just before independence to retrieve all the Zimbabwe birds, of which four and a half were returned from South Africa in 1981 (Matenga 1998: 60). As Matenga put it,

> The determination by the Zimbabwean Government to reclaim the birds stemmed from a desire to rehabilitate Great Zimbabwe as a cultural symbol of the African people. The desire was inspired by the belief that the potency of Great Zimbabwe as the guardian spirit of the nation lies in its possession of sacred artefacts such as the conical tower and the Zimbabwe birds. It was imperative to bring back the bird emblems in order to re-equip and revive the shrine of Great Zimbabwe. It seemed the spirit of the deceased inhabitants of Great Zimbabwe, who had bequeathed to us a wonderful heritage and a name to nationhood, were not going to rest until the birds were recovered. (Matenga 1998: 57)

Clearly this event can be seen as the fulfilment of a 'national imperative' that dates back to the early years of independence. Little surprise then that it was invested with a huge amount of state pomp, televised on national TV and presided over by President Mugabe himself. However, as Ranger describes (2004: 226), the ceremony generated a great deal of criticism and ridicule, both from within Zimbabwe, and in the South African press. This criticism focused on both the inappropriate amount of resources (given Zimbabwe's economic situation) spent on the occasion, and also the perception that the event was an attempt by ZANU PF to monopolise 'national heritage' for its own purposes. Fidelis Mashavakure wrote to *The Standard* (6 June 2003) and, although acknowledging the bird's 'historical significance to … Zimbabweans', asked

> How could the government arrange a multimillion dollar ceremony to deity a mere artefact when the country is facing a myriad of problems? It was astonishing to see women religiously kneeling down in honour of a stone carving. It was even more astonishing to see the president lead the gathering in sloganeering over the carving. Does the carving belong to Zanu PF or Zimbabwe? (*The Standard* 6 June 03)

A satirical article printed in the South African newspaper, the *Sunday Times* took its ridicule much further, saying that 'Mugabe had a "bird in his head" ' (Ranger 2004: 26) which caused the enraged Minister of Information, Jonathon Moyo, to lodge an official complaint with the government of South Africa (*Sunday Mail* 25 May 2003). Mahoso described the offending article as 'a racist attack on Zimbabwe's efforts to recover its heritage and symbols looted in the 1890s and scattered all over the world' (*Sunday Mail* 1 June 2003). As he continued, Mahoso also succinctly encapsulated how the ceremony could indeed be incorporated into ZANU PF's discourse of patriotic history, as an expression of 'deep African memory'.

> The white racist columnist, Hogarth, wrote a column called 'Our Bob's got birds on the brain', which was savage attack on the entire process of 'remembering', that is the process of remobilising African memory by reconnecting symbols, communities, movements and people as the South's answer to Northern driven globalisation, reviving their memory of a world without apartheid. (*Sunday Mail* 1 June 2003)

Such reactions illustrate Ranger's point (2004: 227) that this ceremony was a fulfilment of not only NMMZ's agenda, but also that of ZANU PF, the Foreign Minister Stan Mudenge, and President Mugabe himself. Furthermore, 'in moments like the Zimbabwe bird ceremony or the declaration of the Matopos as a World Heritage site patriotic history and academic archaeology fit together very well' (Ranger 2004: 228). But if this ceremony does reflect an overlap of NMMZ's heritage agenda with that of the ruling party, then another event the following year, again marking the return of the half-bird and also held at Great Zimbabwe, illustrates how this convergence of interests can also be engineered towards including a broader constituency of 'traditional leaders' whilst simultaneously working in favour of particular, localised and individual political interests.

In May 2004, the National Assembly of Chiefs met for two days at Great Zimbabwe National Monument. The assembly 'coincided with the official hand over to the Masvingo chiefs of the Zimbabwe Bird whose lower half had been stolen … in the 1880s' (*The Herald* 8 May 2004). Coming in the run up to both the ZANU PF congress later that year, and the parliamentary elections of the next, the government newspaper *The Herald* (8 May 2004) reported the chiefs' endorsement of the ZANU PF leadership, the government and its land reform programme. In their resolutions the chiefs also stated their desire to be given a role in the management of resettlement areas, as well as welcoming the increases in their monthly allowances, their new access to a car loan scheme, and the expansion of their judicial role which had been announced during the meeting (*The Herald* 6 May 2004, 8 May 2004; *Zimbabwe Independent* 14 May 2004). The Minister of Home Affairs, Kembo Mohadi urged the chiefs 'to work towards removing the relics of colonialism' adding that 'it was important to preserve historical sites [of the liberation war] like Nyadzonia Camp in Mozambique and Mkushi in Zambia because they acted as a reminder of the country's grim past' (*The Herald* 8 May 2004). Furthermore, 'since we do not have a true history of the first and second chimurenga, chiefs should ... work together with the University of Zimbabwe in the rewriting of the country's history' (*The Herald* 8 May 2004).

This meeting of chiefs from across the country was plainly engineered in favour of the ruling party and its patriotic history.[14] But the concurrence of the meeting, its location, and the event of the returned half-bird's 'hand over' to 'the Masvingo chiefs', was not merely window dressing in the ruling party's continuing efforts to appeal to a particular constituency of 'traditional leaders'. It was also an important stage in the ongoing realisation of certain local agendas and political ambitions particularly those of Chief Charumbira, who has become increasingly prominent in his various new roles as deputy Minister of Local Government, Secretary-general of the Council of Chiefs, and now, it seems, as head of the Masvingo chiefs.[15] Various accounts that I heard when I returned to the area in July 2004 (Fieldnotes, 21 July 2004, 22 July 2004, 25 July 2004) describe the central role played by Chief Charumbira during the events marking the return of the bird. One person described what happened as follows

> The bird was put into the museum and covered with a cloth. There were so many chiefs there from right across the country. The local chiefs Nemanwa and Mugabe were looking at each other to see who the bird was going to be given to. But it was given to Chief Charumbira to reveal the bird, who then stood up, uncovered the bird and made a speech in which he said that the bird did not belong to one chief, or another, rather it belong to all the chiefs together. He uncovered the bird in his role as chairman of the Masvingo chiefs. Then the bird was given to the Minister of Home Affairs, and then onto NMMZ to look after properly. (Fieldnotes, 25 July 2004)

Of course, NMMZ remains the custodian of the bird and Great Zimbabwe as whole, but the sequence of the 'hand over' nevertheless became the focus of local rumour and hearsay. The person I quoted above also voiced a common perception that Chief Charumbira was, in fact, acting less as an impartial, 'chairman of the Masvingo chiefs', and more in the interests of a particular Charumbira claim over the ruins which is based on his authority as chief over Headman Nemanwa. Therefore,

> because Nemanwa and Charumbira *vano batana* ['help each other' or 'are on the same side'], the next day, in secret, Charumbira went there with Nemanwa, and they took the bird up the hill, and it was 'given' to Nemanwa, without Mugabe knowing. (Fieldnotes, 25 July 2004)

Not surprisingly, there was some resentment among members of other local clans, some of whom questioned Charumbira's authority as 'chairman of the Masvingo chiefs'. VaHaruzvivishe suggested that 'someone like Chief Masungunye' would be more appropriate in that role, given his senior position above all the VaDuma chiefs

(Fieldnotes 21 July 2004). Moreover VaHaruzvivishe also denounced the absence of any spirit mediums at the handover, stating that Ambuya VaZarira should have been invited. But his biggest complaint was, of course, that the bird should have gone back to Mugabe, not to Nemanwa or Charumbira.

> I ask VaHaruzvivishe about the return of the *shiri* ['bird' i.e. Zimbabwe bird]. Immediately he says that they had said they were going to return the bird to the people from whom it was taken, but they gave it to Nemanwa. If they read the books they would know that the bird was taken from Chief Mugabe, and so that bird should have come back to Mugabe, not Nemanwa. (Discussion with VaHaruzvivishe, Fieldnotes, 21 July 2004)

Although the Nemanwa elders who I spoke to in July 2004 seemed less disgruntled, they also continue to make their own claims on Great Zimbabwe, and to the status of chief in their own right. I heard rumours of a story in which it is claimed that the day after the bird was handed over, it was found by NMMZ workers standing outside the museum, speaking and requesting that beer be brewed by a particular house of the Nemanwa clan, so that 'it can stay well in the museum' (Fieldnotes, 22 July 2004).

Such stories reiterate how state-sponsored events at Great Zimbabwe inevitably feed back into the situated history-scapes constantly being deployed, and reworked in the ongoing local contests over the site. Similarly, these events, and the multiple narratives that are emerging from them, illustrate how the minutia of ZANU PF's renewed nationalist project and patriotic history are often embedded and situated within a complex, multiplicity of longstanding and ongoing struggles, contests and disputes over the past, place and landscape. Regardless of whether patriotic history does indeed offer the radical alternative to the construction of a national past which its rhetoric often promises, in this respect, it is only the latest stage in the ongoing 'movement' and 'becoming' of Great Zimbabwe. As Sinamai has pointed out, Great Zimbabwe remains, like the other Zimbabwe sites, a 'manifestation of power' at the centre of 'multifaceted contests' (Sinamai 2003: 3) which transcend, yet reverberate within, different layers of changing meanings and values. And although it contributes to these ongoing 'multifaceted contests', patriotic history has yet to demonstrate whether it does indeed challenge the continued alienation and marginalisation of alternative voices about Great Zimbabwe's past and management. There has been little mention of local communities' various versions of its past in any of patriotic history's incarnations, and the wider demands of 'traditionalists', spirit mediums and war veterans, from across Masvingo province and beyond, for a national ceremony to thank the ancestors at Great Zimbabwe, have so far remained unanswered. Neither the unity galas nor the bird ceremonies come close to fulfilling those demands, and widely canvassed plans for a national ceremony to rebury the remains taken from mass graves in Mozambique and Zambia have not yet materialised.

Beyond merely failing to challenge Great Zimbabwe's silences, there is also a sense in which current events may be contributing to it. Despite some of patriotic history's more populist efforts, in the context of economic hardship, social turmoil, and widespread hunger, it is to be expected that Great Zimbabwe's relevance for many people will diminish. As Dr T. Mangwenda commented in a letter to the *Daily News* (31 May 2003), 'If a country is ravaged to the point that it cannot provide basics for its citizens, cultural symbols are the first to lose their value and meaning'. Therefore it should be no surprise that 'some young Zimbabweans referred to the other half of the recently returned bird as "just another piece of stone" '. With the decline of the tourist industry, Great Zimbabwe also rapidly loses its value for those people whose livelihoods depend on it. Since the opening of the new Shona village, one of 'traditional' music groups no longer plays at Great Zimbabwe, as it is no longer worth their effort. Similarly Walter Marwizi reported in *The Standard* that during the drought of 2002

> Great Zimbabwe is now just a heap of stones with no benefit for us', cursed the empty-handed Jerina, as she arrived at home. ... 'Will Great Zimbabwe ever rise and be great for us?' the fisherman muttered to himself. ... It was not only Jerina and the fisherman who went home empty-handed. ... In fact it is now the order of the day for drought wracked villagers in Chief Mugabe's area in Masvingo who were earning a living through selling their various wares to tourists who thronged the Great Zimbabwe monuments on a daily basis to explore the mysteries buried at the world acclaimed heritage site. The villagers ... now sing the blues as the monuments have lost their lustre ... Tawanda Magara, a stone carver, said Great Zimbabwe now had a different meaning to him altogether. 'In the past when we saw the Great Zimbabwe monuments we realised that we would always make money since visitors would always come to discover the mystique associated with them. Now we see them just as any other heap of stones. They don't make any difference in our lives.' (*The Standard* 21 June 2002, cited in Ranger 2004: 231)

In light of the accelerated changes and events that have occurred since 2000, both across Zimbabwe's broader political, social and economic landscapes, and at Great Zimbabwe itself, any optimism that the *silence* of Great Zimbabwe could be successfully filled, may seem misplaced. Yet despite all of this, there remains, in my view, cause for optimism. While the narration of the national past has again become a 'battleground of rival attachments'(Lowenthal 1990: 302), this should be taken to indicate, as Matenga commented (Field notes 22 July 2004), how Great Zimbabwe continues to provide inspiration. Although now 'in principle' permitted, NMMZ has maintained its right to exercise a veto on the conduct of ceremonies at the site. But this is not all bad, as it seems to have prevented, at least since the last gala in 2002, a repetition of some of patriotic history's worst excesses. Furthermore, there are growing signs that NMMZ is beginning to rethink its approach to community involvement in heritage management and 'realise that communities and nations are not homogeneous' (Sinamai 2003: 3) and therefore 'managing sites with local communities' will never be straightforward. The notion of 'community', as Pikirayi has commented,[16] has been taken for granted, and needs to be unpacked, for 'it is in understanding all facets of the "community" or "nation" that heritage managers can understand the multifaceted system of values ascribed to these sites' (Sinamai 2003: 3).

It is likely that local clans and individuals within them will continue to put forward and dispute their claims over the site, and that other 'traditionalists' across the country will continue to use Great Zimbabwe in their wider attempts to gain the recognition they feel they deserve. But beyond merely, and reluctantly, conceding that the 'Nemanwa and Mugabe clans will never agree' (Matenga, Fieldnotes, 22 July 2004) NMMZ may be soon able to embrace and celebrate these ongoing contests, and the constantly changing history-scapes through which they are often articulated, as 'signs of success, demonstrating genuine involvement ... and concern with the very essence of contemporary cultural activity' (Ucko 1994: 247). The *silence* of Great Zimbabwe remains to be filled, and if a means of discussing the past, and what to do with it, emerges which can embrace its contested and dynamic nature, and the multiplicity of changing histories, attachments and meanings inevitably associated with any place or landscape, then a substantial leap will have been made towards a truly 'original' and 'authentic' history.

This optimism is not mine alone. Undoubtedly, some of the initial enthusiasm generated by the Chisikana spring ceremony of 2000 has waned. In July 2004 I heard comments suggesting that the Chisikana ceremony had not been conducted properly – that representatives of 'big *masvikiro* of the whole country' should have been there, or that the 'water should have just burst out of the covered spring by itself' rather than with the work of pick axes. But the expectations that arose out of that event have not vanished entirely. For many people the possibility remains that if *chikaranga* is respected, and if the ancestors, the *mhepo dzenyika* – the winds of the land – and their *masvikiro* are again consulted then the *Voice* and mysterious sounds might again be heard on the hill at Great Zimbabwe. As VaHaruzvivishe explained

> About the spring, things are better now, because the water is running from there. But it is still cloudy, it has not settled and begun running clear yet. When things are sorted out the soil is settled – when the dust has settled – then the spring might run clear and then I still think that the *Voice* could return and again be heard from the rocks on that hill in *masvingo* there. (Discussion with VaHaruzvivishe, Fieldnotes, 21 July 2004)

Notes

1. Although the 'multiple origins' of what has been labelled 'Zimbabwe's crisis' can be traced far before 2000, the constitutional referendum of February that year has been taken by many commentators to mark a 'particular watershed in Zimbabwe's post-independence political history' (Hammer and Raftopoulos 2003: 1).

2. Initially characterised as 'demonstrations' and 'invasions', the land occupations were later 'turned into' the government's 'fast-track' land reform programme (see Chaumba et al. 2003a and 2003b). By the time of the official completion of the programme in 2002 the government had acquired 11 million hectares of land for redistribution (ciir.org/content/nes/zim).

3. For more on the 'new politics' of citizenship and 'exclusion' see also Rich Dorman 2003; Raftopoulos 2003; Rutherford 2003.

4. E.g. AIPPA (Access to information and Protection of Privacy Act, 2002); POSA (Public Order and Security Act, 2002); ZEC ACT (Zimbabwe Electoral Commission Act, 2004); and most recently, the NGO Act (Non Governmental Organsiations Act, still to be signed into law).

5. Sikhumbizo Ndiweni, ZANU PF information and publicity Secretary for Bulawayo in the *Bulawayo Chronicle* 26 April 2001 (cited in Ranger 2004: 218).

6. Professor Mahoso in the *Sunday Mail* (16 March 2003) cited in Ranger (2004: 222).

7. One consistently reported feature of recent political events in Zimbabwe is a dramatic reconfiguration of citizenship; of who is 'included' and who is 'excluded'

from the nation (See also Raftopoulos 2003; Rich Dorman 2003; Rutherford 2003).This has often been articulated through a narrow language of 'sell-outs' and revolutionaries (Ranger 2004: 232). It is a re-emerged language which is often associated with memories of violence during the liberation struggle and the *gukurahundi* of the 1980s, and has strong connotations with accusations of witchcraft (Worby 2003: 75).

8.   These include the oral history project of the first and second *chimurenga* which has seen a partnership emerge between NMMZ, the National Archives, and the University of Zimbabwe's History Department (Ranger 2004: 229); a plan to build a new museum at the National Heroes' Acre; and of course the renewed efforts to finally build the widely mooted Great Zimbabwe University (see e.g. *The Masvingo Star* 23–29 July 2004).

9.   As Jonathon Moyo found himself out of favour with the ruling party old guard towards the end of 2004, after having convened the unsanctioned 'Tsholotsho meeting', some of the 'populist' activities he most championed (such as the cultural galas and the PaxAfro albums) were criticised, and his use of tax payers' funds questioned (see 'Questions for Oral Answers', Parliament of Zimbabwe, *Votes and Proceedings of the Parliament of Zimbabwe* 20 October 2004; 'We're not broke say Zim Propogandists', *Mail & Guardian* 25 October 2004; 'Minister blows £1.5 billion on musical galas', *Zim Online (SA)* 2 November 2004; 'Moyo losing Tsholotsho', *Zimbabwe Independent* 10 December 2004). This championing of some aspects of 'popular' culture and the arts has been seen as part of a process of assimilating and co-opting potential arenas of criticism into the government's propaganda machine. While the music of Thomas Mapfumo, Leonard Zhakata, Raymond Mjongwe, among others, has been banned from the airwaves, other performing individuals and groups have been commissioned by the Department of Information and Publicity (see 'Has Cont Mhlanga sold out?' *The Standard* 28 July 2002; 'The State of the Arts in Zimbabwe Theatre. Producers always try to balance', P. Zenenga 4 November 2002, The Nordic Africa Institute (http://www.nai.uu.se/forsk/current/stateof thearts/theatre/zenengasve.html); 'Zimbabwean artists battle not to offend political gods', *Zim Online* 2 October 2004).

10.   For more newspaper reports on these horrific accidents see '37 perish in bus disaster' *Daily News* 11 July 2002; 'Zimbabwe declares disaster after crash' BBC News website, 10 June 2002; 'Thirty-four people killed in bus accident in Zimbabwe in another crash' and 'Zimbabwe Bus crash kills 11', *Associated Press* 11 July 2002.

11.   Although, at least until late 2004, 'cultural galas' have continued to be held at regular intervals across the country to commemorate different events. In October 2004 Jonathan Moyo told Parliament that his ministry had spent $1.5 billion in the previous four months on a gala held in July to commemorate the life of Joshua Nkomo, another in August for the late Simon Muzenda, and a third in October at Chimoio in Mozambique to 'celebrate unity between Zimbabwe and its eastern neighbour'. Another was planned for Unity Day in December 2004. See 'Minister blows $1.5 billion on musical galas' *Zim Online (SA)* 2 November 2004.

12.   Who included the Deputy officer commanding Manicaland, Assistant Commissioner Davis Chifamba, and the Masvingo district administrator,

Mr Makanzwei Jecheche (*The Herald* 14 June 2003), as well as members of the Department of Information and Publicity in the Office of the President (*Daily News* 14 May 2003).

13. One bird remains in South Africa, at Cecil Rhodes's former South African residence, Groote Schuur in Cape Town (Matenga 1998: 61). In July 2004 I was interested to note that in the museum at Great Zimbabwe, not only was the newly returned bird's spot in the display highlighted with 'mini-Zimbabwean flags' and some accompanying, explanatory text, but also the spot of this bird which remains in South Africa (Fieldnotes, 21 July 2004). A brief perusal of the visitor's comments book suggests that any effort by NMMZ to secure the return of this last remaining bird would be met with great approval by visitors to the site.

14. For his part, the Minister of Local Government, Public works and National Housing, Ignatious Chombo 'urged the chiefs to assist in the voter registration exercise by mobilising the communities which live under them', and closing the conference, 'praised the chiefs for holding ... one of the most successful chiefs assemblies ever'(*The Herald* 8 May 2004).

15. Chief Charumbira now frequently appears in national newspapers. In this case he appeared in all the three articles which I saw about the chiefs' assembly of May 2004 (*The Herald* 6 May 2004, 8 May 2004; *Zimbabwe Independent* 14 May 2004). More recently still, after the parliamentary elections of March 2005, Chief Charumbira was elected, unopposed, as president of the Zimbabwe Council of Chiefs after the retirement of Chief Jonathon Mangwenda (*Daily Mirror* 7 April 2005).

16. From my notes (8 July 2004) of the final plenary session of an international conference entitled 'Heritage in Southern and Eastern Africa: Imagining and Marketing Public Culture and History', convened by the National Heritage Conservation Commission, and the National Museums Board of Zambia, the British Institute in Eastern Africa and the *Journal of Southern African studies*, in Livingston, Zambia, 5–8 July 2004.

Bibliography

Abraham, D. P. (1966) 'The Roles of Chaminuka and the Mhondoro Cults in Shona Political History', in Stokes, E.T. and Brown, R. (eds), *The Zambesian Past: Studies in Central African History*, Manchester: Manchester University Press.

Abrams, P. (1988) 'Notes on the difficulty of studying the State', *Journal of Historical Sociology*, 1(1): 58–9.

Abu El-Haj, N. (1998) 'Translating truths: nationalism, the practice of archaeology and the remaking of the past and present in contemporary Jerusalem', *American Ethnologist*, 25(2):166–88.

Abu El-Haj, N. (2001) *Facts on the Ground: Archaeological Practice and Territorial Self-Fashioning in Israeli Society*, London: University of Chicago Press.

Alexander, J. (1990) 'Modernisation, tradition and control. Local and national struggles over authority and land: a case study on Chimanimani District Zimbabwe', manuscript, Balliol College, Oxford.

Alexander, J. (1995) 'Things Fall Apart The Centre *Can* Hold: Processes of Post-War Political Change in Zimbabwe's Rural Areas', in Bhebe, N. and Ranger, T. (eds) *Society in Zimbabwe's Liberation War*, London: James Currey.

Alexander J. (2003) 'Squatters, Veterans and the State in Zimbabwe', in Hammer, A., Raftopoulos, B. and Jensen, S. (eds) *Zimbabwe's Unfinished Business: Rethinking Land, State and Nation in the Context of Crisis*, Harare: Weaver Press.

Althusser, L. (1971) 'Ideology and Ideological State Apparatus', in *Lenin and Philosophy and Other Essays*, London: New Left Books.

Anderson, B. (1983) *Imagined Communities*, London: Verso.

Asad, T. (1973) *Anthropology and the Colonial Encounter*, London: Ithaca Press.

Asad, T. (1983) 'Anthropological Conceptions of Religion: reflections on Geertz', *Man*, 18: 237–59.

Aquina, Sister Mary (1965) 'Tribes in Victoria Reserve', *NADA*, 9 (2).

Aquiran, Sister Mary (1969–70) 'Karanga history and the Mwari cult' *Culture and Development*, 2 (2) 389–405.

Balibar, E. (1991) 'The Nation Form: History and Ideology', in Balibar, E. and Wallerstein, I. (eds) *Race, Nation, Class: Ambiguous Identities*, London: Verso.

Beach, D. (1973a) 'A historiography of the people of Zimbabwe', *Rhodesian Prehistory*, 4:21–30.

Beach, D. (1973b) 'Great Zimbabwe as a Mwari Cult centre', *Rhodesian Prehistory*, 11: 11–12.

Beach, D. (1973c) 'The Initial Impact of Christianity: The Protestant and The Southern Shona', in Douglas, A.J. (ed.) *Christianity South of the Zambezi*, Gwelo: Mambo Press.

Beach, D. (1980) *The Shona and Zimbabwe 900–1850*, London: Heinemann.

Beach, D. (1984) *Zimbabwe before 1900*, Gweru: Mambo Press.

Beach, D. (1984b) 'Is this the right way to interpret our most important heritage: review article of Dzimbahwe by K. Mufuka', *Teacher's Forum*, February.

Beach, D. (1994a) *A Zimbabwean Past: Shona Dynastic Histories and Oral Traditions*, Gweru: Mambo Press.

Beach, D. (1994b) *The Shona and their Neighbours*, Oxford: Blackwell.

Beach, D. (1998) 'Cognitive Archaeology and Imaginary History at Great Zimbabwe', *Current Anthropology*, February, 39(1).

Beidelman, T.O. (1971) 'Nuer Priests and Prophets: Charisma, Authority and Power among the Nuer', in Beidelman, T.O. (ed.) *The Translation of Culture*, London: Tavistock.

Bender, B. (1998) *Stonehenge: Making Space*, Oxford: Berg.

Bender, B. and Winer, M. (eds) (2001) *Contested Landscapes: Movement, Exile and Place*, Oxford: Berg.

Bent, J.T. (1893) 'The Ruins of Mashonaland', *Geographical Journal*, 2(5).

Bent, J. [1893] (1896) *The Ruined Cities of Mashonaland*, 2nd edition, London: Longman's Green.

Berliner, P.F. (1978) *The Soul of Mbira. Music and Traditions of the Shona People of Zimbabwe*, Berkeley: University of California Press.

Bhabha, H. (1994) *The Location of Culture*, London: Routledge.

Bhebe, N. and Ranger, T. (eds) (1995) *Soldiers in Zimbabwe's Liberation War*, University of Zimbabwe and London: James Curry.

Bloch, M. and Bloch J. (1980) 'Women and the Dialectics of Nature in Eighteenth Century French Thought', in MacCormack and Strathern (eds) *Nature, Culture and Gender*, Cambridge: University Press.

Bond, P. and Manyanya, M. (2003) *Zimbabwe's Plunge. Exhausted Nationalism, Neoliberalism and the Search for Social Justice*, 2nd edition, London: Merlin Press; South Africa: University of Natal Press; Zimbabwe: Weaver Press.

Bourdieu, P. (1991) *Language and Symbolic Power*, Cambridge: Polity Press.

Bourdillon, M.F.C. [1976] (1987a). *The Shona Peoples*, 3rd edition, Gweru: Mambo Press.

Bourdillon, M.F.C. (1987b) 'Guns and rain: taking structural analysis too far?', *Africa*, 57(2).

Bratton, M. (1978) *Beyond Community Development: The Political Economy of Rural Administration in Rural Zimbabwe*, Gweru: Mambo Press.

Brown-Lowe, R. (2003) *The Lost City of Solomon and Sheba: An African Mystery*, London: Sutton.

Bruwer, A. J. (1965) *Zimbabwe: Rhodesia's Ancient Greatness*, Johannesburg: Hugh Keartland.

Burke, E.E. (1969) *The Journals of Carl Mauch 1869–1872*, Salisbury: National Archives of Rhodesia.

Campbell, H. (2003) *Reclaiming Zimbabwe. The Exhaustion of the Patriarchal Model of Liberation*, South Africa, New York: Africa World Press.

Caton-Thompson, G. (1931) *The Zimbabwe Culture: Ruins & Reactions*, Oxford: Clarendon Press.

Chambers, R. (1997) *Whose Reality Counts? Putting the First Last*, London: Intermediate Technology.

Chatterjee, P. (1986) *Nationalist Thought and the Colonial World: A Derivative Discourse?*, Minneapolis: University of Minnesota Press.

Chatterjee, P. (1993) *The Nation and Its Fragments: Colonial and Post-colonial Histories*, New Jersey: Princeton University Press.

Chatterjee P. (1996) 'Whose Imagined Community?', in Balakrishnan, G. (ed.) *Mapping the Nation*, New Left Review, London: Verso.

Chaumba, J., Scoones, I. and Wolmer, W. (2003a) 'From *jambanja* to planning: the reassertion of technocracy in land reform in south-eastern Zimbabwe?', *Journal of Modern African Studies*, 41(4): 533–54.

Chaumba, J., Scoones, I. and Wolmer, W. (2003b) 'New politics, new livelihoods: agrarian change in Zimbabwe', *Review of African Political Economy*, 98: 585–608.

Chigwedere, A. (1980) *From Mutapa to Rhodes*, London: Macmillan.

Chiguredere, A. (1982) *The Birth of Bantu Africa*, Harare: Books for Africa.

Chiguredere, A. (1985) *The Karanga Empire*, Harare: Books for Africa.

Chiguredere, A. (1998) *The Roots of the Bantu*, Marondera: Mutapa Publishing House.

Chiguredere, A. (2001) *British Betrayal of the Africans. Land, Cattle and Human Rights. Case for Zimbabwe*. Marondera: Mutapa Publishing House.

Christian, W.A. (1972) *Person and God in a Spanish Valley*, London: Academic Press.

Chung, F. (1995) 'Education and the War', in Bhebe, N. and Ranger, T. (eds) *Society in Zimbabwe's Liberation War*, London: James Curry.

Clifford, J. (2004) 'Looking Several Ways. Anthropology and Native Heritage in Alaska', *Current Anthropology*, 45(1): 5–30.

Cohen, A. (1993) *Masquerade Politics: Explanations in the Structure of Urban Cultural Movements*, Oxford: Berg.

Collet, D.A. (1992) *The Master Plan for the Conservation and Resources Development of Archaeological Heritage*, Harare: NMMZ.

Collins. (1999) *PaperBack English Dictionary*, 4th edition, Glasgow: HarperCollins.

Colquhoun, A. (1914) 'A visit to King Solomon's Mines', *United Empire*, 5(6).

Connah, G. (1987) *African Civilisations*, Cambridge: Cambridge University Press.

Daneel, M. (1970) *God of the Matopo Hills*, The Hague: Mouton.

Daneel, M. (1995) (as Mafuranhunzi Gumbo) *Guerrilla Snuff*, Harare: Baobab Books.

Daneels M. (1998) *African Earthkeepers Volume 1 Interfaith Mission in Earth-care*, Pretoria: Unisa Press.

Daniels, S. and Cosgrove, D. (eds) (1988) *The Iconography of Landscape*, Cambridge: Cambridge University Press.

Dart, R. (1925) 'The historical succession of cultural impacts upon South Africa', *Nature*, 115: 425–29.

Dart, R. (1955) 'Foreign influences of the Zimbabwe and pre-Zimbabwe eras', *NADA*, 32: 19–30.

Delineation Reports (1965) Report drawn up by the District Delineation Officer B.P. Kaschula for the Native Affairs Department, National Archives of Zimbabwe, S2929/8/5.

Derman, B. (2003) 'Culture of development and indigenous knowledge: the erosion of traditional boundaries', *Africa Today*, 50(2).

Descola, P. and Palsson, D. (1996) 'Introduction', in Descola, P. and Palsson, D. (eds) *Nature & Society: Anthropological Perspectives*, London: Routledge.

Diaz-Andreu, A. (1995) 'Archaeology and Nationalism in Spain', in Kohl, P.L. and Fawcett, C. (eds) *Nationalism, Politics and the Practise of Archaeology*, Cambridge: Cambridge University Press.

Dorman, S.R. (2003) 'NGOs and the constitutional debate in Zimbabwe: from inclusion to exclusion', *Journal of Southern African Studies*, 50(2).

Durkheim, E. (1915) *Elementary Forms of the Religious Life*, London: Hollen Street Press.

Eliade, M. (1954) *The Myth of the Eternal Return*, New Jersey: Princeton University Press.

Evans- Pritchard (1956) *Nuer Religion*, Oxford: Oxford University Press.

Ferguson, J. (1990) *The Anti-politics Machine*, Cambridge: Cambridge University Press.

Fienup-Riordan, A. (1996) *The Living Tradition of Yup'ik Masks*: Agayuliyraput *(Our Way of Making Prayer)*, Seattle: University of Washington Press.

Fienup-Riordan, A. (2000) *Hunting Tradition in a Changing World: Yup'ik Lives in Alaska Today*, New Brunswick: Rutgers University Press.

Fontein, J. (1997) 'Great Zimbabwe: A Sacred Site or a National Monument?', unpublished dissertation for degree of MA (Hons) Social Anthropology, University of Edinburgh.

Fontein, J. (2000) 'UNESCO, Heritage and Africa: An Anthropological Critique of World Heritage', Occasional Paper No. 80, Centre of Africa Studies, University of Edinburgh.

Fontein, J. (2003a) 'The Silence of Great Zimbabwe: Contested Landscapes and the Power of Heritage', unpublished PhD thesis, University of Edinburgh, May.

Fontein, J. (2003b) 'Great Zimbabwe and Patriotic History', unpublished manuscript, December.

Fontein, J. (2005) 'An Ethnographic Study of the Politics of Land, water and "tradition" around Lake Kyle/Mutirikwi in southern Zimbabwe', in Templehoff, J.W.N. (ed.) *African water Histories: Trans-disciplinary Discourses*, Vanderbijle Park: Northwest University, pp. 285–315.

Forty, A. and Küchler, S. (eds) (1999) *The Art of Forgetting*, Oxford: Berg.

Frederikse, J. (1982) *None but Ourselves*, Johannesburg: Raven Press.

Frobenius, L. (1928) 'The Mystery of Zimbabwe', *The African World*, 26.

Ganter, R. (2003) *Zimbabwe's Heavenly Ruins: A Mystery Explained*, Leicestershire: Upfront.

Garbett, K. (1966) 'Religious aspects of political succession among Valley Korekore (N. Shona)', in Stokes, E. and Brown, R. (eds) *The Zambesian Past*, Manchester: Manchester University Press.

Garbett, K. (1977) 'Disparate regional cults and a unitary field in Zimbabwe', in Werbner, R. (ed.) *Regional Cults*, London: Academic Press, ASA monograph no. 16.

Garbett, K. (1992) 'From conquerors to autochthons: cultural logic, structural transformation, and Korekore regional cults', *Social Analysis*, No. 31: 12–43.

Garlake, P. (1968) 'The wake of imported ceramics in the dating and interpretation of the Rhodesian Iron Age', *Journal of African History*, 9 (1): 13–33.

Garlake, P. (1973) *Great Zimbabwe*, London: Thames & Hudson.

Garlake, P. (1974) *The Ruins of Zimbabwe*, Zambia: Historical Association of Zambia.

Garlake, P. (1978) 'Pastoralism and Zimbabwe', *Journal of African History*, 19(4): 479–93.

Garlake, P. (1982) *Great Zimbabwe: Described and Explained*, Harare: Zimbabwe Publishing House.

Garlake, P. (1982b) 'Museums remain rooted in the past', in *Moto*, July: 31–2.

Garlake, P. (1983) 'Prehistory and ideology in Zimbabwe', in Peel, J.D.Y. and Ranger, T.O. (eds) *Past & Present in Zimbabwe*, Manchester: Manchester University Press.

Garlake, P. [1982] (1985) *Great Zimbabwe: Described and Explained*, Harare: Zimbabwe Publishing House.

Gathercole, P. and Lowenthal, D. (1990) *The Politics of the Past*, London: Unwin Hyman.

Gayre, R. (1972) *Origin of the Zimbabwe Civilisation*, Salisbury: Galaxie Press.

Gelfand, M. (1959) *Shona Ritual: With Special Reference to the Chaminuka Cult*, Cape Town: Juta.

Gelfand, M. (1962) *Shona Religion*, Cape Town: Juta.

Gellner, E. (1983) *Nations and Nationalism*, Oxford: Basil Blackwell.

Gero, J. and Root, D. (1990) Public Presentations and Private Concerns: Archaeology in the Pages of the National Geographic', in Gathecole, P. and Lowenthal, D. (eds) *The Politics of the Past*, London: Unwin Hyman.

Giddens, A. (1990) *The Consequences of Modernity*, Cambridge: Polity.

Gosden, C. and Head, L. (1994) 'Landscape—a usefully ambiguous concept', *Archaeology in Oceania*, 29: 113–16.

Gupta, A. (1992) 'The Song of the Nonaligned World: Transnational Identities and the Reinscription of Space in Late Capitalism', *Cultural Anthropology*, 7(1) 63–79.

Haggard, R. (1885) *King Solomon's Mines*, London: Thomas Nelson and Sons.

Hall, M. (1984) 'The Burden of Tribalism: The Social Context of Southern African Iron Age Studies', *American Antiquity*, 49(1).

Hall, M. (1995) 'Great Zimbabwe and the Lost City', in Ucko, P. (ed.) *Theory in Archaeology: A World Perspective*, London: Routledge.

Hall, R.N. (1902a) 'letter from "Havilah Camp," ' *Rhodesia*, 12 July, see National Archives of Zimbabwe, S142/13/5.

Hall, R.N. (1902b) 'Sunday at Zimbabwe', *Rhodesian Times*, 27 September see National Archives of Zimbabwe, S142/13/5.

Hall, R.N. (1905) *Great Zimbabwe, Mashonaland, Rhodesia. An Account of Two Years Examination Work in 1902–4 on Behalf of the Government of Rhodesia*, London: Metheun.

Hall, R.N. (1909) *Prehistoric Rhodesia*, London: Unwin.

Hall, R.N. and Neal, W. (1902) *The Ancient Ruins of Rhodesia*, London: Methueun.

Hammer, A. and Raftopoulos, B. (2003) 'Zimbabwe's Unfinished Business: Rethinking Land, State and Nation', in Hammer, A., Raftopoulos, B. and Jensen, S. (eds) *Zimbabwe's UnFinished Business. Rethinking Land, State and Nation in the Context of Crisis*, Harare: Weaver Press.

Hodder, I. (1995) *Interpreting Archaeology*, London: Routledge.

Holl, A. (1990) 'West African Archaeology: Colonialism and Nationalism', in Robertshaw, P. *A History of African Archaeology*.

Hove, C. and Trojanov, A. (1996) *Guardians of the Soil: Meeting Zimbabwe's Elders*, Munich: Frederking & Raler Verlong.

Hromnik, C. (1981) *Indo-Africa: Towards a New Understanding of the History of Sub-Saharan Africa*, Cape Town: Juta.

Huffman, T.N. (1972) 'The rise and fall of Zimbabwe', *Journal of African History*, 13(3): 353–66.

Huffman, T.N. (1981) 'Snakes and Birds: Expressive Space at Great Zimbabwe', *African Studies*, 40.2.81.

Huffman, T.N. (1984a) ' "Where you are the girls gather to play": The Great Enclosure at Great Zimbabwe', in Hall M. (eds) *Frontiers: Southern African Archaeology Today*, British Archaeological Reports International Series 207.

Huffman, T.N. (1984b) 'Expressive Space in the Zimbabwe Culture', *Man*, 19(2): 593–612.

Huffman, T.N. (1987) *Symbols in Stone: Unravelling the Mystery of Great Zimbabwe*, Johannesburg: Witwatersrand University Press.

Huffman, T.N. (1996) *Snakes & Crocodiles: Power and Symbolism in Ancient Zimbabwe*, Johannesburg: Witwatersrand University Press.

ICOMOS (1985) 'Report to World Heritage Committee on Nomination of Great Zimbabwe to the World Heritage List', June 25.

Ingold, I. (1992) 'Culture and the Perception of the Environment', in Croll, E. and Parkin, D. (eds) *Bush Base: Forest Farm: Culture, Environment, Development*, London: Routledge.

Ingold, I. (1993) 'The Temporality of the Landscape', *World Archaeology*, 25(3).

Jeffreys, M.D. (1954) 'Zimbabwe and Galla Culture', *South African Archaeological Bulletin*, 9:152.

Johanis, G. (1990) ' "Who built Great Zimbabwe?" Oral history and academic credibility' *Moto*, 92/93, September/October.

Johnstone, H.H. (1909) 'Prehistoric Rhodesia: Review of R. Hall's *Prehistoric Rhodesia*', *Geographical Journal*, 34: 502–64.

Kaiser, T. (1995) 'Archaeology and ideology in southeast Europe', in Khol, P.L. and Fawcett, C. (eds) *Nationalism, Politics and the Practise of Archaeology*, Cambridge: Cambridge University Press.

Kaplan, F.E.S. (1994) 'Nigerian Museums: Envisaging Culture as National Identity', in Kaplan, F.E.S. (ed.) *Museums and the Making of Ourselves*, London: Leicester University Press.

Kapungu, L.T. (1988) *Rhodesia: The Struggle for Freedom*, New York: Orbis Press.

Katanekwa, N.M. (2004) 'The heritage of Violence and Resistance: SADC Freedom War Heritage', paper presented to *Livingston Conference on Heritage Management in Southern and Eastern Africa*, Livingston, Zambia, July.

Kelly, J.P. and Kaplan, M. (1990) 'History, structure and ritual', *Annual Review of Anthropology*, 19: 119–50.

Kertzer, D.I. (1988) *Ritual, Politics and Power*, London: Yale University Press.

Kohl, P.L. and Fawcett, C. (1995) 'Archaeology in the Service of the State: Theoretical Considerations', in Kohl, P.L. and Fawcett, C. (eds) *Nationalism, Politics and the Practise of Archaeology*, Cambridge: Cambridge University Press.

Kriger, N. (1988) 'The Zimbabwean War of Liberation: struggles within the struggle', *Journal of Southern African Studies*, 14(2).

Kriger, N. (1992) *Zimbabwe's Guerrilla War: Peasant Voices*, Cambridge: Cambridge University Press.

Kriger, N. (1995) 'The Politics of Creating National Heroes', in Bhebe, N. and Ranger, T. (eds) *Soldiers in Zimbabwe's Liberation War*, University of Zimbabwe and London: James Currey.

Kriger, N. (2003) *Guerrilla Veterans in Post-War Zimbabwe*, Cambridge: Cambridge University Press.

Kuklick, H. (1991) 'Contested Monuments', in Stocker, G.W. (ed.) *Colonial Situations*, London: University of Wisconsin Press.

Lan, D. (1985) *Guns and Rain: Guerrillas & Spirit Mediums in Zimbabwe*, London: James Currey.

Layton, R. and Ucko, P. (eds) (1999) *The Archaeology and Anthropology of Landscape*, ondon: Routledge.

Lévis-Strauss L. (1996) 'A "global strategy" to improve the representativeness of the World Heritage List,' paper presented at the *African Cultural Heritage and the World*

Heritage Convention, Second Global Strategy Meeting, Addis-Ababa, Ethiopia 29 July–1 August, in Hirsch, B., Levis-Strauss, L. and Saouma-Forero, G. (eds) Document WHC-97/WS//5.

Linden, I. (1980) *The Catholic Church and the Struggle for Zimbabwe*, London: Longman.

Lowenthal, D. (1985) *The Past is a Foreign Country*, Cambridge: Cambridge University Press.

Lowenthal, D. (1990) 'Conclusion: Archaeologists and Others', in Gathercole, P. and Lowenthal, D. (eds) *The Politics of the Past*, London: Unwin Hyman.

Lowenthal, D. (1998) *The Heritage Crusade and the Spoils of History*, Cambridge: Cambridge University Press.

MacCormack, C. and Strathern, M. (1980) 'Nature & Culture: a critique', in MacCormack and Strathern (eds) *Nature, Culture and Gender*, Cambridge: Cambridge University Press.

Mahachi, G. and Ndoro, W. (1996) 'The socio-politicial context of southern African Iron Age studies with special refrence to Great Zimbabwe', in Pwiti, G. (ed.) *Caves Monuments and Texts: Zimbabwean Archaeology Today*, Uppsala: Department of Archaeology and Ancient History.

Makamuri, B.B. (1991) 'Ecological Religion', in Virtanen, P. (ed.) *Management of Natural Resources in Zimbabwe: Report on the Research and Training Programme on Energy, Environment and Development*, Tampere: University of Tampere.

Malkki, L. (1994) 'Citizens of Humanity: Internationalism and the Imagined Community of Nations', *Diaspora*, 3: 1: 41–68.

Mallows, W. (1986) *The Mystery of Great Zimbabwe: The Key to a Major Archaeological Enigma*, London: Robert Hale.

Masey Report (1909) National Archives of Zimbabwe, file WI/3/4/1.

Matenga, E. (1996) 'Conservation History of the Great Enclosure, Great Zimbabwe with Reference to the Proposed Restoration of a Lintel Entrance', in Pwiti, G. and Soper, R. (eds) *Aspects of African Archaeology*, Harare: University of Zimbabwe Publications.

Matenga, E. (1998) *The Soapstone Birds of Zimbabwe*, Harare: African Publishing Group.

Matenga, E. (2000) 'Traditional Ritual Ceremony held at Great Zimbabwe', *NAMMO Bulletin*, Official Magazine of NMMZ, 1 (9) December: 14–15.

Maxwell, D. (1999) *Christians and Chiefs in Zimbabwe: A Social History of the Hwesa People, 1870s–1990s*, Edinburgh: Edinburgh University Press.

Mazarire, G. (2003) 'Changing landscape and oral memory in South-Central Zimbabwe: towards a historical geography of Chishanga, c. 1850–1990', *Journal of Southern African Studies*, 29(3): 701–15.

McGregor, J. (2004) 'The Social life of Ruins: Sites of Memory and the Politics of a Zimbabwean Periphery', paper presented at the international conference entitled *Heritage in Southern and Eastern Africa*, Convened by NHCC, NMB, JSAS and BIEA in Livingston, Zambia, 5–8 July.

Middleton, J. (1960) *Lugbara Religion*, Oxford University Press.

Middleton, J. (1971) 'Prophets and Rainmakers: The Agents of Social Change Amongst the Lugbara', in Beidelman, T. (ed.) *The Translation of Culture*, London: Tavistock.

Middleton, J. (1982) 'Lugbara Death', in Bloch M. and Parry, J. (eds) *Death and the Regeneration of Life*, Cambridge: Cambridge University Press.

Mitchell, T. (1996) 'Ritual', in Barnard, A. and Spencer, J. (eds) *The Encyclopedia of Social and Cultural Anthropology*, London: Routledge.

Morphy, H. (1995) 'Landscape and the Reproduction of the Ancestral Past', in Hirsch, E. and O'Hanlon, M. (eds) *The Anthropology of Landscape*, Oxford: Clarendon Press.

Moyo, S. (2001) 'The land occupations movement and democratisation in Zimbabwe: contradictions of neoliberalism', *Millenium Journal of International Studies*, 30 (2).

Mtetwa, R.M.G. (1976) 'The "Political" and Economic History of the Duma people of South-Eastern Rhodesia from early 18th Century to 1945', unpublished PhD thesis, History Department, University of Rhodesia.

Mudenge, I.S.G. (1988) *A Political History of Munhumutapa c 1400–1902*, Harare: Zimbabwe Publishing House.

Mudenge, I.S.G. (1998) 'Foreword' in Matenga, E. (ed.) *The Soapstone Birds of Zimbabwe*, Harare: African Publishing Group.

Mufuka, K. (1983) *DZIMBAHWE: Life and Politics in the Golden Age 1100–1500 A.D.*, Harare: Zimbabwe Publishing House.

Mufuka, K. (1984) of *DZIMBAHWE: Life and Politics in the Golden Age 1100–1500 A.D.*, 2nd edition, Harare: Zimbabwe Publishing House.

Mufuka, K. (2002) 'ZANU PF's paranoia against whites explained—American Notes', *The Standard*, 8 September, available at www.zimbabwesituation.com/dec_2002_archive.html.

Moyo, J. (2001). 'The land occupations movement and democratisation in Zimbabwe: Contradictions of neoliberalism', in *Millenium: Journal of International Studies* 30(2).

National Archives of Zimbabwe (1984) *The Zimbabwe Epic* edited by Douglas R.G.S. Harare: National Archives.

Ncube, M. and Ranger, T. (1995) 'Religion and War in Southern Matabeleland', in Bhebe, N. and Ranger, T. (eds) *Soldiers in Zimbabwe's Liberation War*, University of Zimbabwe and London: James Currey.

Ndlovu, J. (2003) 'Breaking the Taboo. Mzilikazi and National Heritage', Lozikeyi Lecture, Bulawayo National Gallery, 7 August.

Ndoro, W. (1996) 'The Evolution of a Management Policy at Great Zimbabwe', in Pwiti, G. (ed.) *Caves, Monuments and Texts; Zimbabwean Archaeology Today*, Uppsala: Department of Archaeology and Ancient History.

Ndoro, W. (2001) *Your Monument, Our Shrine: The Preservation of Great Zimbabwe*, Studies in African Archaeology 19, Uppsala: Societas Archaeogica Upsaliensis.

Ndoro, W. and Pwiti, G. (1997) 'Marketing the Past: The Shona Village at Great Zimbabwe', *Conservation and Management of Archaeological sites*, 2(1).

Ndoro, W. and Puriti, J. (2001) 'Heritage management in Southern Africa: local, national and international discourse', *Public Archaeology* 2(1).

Nehowa, O. (1993) 'Technology and the Monument—Why?', *Conservation News*, Newsletter of NMMZ, 1(2), July.

Ngara, E. (1992) *Songs from the Temple*, Gweru: Mambo Press.

Nora, P. (1989) 'Between memory and history: *Les lieux de memoires*', *Representations*, 26: 7–25.

Nthoi, L. (1995) 'Perspectives of Religion. A Study of the Mwali Cult of Southern Africa', PhD thesis, University of Manchester.

Nyathi, L. (2003) 'The Matopos Hills Shrines: A Comparative Study of the Dula, Njelele and Zhame Shrines and Their Impact on the Surrounding Communities', History Honours dissertation, University of Zimbabwe.

Obeyesekere, G. (1963) 'The great tradition and the little in the perspective of Sinhalese Buddhism', *The Journal of Asian Studies*, 22: 139–53.

Ortner, S. (1995) 'Resistance and the problem of ethnographic refusal', *Comparative Studies in Society and History*, 37(1): 173–93.

Paine, R. (1994) 'Masada: A History of a Memory', *History and Anthropology*, 6 (4): 371–409.

Pakenham, T. (1991) *The Scramble for Africa*, New York: Avon Books.

Parfit, I. (1992) *Journey to the Vanished City: The Journey for a Lost Tribe of Israel*, London: Hodder & Houghton.

Paver, B.G. [1950] (1957) *Zimbabwe Cavalcade*, London: Cassell & Company.

PhotoSafari (Pvt.) Ltd. (1999) *Great Zimbabwe*, tourist pamphlet.

Pikirayi, I. (2001) *The Zimbabwe Culture: Origins and Decline of Southern Zambezian States*, Oxford: Alta Mira Press.

Pocock, D. (1973–74) 'Nuer religion; a supplementary view', *Journal of the Anthropology Society of Oxford*, 5: 69–79.

Pongweni, A. (1982) *Songs that Won the Liberation War*, Harare: College Press.

Posselt, W. (1924) 'Early Days of Mashona Land', *NADA*, 2: 70.

Pressouyre, L. (1996a) (English trans. – original in French 1993) *The World Heritage Convention, Twenty Years Later*, Paris: UNESCO Publishing.

Pressouyre, L. (1996b) 'Cultural Heritage and the 1972 Convention: Definition and Evolution of a Concept', paper presented at the *African Cultural Heritage and the World Heritage Convention, Second Global Strategy Meeting, Addis-Ababa, Ethiopia 29 July – 1 August*, Hirsch, B., Levi-Strauss, L. and Saouma-Forero, G., Document WHC – 97/WS//5.

Puriti, G. (1996b) *Continuity and Change: An Archaeological Study of Farming Communities in Northern Zimbabwe, AD 500–1700*, Studies in African Archaeology 13, Uppsala: Societas Archaeogica Upsaliensis.

Pwiti, G. (1996) 'Let the ancestors rest in peace? New challenges for cultural heritage management in Zimbabwe', *Conservation and Management of Archaeological Sites*, (1)3.

Pwiti, G. and Muvenge, G. (1996) 'Archaeologists, Tourists and Rainmakers: Problems in the Management of Rock Art Sites in Zimbabwe. A Case Study of Domboshava National Monument', in Pwiti, G. and Soper, R. (eds) *Aspects of African Archaeology*, Harare: University of Zimbabwe Publications.

Pwiti, G. and Ndoro, W. (1999) 'The legacy of colonialism: perceptions of the cultural heritage in southern Africa with special reference to Zimbabwe', *African Archaeological Review*, (2)1.

Puriti, G. and Ndoro, W. (2000) *The Traditional Village at Great Zimbabwe National Monument – Final Report*, August, University of Zimbabwe, History Department, Archaeology Unit, Report for NMMZ, NMMZ file H4.

Raftopoulos, B. (2003) 'The State in Crisis: Authoritarian Nationalism, Selective Citizenship and Distortions of Democracy in Zimbabwe', in Hammer, A., Raftopoulos, B. and Jensen, S. (eds) *Zimbabwe's UnFinished Business. Rethinking Land, State and Nation in the Context of Crisis*, Harare: Weaver Press.

Ramaswamy, S. (1994) 'Nationalist Thought and the Colonial World,' review article, *Journal of Asian Studies*, 53.

Randal-MacIver, D. (1906) *Medieval Rhodesia*, London: Macmillan & Co.

Ranger, T. (1967) *Revolt in Southern Rhodesia 1896–7: A Study in African Resistance*, London: Heineman.

Ranger, T. (1979) 'Rhodesia's politics of Tribalism', *New Society*, 6 September.

Ranger, T. (1982) 'The death of Chaminuka: spirit mediums, nationalism and the guerilla war in Zimbabwe', *African Affairs*, 324.

Ranger, T. (1983) 'The Invention of Tradition in Colonial Africa', in Hobsbawm, E. and Ranger, T. (eds) *The Invention Of Tradition*, Cambridge: Cambridge University Press.

Ranger, T. (1985) *Peasant Consciousness and Guerrilla War in Zimbabwe*, London: James Currey.

Ranger, T. (1987) 'Taking hold of the land: holy places and pilgrimages in 20th century Zimbabwe', *Past & Present*, 117.

Ranger, T. (1989) 'Whose heritage? The case of the Matobo National Park', *Journal of Southern African Studies*, 15(2).

Ranger, T. (1996) 'Great Spaces Washed with Sun: the Matopos and Uluru Compared', in Darian-Smith K. et al. (eds) *Text, Theory, Space*, London: Routledge.

Ranger, T. (1999) *Voices from the Rocks: Nature, Culture and History in the Matopos Hills of Zimbabwe*, Oxford: James Currey.

Ranger T. (2002) 'The Zimbabwe elections: a personal experience', *Transformation*, July: 60.

Ranger, T. (2003) 'Women and Environment in African Religion: The Case of Zimbabwe', in Beinart, W. and McGregor, J. (eds) *Social History & African Environments*, Oxford: James Currey.

Ranger, T. (2004a) 'Historiography, patriotic history and the history of the nation: the struggle over the past in Zimbabwe', *Journal of Southern African Studies*, 30(2) June.

Ranger, T. (2004b) 'Living Ritual and Indigenous Archaeology: The Case of Zimbabwe', paper presented at *UCLA Conference on Archaeology and Ritual*, January/February.

Ranger, T., Alexander, J. and McGregor, J. (2000) *Violence and Memory: One Hundred Years in the 'Dark Forests' of Matabeleland*, Oxford: James Currey.

Ratnagar, S. (2004) 'Archaeology at the heart of a political confrontation', *Current Anthropology*, 45(2): 239–59.

Robinson, K. (1961) 'Zimbabwe Pottery', *Occasional Papers of National Museums of Southern Rhodesia*, 23A: 193–220.

Robinson, K. (1966) 'The Archaeology of the Rozvi', in Stokes, E. and Brown, R. (eds) *The Zambesian Past*, Manchester: Manchester University Press.

Robinson, K., Summers, R. and Whitty, A. (1961) 'Zimbabwe Excavations 1958' *Occasional Papers of the National Museums of Southern Rhodesia*, 23A.

Rodrigues, M. and Mauelshagen, B. (1987) *Preservation of Great Zimbabwe and Khami Ruins*, Series No. FMR/CC/CH/87/217/, Paris: UNDP.

Rössler, M. (1998) 'The World Heritage Convention', *Naturopa*, 86: 19.

Rutherford, B. (2003) 'Belonging to the Farm(er): Farm Workers, Farmers, and the Shifting Politics of Citizenship', in Hammer, A., Raftopoulos, B. and Jensen, S. (eds) *Zimbabwe's UnFinished Business. Rethinking Land, State and Nation in the Context of Crisis*, Harare: Weaver Press.

Said, E. (1978) *Orientalism*, London: Routledge.

Said, E. [1994] (2002) 'Invention, Memory and Place', in Mitchell, W.J.T. (ed.) *Landscape and Power*, 2nd edition, London: University of Chicago Press.

Saouma-Forero, G. (1999) 'The Global Strategy in Africa', unpublished UNESCO, information document – mes doc\assemb.gen\global strategy in Africa.

Sassoon, H. (1982) *Preservation of Great Zimbabwe*, Technical Report, Series FMR/CCCT/CT/82/156, Paris: UNESCO.

Saul, J. (1979) *The State and Revolution in East Africa*, London: Heineman.

Schoffeeers, J.M. (ed.) (1978) *Guardians of the Land*, Gwelo: Mambo Press.

Shamuyarira, N. (1965) *Crisis in Rhodesia*, London: Deutsch.

Shore, C. and Wright, S. (eds) *The Anthropology of Policy*, London: Routledge.

Sinamai, A. (2003) 'Cultural Shifting-Sands: Changing Meanings of Zimbabwe Sites in Zimbabwe, South Africa and Botswana', paper presented to ICOMOS Conference, Victoria Falls, available at http://www.international.icomos.org/victoriafalls2003/papers/c3-7%20-%70sinamaipdf, accessed on 24/02/2004.

Sinclair, P. (1987) *Space, Time, and Social Formation: A Territorial Approach to the Archaeology and Anthropology of Zimbabwe and Mozambique, c. 0–1700 AD*, Uppsala: Societas Archaeologica Upsaliensis.

Sinclair, P., Pikirayi, I., Pwiti, G. and Soper, R. (1993) 'Urban Trajectories on the Zimbabwean Plateau', in Shaw, T., Sinclair, P., Andan, B. and Okopo, A. (eds) *The Archaeology of Africa: Foods, Metals & Towns*, London: Routledge.

Sithole, M. (1980) 'Ethnicity and factionalism in Zimbabwean nationalist politics 1957–79', *Ethnic & Racial Studies*, 3(1).

Sithole, M. (1984) 'Class and factionalism in the Zimbabwean nationalist movement' *African Studies Review*, 3(2).

Sithole, M. (1985) 'The salience of ethnicity in African politics: the case of Zimbabwe', *Journal of Asian and African Studies*, 20.

Smith L.T. (1999) *Decolonising Methodologies*, London: Zed Books.

Spierenburg, M. (2003) *Strangers, Spirits and Land Reforms. Conflicts about Land in Dande, Northern Zimbabwe*, PhD, University of Amsterdam.

Sprague De Camp, L. (1968) 'Secret of Zimbabwe — search for King Solomon's gold', *Science Digest*, February.

Steedly, M. (1993) *Hanging Without a Rope: Narrative Experience in Colonial and Postcolonial Karoland*, Princetown: Princetown University Press.

Stirrat, R.L. (1984) 'Sacred Models' in *Man*, 19: 2.

Strawbridge, A. (1982) 'Althusser's theory of ideology and Durkheim's account of religion: an examination of some striking parallels', in *Sociology Review*, 30, 125–40.

Summers, R. (1955) 'The dating of the Zimbabwe ruins', *Antiquity*, 29.

Summers, R. (1963) *Zimbabwe: A Rhodesian Mystery*, Johannesburg: Nelson.

Summers, R. (1963b) 'Mystery, Myth & Method at Zimbabwe', paper presented at *History of Central African People's Conference*, Rhodes-Livingston Institute, Lusaka, Zambia, 28 May–1 June (National Archives of Zimbabwe, file Gen-f/his).

Summers, R. and Thompson, Blake (1956) 'Mlimo and Mwari', *NADA*, 33: 53.

Tilley, C. (1994) *The Phenomenology of Landscape*, Oxford: Berg.

Turner, V. (1969) *The Ritual Process*, Harmondsworth: Penguin.

Ucko, P.J. (1981) 'Report on a proposal to initiate "Culture Houses" in Zimbabwe', unpublished report to the Government of Zimbabwe.

Ucko, P. J. (1990) 'Foreword', in Gathecole, P. and Lowenthal, D. (eds) *The Politics of the Past*, London: Unwin Hyman.

Ucko, P. J. (1994) 'Museums and sites: cultures of the past within education – Zimbabwe, some ten years on', in Stone, P.G. and Molyneaux, B.L. (eds) *The Presented Past*, London: Routledge.

Ucko, P. J. (2000) 'Enlivening a "dead" past', *Conservation and Management of Archaeological Sites*, 4: 67–92.

UNESCO (1972) *Convention Concerning the Protection of World Cultural and Natural Heritage*, adopted by the General Conference at its seventeenth session, 16 November, Paris.

UNESCO (1997) *Records of the General Conference*, Volume 1, *Resolutions*, Twenty-ninth Session, 21 October – 12 November, General Conference, Paris: UNESCO.

UNESCO World Heritage Centre (1995) *Report of the African Cultural Heritage and the World Heritage Convention, First Global Strategy Meeting, Harare, Zimbabwe, 11–13 October 1995*, Munjeri, D., Ndoro, W., Sibanda, C., Saouma-Forero, G., Lévi-Strauss, L. and L.Mbuyamba (eds).

UNESCO World Heritage Centre (1997) *Operational Guidelines for the Implementation of the World Heritage Convention*, Document WHC-97/WS/1 Rev, revision adopted by the World Heritage Committee at its 21st Session, December.

UNESCO World Heritage Centre (1997) *Report of the African Cultural Heritage and the World Heritage Convention, Second Global Strategy Meeting, Addis-Ababa, Ethiopia. 29 July–1 August*. Document WHC -97/WS/5, (eds) B. Hirsch, L. Levi-Strauss and G. Saouma-Forero.

UNESCO World Heritage Centre (1997) *Format for the Nomination of Cultural and Natural Properties for Inscription on the World Heritage List*, Document WHC-97/WS/6.

UNESCO World Heritage Centre (1998a) Document WHC.99/2, March 1999, revision adopted by the World Heritage Committee at its 22th Session, December 1998.

UNESCO World Heritage Centre (1998b) *Report of the World Heritage Global Strategy Natural and Cultural Expert Meeting, 25–29 March 1998, Amsterdam, The Netherlands*, Document WHC-98/conf.203/INF.7.

UNESCO World Heritage Centre (1999a) *Brief Descriptions of Sites Inscribed on the World Heritage List*, Document WHC.99/15, January.

UNESCO World Heritage Centre (1999b) *Properties Included in the World Heritage List* Document WHC.99/3, January.

UNESCO World Heritage Centre (1999c) *Report of the Twenty-second Session of the World Heritage Committee, Kyoto, Japan, 30 November–5 December 1998*, Document WHC-98/Conf.203/18, 29 January.

UNESCO World Heritage Centre (2002) *Decisions of the 26th Session of the World Heritage Committee* 24–29 June, Document WH-02/conf.202/25.

Van Gennep (1909) *Les Rites de passage*, Paris: Payot.

Vansina, J. (1965) *Oral Tradition: A Study in Historical Methodology*, London: Routledge.

Vansina, J. (1980) 'Memory and Oral Tradition', in Miller, J.C. (ed.) *African Past Speaks*, 262–79.

Vansina, J. (1985) *Oral Tradition as History*, Madison: University of Wisconsin Press.

Wainright, A.K. (1949) 'The founders of the Zimbabwe civilization', *Man*, 49: 80.

Walker, P. and Dickens, J. (1992) *An Engineering Study of Dry Stone Monuments in Zimbabwe*, Vol. 1 & 2, Nottingham: Loughborough University.

Walsh, K. (1992) *The Representation of the Past: Museums and Heritage in the Post-modern World*, London: Routledge.

Werbner, R. (1989) 'Regional Cult of God Above. Achieving and Defending the Macrocosm', in Werbner, R. (ed.) *Ritual Passage, Sacred Journey*, Washington: Smithsonian.

Werbner, R. (1991) *Tears of the Dead: The Social Biography of an African Family*, Edinburgh: Edinburgh University Press.

Werbner, R. (1998) 'Smoke from the Barrel of a Gun: Postwars of the Dead, Memory and Reinscription in Zimbabwe', in Werbner R. (ed) *Memory and the Postcolony: African Anthropology and the Critique of Power*, London: Zed Books.

White, L. (2003) *The Assassination of Herbert Chitepo: Texts and Politics in Zimbabwe*, Bloomington: Indianna University Press.

Whitty, A. (1961) 'Architectural style of Zimbabwe', *Occasional Paper of National Museums and Monuments of Southern Rhodesia*, 3: 289–305.

Willet, F. (1990) 'Museums: Two Case Studies of Reaction to Colonialism', in Gathercole, P. and Lowenthal, D. (eds) *The Politics of the Past*, London: Unwin Hyman.

Willoughby, J. (1893) *Further Excavations at Zimbabye*, London: George Phillip.

Wilmot, A. (1896) *Monomutapa (Rhodesia)*, London: Unwin.

Worby, E. (2003) 'The End of Modernity in Zimbabwe? Passages from Development to Sovereignty', in Hammer, A., Raftopoulos, B. and Jensen, S. (eds) *Zimbabwe's UnFinished Business. Rethinking Land, State and Nation in the Context of Crisis*, Harare: Weaver Press.

Zawaira, A. and Malorera, C. (1983) 'Foreword', in Mufuka, K. (ed.) *DZIMBAHWE: Life and Politics in the Golden age 1100–1500 AD*, Harare: Zimbabwe Publishing House.

Zerubavel, Y. (1995) *Recovered Roots: Collective Memory and the Making of Israeli National Tradition*, Chicago: Chicago University Press.

Zimunya, M.B. (1982) *Thought Tracks*, Essex: Longman.